Venom and Toxin as Targeted Therapy

Venom and Toxin as Targeted Therapy

Special Issue Editor

Hang Fai (Henry) Kwok

MDPI • Basel • Beijing • Wuhan • Barcelona • Belgrade

Special Issue Editor
Hang Fai (Henry) Kwok
Faculty of Health Sciences
University of Macau
Avenida de Universidade
Macau SAR

Editorial Office
MDPI
St. Alban-Anlage 66
4052 Basel, Switzerland

This is a reprint of articles from the Special Issue published online in the open access journal *Toxins* (ISSN 2072-6651) from 2017 to 2019 (available at: https://www.mdpi.com/journal/toxins/special_issues/Venom_therapy)

For citation purposes, cite each article independently as indicated on the article page online and as indicated below:

LastName, A.A.; LastName, B.B.; LastName, C.C. Article Title. *Journal Name* **Year**, *Article Number*, Page Range.

ISBN 978-3-03921-189-0 (Pbk)
ISBN 978-3-03921-190-6 (PDF)

Contents

About the Special Issue Editor

Hang Fai (Henry) Kwok is an associate professor and a histopathology core consultant in the Faculty of Health Sciences at the University of Macau. He is also a visiting scientist at the Cancer Research UK (CRUK) Cambridge Institute at the University of Cambridge.

Born and raised in Hong Kong, Prof. Kwok moved to the United Kingdom in 1999 to study and work. Before joining the University of Macau in 2014, he was a senior research fellow in the Department of Oncology and the CRUK Cambridge Institute at the University of Cambridge. His research interests are in the areas of protease biochemistry with biologics development (e.g., antibodies, peptides/toxins, and small molecules) with a focus on pursuing novel therapeutic and prognostic approaches in the treatment of diseases, such as cancer, diabetes, and rheumatoid arthritis. In addition, Prof. Kwok is also interested in the discovery and characterization of novel bioactive molecules from sources in nature, including amphibian defensive skin secretions and reptile, scorpion, and insect venoms, to exploit their anticancer and antibacterial/antifungal therapeutic potential.

toxins

MDPI

Editorial

Venom Toxins as Potential Targeted Therapies

Hang Fai Kwok [1,2]

[1] Cancer Centre, Faculty of Health Sciences, University of Macau, Taipa, Macau, China; hfkwok@um.edu.mo
[2] Institute of Translational Medicine, Faculty of Health Sciences, University of Macau, Taipa, Macau, China

Received: 9 May 2019; Accepted: 10 June 2019; Published: 13 June 2019

Targeted therapy has been a very hot research topic in the last decade. It focuses on specific medications for treatment of particular diseases, such as cancer, diabetes, heart disease, etc. One of the most exciting recent developments in targeted therapies is the isolation of disease-specific molecules from natural resources, such as animal venoms and plant metabolites/toxins, to use as templates for new drug motif design.

This Special Issue of Toxins includes three recent advanced research studies related to bee venoms as potential medicinal therapy in different aspects [1–3]. Furthermore, recent advances in bioactive molecules finding from frog skins, mushroom and venom/toxin/immunotoxins for targeted cancer therapy and immunotherapy are discussed [4–8]. The discussion on using novel disease-specific venom-based protein/peptide/toxin along with currently available FDA approved drugs as combinatorial treatment, such as a family of novel types of antimicrobial agents, were also encouraged to be discussed in these contexts. For examples, an overview of some selected promising snake and ant venom-based peptides/toxins potentially able to address the forthcoming challenges in this field were included [9,10]. In addition, four detailed review articles openly discuss the venom proteins/peptides in different species of snake venoms and toad toxins [11–13]; moreover, plant toxins from Bouganin naturally targeted, as immunotoxins [14], mammalian receptors and demonstrated high specificity and selectivity towards defined ion channels of cell membranes and receptors.

To sum up, all research and review articles proposing novelties or overviews, respectively, were successfully and carefully selected in this Special Issue after rigorous revision by the expert peer reviewers. As a guest editor, I would like to express my deep appreciation to all the selfless and fair reviewers.

Acknowledgments: The editor is grateful to all the important authors who contributed to this special issue "Venom and Toxin as Targeted Therapy". They are also mindful that without the rigorous and selfless evaluation of the submitted manuscripts by external peer reviewers/expertise, this special issue could not have happened. Moreover, the editor (Kwok H. F.) thanks for the support from the Science and Technology Development Fund of Macau SAR (FDCT) [019/2017/A1] and the Faculty of Health Sciences (FHS) University of Macau. Finally, the valuable contributions, organization, and editorial support of the MDPI management team and staff are greatly appreciated.

Conflicts of Interest: The authors declare no conflict of interest.

References

1. Shin, D.; Choi, W.; Bae, H. Bee venom phospholipase A2 alleviate house dust mite-induced atopic dermatitis-like skin lesions by the CD206 mannose receptor. *Toxins* **2018**, *10*, 146. [CrossRef] [PubMed]
2. Kim, Y.; Lee, Y.W.; Kim, H.; Chung, D.K. Bee Venom Alleviates Atopic Dermatitis Symptoms through the Upregulation of Decay-Accelerating Factor (DAF/CD55). *Toxins* **2019**, *11*, 239. [CrossRef] [PubMed]
3. Frangieh, J.; Salma, Y.; Haddad, K.; Mattei, C.; Legros, C.; Fajloun, Z.; El Obeid, D. First Characterization of The Venom from *Apis mellifera syriaca*, A Honeybee from The Middle East Region. *Toxins* **2019**, *11*, 191. [CrossRef] [PubMed]
4. Najar, D.; Haraldsson, B.; Thorsell, A.; Sihlbom, C.; Nyström, J.; Ebefors, K. Pharmacokinetic Properties of the Nephrotoxin Orellanine in Rats. *Toxins* **2018**, *10*, 333. [CrossRef] [PubMed]

5. Tan, Y.; Chen, X.; Ma, C.; Xi, X.; Wang, L.; Zhou, M.; Burrows, J.; Kwok, H.; Chen, T. Biological Activities of Cationicity-Enhanced and Hydrophobicity-Optimized Analogues of an Antimicrobial Peptide, Dermaseptin-PS3, from the Skin Secretion of *Phyllomedusa sauvagii*. *Toxins* **2018**, *10*, 320. [CrossRef] [PubMed]
6. Rust, A.; Shah, S.; Hautbergue, G.; Davletov, B. Burkholderia Lethal Factor 1, a Novel Anti-Cancer Toxin, Demonstrates Selective Cytotoxicity in MYCN-Amplified Neuroblastoma Cells. *Toxins* **2018**, *10*, 261. [CrossRef] [PubMed]
7. Lin, Y.; Hu, N.; He, H.; Ma, C.; Zhou, M.; Wang, L.; Chen, T. A Hylarana latouchii Skin Secretion-Derived Novel Bombesin-Related Pentadecapeptide (Ranatensin-HLa) Evoke Myotropic Effects on the in vitro Rat Smooth Muscles. *Toxins* **2019**, *11*, 204. [CrossRef] [PubMed]
8. Müller, F.; Cunningham, T.; Beers, R.; Bera, T.; Wayne, A.; Pastan, I. Domain II of Pseudomonas Exotoxin Is Critical for Efficacy of Bolus Doses in a Xenograft Model of Acute Lymphoblastic Leukemia. *Toxins* **2018**, *10*, 210. [CrossRef] [PubMed]
9. Monteiro, D.; Selistre-de-Araújo, H.; Tavares, D.; Fernandes, M.; Kalinin, A.; Rantin, F. Alternagin-C (ALT-C), a Disintegrin-Like Cys-Rich Protein Isolated from the Venom of the Snake *Rhinocerophis alternatus*, Stimulates Angiogenesis and Antioxidant Defenses in the Liver of Freshwater Fish, *Hoplias malabaricus*. *Toxins* **2017**, *9*, 307. [CrossRef] [PubMed]
10. Guzman, J.; Téné, N.; Touchard, A.; Castillo, D.; Belkhelfa, H.; Haddioui-Hbabi, L.; Treilhou, M.; Sauvain, M. Anti-Helicobacter pylori Properties of the Ant-Venom Peptide Bicarinalin. *Toxins* **2018**, *10*, 21. [CrossRef] [PubMed]
11. Li, L.; Huang, J.; Lin, Y. Snake venoms in cancer therapy: Past, present and future. *Toxins* **2018**, *10*, 346. [CrossRef] [PubMed]
12. Qi, J.; Zulfiker, A.; Li, C.; Good, D.; Wei, M. The development of toad toxins as potential therapeutic agents. *Toxins* **2018**, *10*, 336. [CrossRef] [PubMed]
13. Abidin, Z.; Asnawi, S.; Lee, Y.Q.; Othman, I.; Naidu, R. Malaysian Cobra Venom: A Potential Source of Anti-Cancer Therapeutic Agents. *Toxins* **2019**, *11*, 75. [CrossRef] [PubMed]
14. Bortolotti, M.; Bolognesi, A.; Polito, L. Bouganin, an Attractive Weapon for Immunotoxins. *Toxins* **2018**, *10*, 323. [CrossRef] [PubMed]

toxins

MDPI

Article

Bee Venom Phospholipase A2 Alleviate House Dust Mite-Induced Atopic Dermatitis-Like Skin Lesions by the CD206 Mannose Receptor

Dasom Shin, Won Choi and Hyunsu Bae *

Department of Science in Korean Medicine, College of Korean Medicine, Kyung Hee University, 26 kyungheedae-ro, dongdaemoon-ku, Seoul 02447, Korea; ssd060@naver.com (D.S.); wonones@naver.com (W.C.)

* Correspondence: hbae@khu.ac.kr; Tel.: +82-2961-9316

Received: 8 February 2018; Accepted: 31 March 2018; Published: 2 April 2018

Abstract: Atopic dermatitis (AD) is a chronic inflammatory skin disease characterized by highly pruritic, erythematous, and eczematous skin plaques. We previously reported that phospholipase A2 (PLA2) derived from bee venom alleviates AD-like skin lesions induced by 2,4-dinitrochlorobenzene (DNCB) and house dust mite extract (*Dermatophagoides farinae* extract, DFE) in a murine model. However, the underlying mechanisms of PLA2 action in actopic dermatitis remain unclear. In this study, we showed that PLA2 treatment inhibited epidermal thickness, serum immunoglobulin E (IgE) and cytokine levels, macrophage and mast cell infiltration in the ear of an AD model induced by DFE and DNCB. In contrast, these effects were abrogated in CD206 mannose receptor-deficient mice exposed to DFE and DNCB in the ear. These data suggest that bvPLA2 alleviates atopic skin inflammation via interaction with CD206.

Keywords: atopic dermatitis (AD); house dust mite extract (DFE); 2,4-dinitrochlorobenzene (DNCB); bee venom phospholipase A2 (bvPLA2); skin inflammation; CD206; mannose receptor

Key Contribution: This study highlights the potential utility of bvPLA2 to alleviate atopic skin inflammation through interaction with CD206.

1. Introduction

Atopic dermatitis (AD), also known as atopic eczema, is a chronic inflammatory skin disease associated with intense pruritus [1]. The details of mechanisms underlying AD process are still unclear, although large studies investigated therapeutic targets and pathophysiology [2]. AD affects nearly 20% of children and almost 7% of adults [3,4].

Although it is still disputed, it is generally suggested that AD is a chronic inflammatory disease associated with increased level of proinflammatory cytokines, serum immunoglobulin E (IgE) in the mast cells of the skin [5,6]. Until now, topical corticosteroids have been used as a first-line treatment for AD [7]. Although further studies are needed for validation, novel therapeutic approaches have been discovered, including ceramides and other components involved in epidermal differentiation and maintenance of epidermal barriers [8]. However, most AD therapies have been reported to have side effects, especially in children [9]. Recent advances in herbal medicine suggest potential therapies for inflammatory skin diseases [10].

Bee venom, which has long been used in alternative medicine, has been used to treat asthma, multiple sclerosis, cancer, and rheumatoid arthritis, as a powerful immuno-modulatory agent [11–14]. Bee venom is composed of various peptides and proteins such as melittin, phospholipase A2, apamin, and mast cell degranulating peptide [15–17]. The bee venom phospholipase A2 (bvPLA2), one of the major components of bee venom, plays a central role in the regulation of phospholipid metabolism, signal transduction, inflammatory and immune response [18–20]. We recently reported that bvPLA2 acts by activating regulatory T cell (Tregs) in a murine model of allergic asthma and cisplatin-induced nephrotoxicity via CD206 mannose receptor [21,22]. In addition, bvPLA2 suppresses AD-like skin lesions in Balb/c mouse model induced by *Dermatophagoides farinae* extract (DFE) and 2,4-dinitrochlorobenzene (DNCB) [10]. Based on our previous studies, we hypothesized that bvPLA2 has an inhibitory effect on AD-like skin lesions through the CD206 receptor. AD signs and symptoms were evaluated in wild-type or CD206-deficient mice challenged on the ear with DFE and DNCB.

2. Results

2.1. *bvPLA2 Treatment Alleviated DFE/DNCB-Induced Ear Thickness and AD-Like Symptoms*

To investigate the effect of bvPLA2 in DFE/DNCB-induced AD-like skin lesion, bvPLA2 was used to treat mice subjected to DFE/DNCB treatment. As shown in Figure 1, DFE/DNCB induced significant skin swelling on both ears in wild-type (WT) and CD206-deficient ($CD206^{-/-}$) mice. However, bvPLA2 effectively suppressed AD-related skin swelling in WT mice, but not $CD206^{-/-}$ mice.

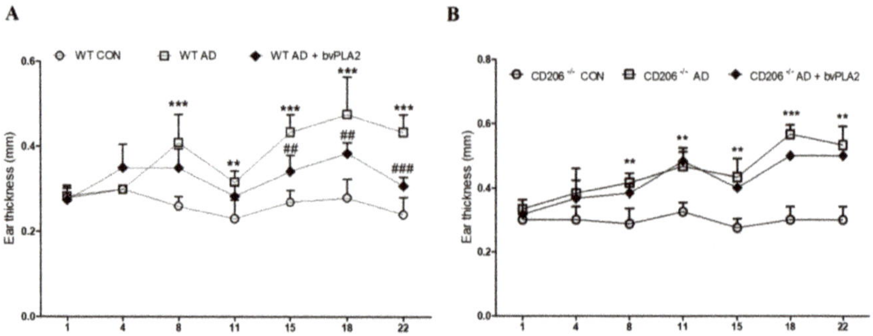

Figure 1. Effect of bvPLA2 on ear thickness of DFE/DNCB-induced AD mice depending on CD206 receptor. Images of the ear skin lesions in groups of mice taken on day 28. Ear thickness was measured using a dial thickness gauge 24 h after DFE/DNCB application for 4 weeks. DFE/DNCB caused ear swelling in wild-type (WT) and $CD206^{-/-}$ mice but bvPLA2 effectively suppressed AD-related ear swelling only in WT mice (**A**), but not $CD206^{-/-}$ mice (**B**). CON: normal control, AD: DFE/DNCB, and AD + bvPLA2: DFE/DNCB + bvPLA2 (80 ng/ear, 20 μL). The data are displayed as the mean ± SD. The statistical analyses were conducted with one-way ANOVA followed by Newman–Keuls multiple comparison tests (*** $p < 0.001$, ** $p < 0.01$ vs. con and ### $p < 0.001$, ## $p < 0.01$ vs. AD; n = 3–5).

The ear thickness was evaluated using dial thickness gauge during 4 weeks (Figure 1). In WT mice, DFE/DNCB treatment significantly increased the ear thickness compared with the control group while bvPLA2 improved ear thickness in DFE/DNCB-induced AD mice (Figure 1A). On the other hand, in $CD206^{-/-}$ mice, no significant effect was observed after bvPLA2 treatment (Figure 1B).

2.2. *bvPLA2 Reduced Serum IgE in DFE/DNCB-Treated Mice*

We previously reported the therapeutic effect of bvPLA2 in DFE/DNCB-induced AD mice [10]. In this study, we measured the total IgE from serum following bvPLA2 treatment of WT and $CD206^{-/-}$ mice. As shown in Figure 2A, DEF/DNCB treatment induced a dramatic increase in serum IgE in mice. Interestingly, bvPLA2 treatment effectively inhibited IgE increase in WT mice challenged with DFE and DNCB. In contrast, bvPLA2 treatment showed no significant difference in $CD206^{-/-}$ mice after DFE/DNCB treatment.

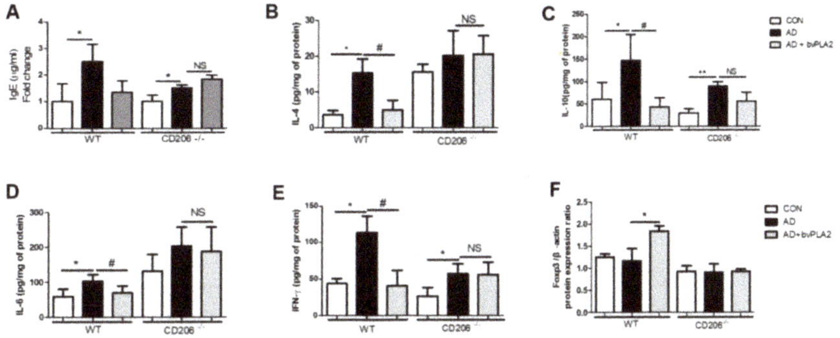

Figure 2. Effects of bvPLA2 on expression of serum total immunoglobulin E (IgE) and cytokines in the ear tissues of DFE/DNCB-induced AD mice via CD206. The total serum IgE levels and the concentration of Th1 and Th2 cytokines were measured by ELISA and Foxp3 expression was observed by western blot. (**A**) IgE; (**B**) Interleukin-4 (IL-4); (**C**) Interleukin-10 (IL-10); (**D**) Interleukin-6 (IL-6); (**E**) Interferon- γ (IFN-γ) and (**F**) Foxp3. The data are shown as the means ± SD. The statistical analyses were conducted with a one-way ANOVA followed by Newman–Keuls multiple comparison tests. CON: normal control, AD: DFE/DNCB treatment, and AD+bvPLA2: DFE/DNCB treatment + bvPLA2 treatment. (** $p < 0.01$, * $p < 0.05$ vs. Control and # $p < 0.05$, NS vs. AD; n = 3–5).

2.3. bvPLA2 Abrogated AD-Related Th1 and Th2 Cytokine Production via CD206

We evaluated the expression of Th1- and Th2-cytokines using ELISA (Figure 2B–E). Similar to our previous study showing that bvPLA2 was effective in decreasing Th2 cytokine expression in an asthma mouse model [22], treatment of WT mice with bvPLA2 following DFE/DNCB exposure significantly decreased the expression of Th1 and Th2 cytokines including IFN-γ, IL-4, IL-6, and IL-10. The cytokine expression was remarkably blocked by bvPLA2 in WT mice treated with DFE/DNCB. However, bvPLA2 treatment of CD206$^{-/-}$ mice induced no significant variation in expression of Th1 or Th2 cytokines following exposure to DFE/DNCB.

2.4. bvPLA2 Is Associated with Treg Induction through the CD206 Receptor

Previous reports have shown that bvPLA2 treatment effectively increased Foxp3-expressing $CD4^{+}CD25^{+}$Tregs in AD-like skin lesions induced by DFE/DNCB [11]. To confirm that bvPLA2 induced Treg through CD206, the expression of Foxp3 in the ear tissues were measured using western blot. In WT mice, bvPLA2 treated group showed a significant increase in Foxp3 protein expression compared to AD group, while in CD206$^{-/-}$ mice, bvPLA2 was not effective in inducing Treg (Figure 2F).

2.5. bvPLA2 Decreased DFE/DNCB-Induced Epidermal and Dermal Thickness and Infiltration of Inflammatory Cells Depending on CD206

To examine histological changes and macrophage infiltration in the ear tissues, ear tissues were stained with Hematoxylin and Eosin (H&E) (Figure 3) and immunohistochemistry (IHC) (Figure 4). Similar to our previous study, the AD group of WT mice exhibited an increase in epidermal thickness, fibrosis in the dermis, and accumulation of inflammatory cells in the dermis. Notably, bvPLA2 treatment induced a dramatic decrease in epidermal and dermal hyperplasia (Figure 3). However, bvPLA2 treatment in CD206$^{-/-}$ mice showed no histological changes compared with AD in CD206$^{-/-}$ mice (Figure 3). In addition, epidermis and dermis thickness were not affected by bvPLA2 treatment in CD206$^{-/-}$ mice (Figure 3B). Similarly, macrophage infiltration of bvPLA2 in WT reduced compared with AD in WT. However, bvPLA2 treatment in CD206$^{-/-}$ mice did not changed macrophage infiltrations in ear tissues (Figure 4).

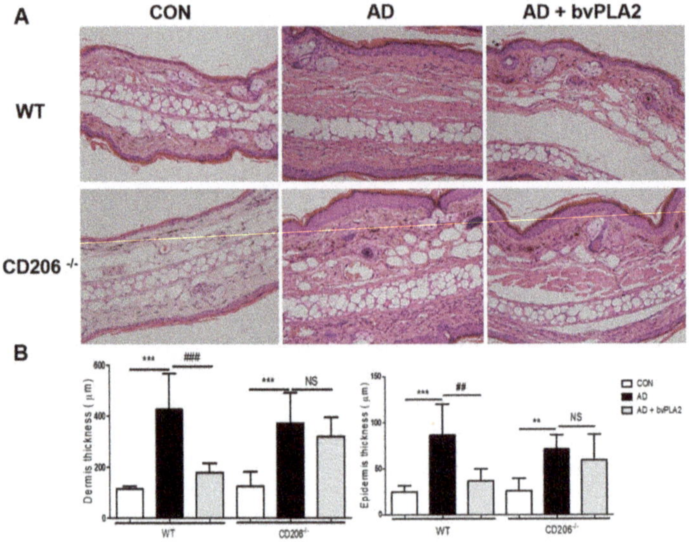

Figure 3. Protective effect of bvPLA2 on the histology of the ear in DFE/DNCB-induced mice via CD206. The ear tissue sections obtained from wild-type (WT) mice (**A**) or CD206-deficient mice (CD206$^{-/-}$ mice) (**A**) were stained with Hematoxylin and Eosin (H&E) (magnification ×200). The thicknesses of dermis and epidermis were quantified based on the H&E stained sections (**B**). CON: normal control, AD: DFE/DNCB treatment, and AD + bvPLA2: DFE/DNCB treatment + bvPLA2 treatment. (** $p < 0.01$, *** $p < 0.001$ versus Control; NS, ## $p < 0.01$, ### $p < 0.001$, versus AD).

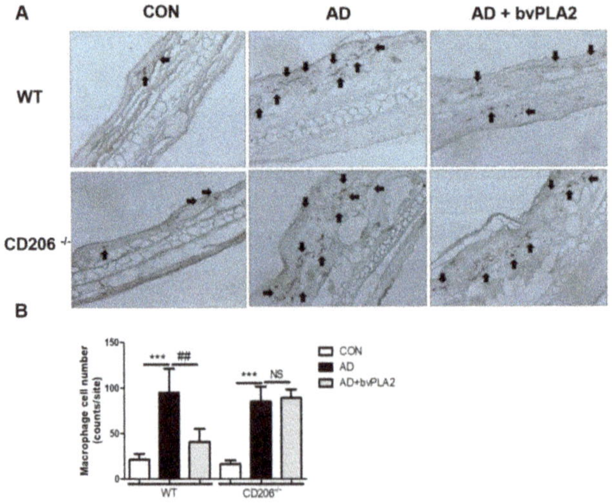

Figure 4. Protective effect of bvPLA2 on macrophage infiltration in DFE/DNCB-induced mice via CD206. The ear tissue sections obtained from wild-type (WT) mice (**A**) or CD206-deficient mice (CD206$^{-/-}$ mice) (**A**) were stained with IHC (magnification ×200). Arrows indicate infiltrated macrophages. The macrophage cell number count graph on the IHC stained sections (**B**). CON: normal control, AD: DFE/DNCB treatment, and AD + bvPLA2: DFE/DNCB treatment + bvPLA2 treatment. (*** $p < 0.001$ versus Control; NS, ## $p < 0.01$ versus AD).

2.6. bvPLA2 Blocked Mast Cell Infiltration in DFE/DNCB-Induced AD via CD206

DFE/DNCB treatment induced apparent mast cell infiltration both in wild-type (WT) and CD206-deficient (CD206$^{-/-}$) mice compared with control group as shown in Figure 5. bvPLA2 treatment of WT mice, but not CD206$^{-/-}$ mice, suppressed the increase in mast cell infiltration, as shown by histological analysis of mast cells with toluidine blue (TB) staining (Figure 5) suggesting that bvPLA2 played an anti-inflammatory role through CD206 receptor. In this study, we observed a strong inhibitory effect of bvPLA2 on mast cell infiltration into AD-induced skin lesions (Figure 5).

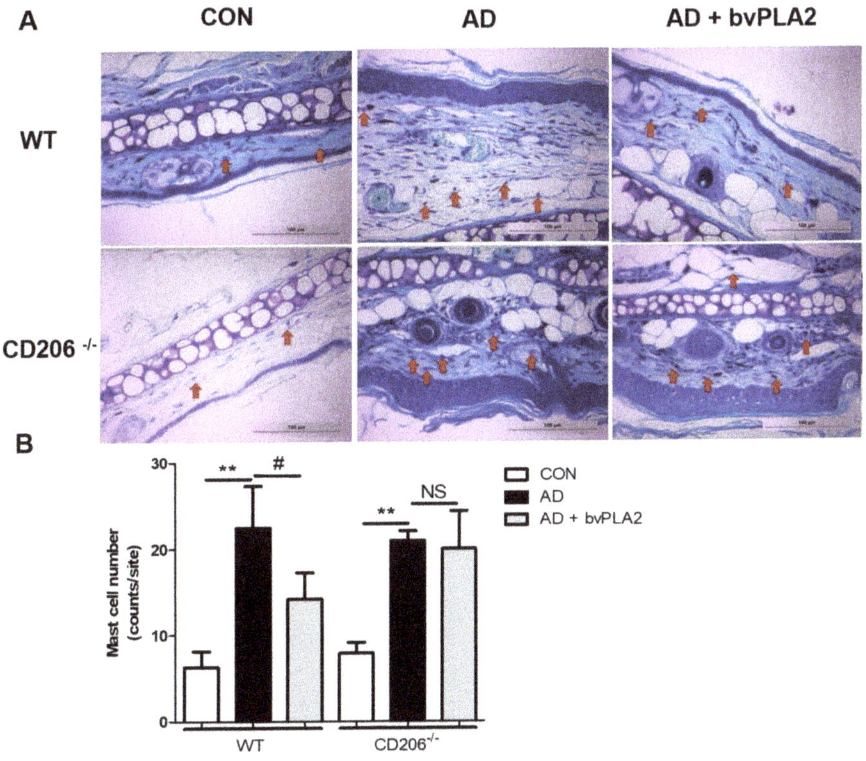

Figure 5. Inhibitory effect of bvPLA2 on DFE/DNCB-induced mast cell infiltration in skin lesions depending on CD206. Mast cell infiltration in skin lesions from wild-type (WT) (**A**) or CD206-deficient (CD206$^{-/-}$) mice (**A**) was evaluated by staining with toluidine blue (TB) (magnification ×200). Mast cell number counts graph (**B**). bvPLA2 was used to treat WT or CD206$^{-/-}$ mice exposed to DFE and DNCB to study whether the suppression of AD-like skin lesions was dependent on CD206. Arrows indicate infiltrated mast cells. CON: normal control, AD: DFE/DNCB treatment, and AD + bvPLA2: DFE/DNCB treatment + bvPLA2 treatment. (** $p < 0.01$ versus Control; NS, # $p < 0.05$, NS versus AD).

3. Discussion

AD is one of the most common inflammatory skin diseases of unknown etiology [23]. The disease is characterized by the induction of Th2 immune response and IgE hypersensitivity, and symptoms of pruritus and eczematous skin lesions [24]. Although it is still controversial, it is generally known that AD is triggered immunologically by food, aeroallergens, and *Staphylococcus aureus* [25]. Since childhood diseases such as allergic rhinitis, allergic dermatitis, and asthma are closely related to AD, AD patients are strongly susceptible to these diseases [26].

Our previous studies showed that bvPLA2 effectively alleviated AD-like skin lesions in DFE/DNCB-treated mice, suggesting the therapeutic effect of bvPLA2 on AD [10]. In this study, we investigated the cellular metabolism underlying this protective effect of bvPLA2 using a murine model of AD. We found that bvPLA2 (1) comprehensively decreased epidermal and dermal thickness; (2) inhibited upregulation of IgE level; (3) suppressed inflammatory cytokines; (4) increased Treg and (5) blocked infiltration of macrophage and mast cells in AD mice. However, these protective effects of bvPLA2 on AD-like skin lesions were completely abolished in DFE/DNCB-treated CD206-deficient mice suggesting that the benefit of bvPLA2 in AD was mediated via interaction with CD206 mannose receptor. PLA2, a lipolytic enzyme that cleaves the sn-2 acyl bond of the phospholipid, is classified into 15 distinct groups and four main types, depending on their structure [27]. We previously compared PLA2s from various sources in Treg inducing effects and found that PLA2 from bee venom was the most effective among others from cow, pig and snake (unpublished data). Thus, we suggest that an increase in Treg is a specific effect of bee venom PLA2.

It is known that the acute phase of AD is mediated via secretion of Th2 cytokines (IL-4, IL-10, and IL-13) and chronic AD via the secretion of Th1 cytokines (IFN-γ, and IL-6) [28,29]. IL-4 plays a key role in Th2 cell differentiation, IgE synthesis, and eosinophil recruitment. IL-13 also induces isotype switching to IgE synthesis. IL-13, which acts in both acute and chronic stages, also induces isotype switching in IgE synthesis [30]. Total IgE overexpression in serum is one of the major characteristics of AD. In our previous study, we demonstrated that bvPLA2 decreased the expression of serum total IgE in ovalbumin-induced asthma mice model [22]. We also observed a strong inhibitory effect of bvPLA2 on Th1 and Th2 cytokine production in DFE/DNCB-induced Balb/c mice. Despite advances in the development of drug targets in AD, many patients still experience skin challenges mostly due to drug-induced side effects [10].

In this study, we demonstrated a strong inhibitory effect of bvPLA2 on AD-like skin lesions in a murine mice model. However, in the CD206-deficient mouse model, $bvPLA_2$ displayed no significant effects on AD-like skin lesions including regulation of serum IgE and inflammatory cytokines. Our histological findings (morphological changes and macrophage infiltration) and mast cell infiltration results also demonstrated the effects in WT mice, but not in CD206-deficient mice, with AD-like skin lesions.

In this study, we investigated the effects of bvPLA2 on AD-like skin lesions using a C57BL/6 mouse model treated with DFE/DNCB. Treatment with bvPLA2 showed significant anti-inflammatory effects on AD-like skin lesions, which were absent in CD206-deficient mice. The study elucidated some of the cellular mechanisms underlying the role of bvPLA2 in the pathogenesis of AD.

4. Materials and Methods

4.1. Animal

Male 7 to 8-week old C57BL/6 and $CD206^{-/-}$ (B6.129P2-^{Mrc1tm1Mnz}/J, Stock No: 007620) mice were purchased from The Jackson Laboratory (Bar Harbor, ME, USA). All mice were housed under pathogen-free conditions with air conditioning on a 12-h light/dark cycle with free access to food and water during the experimental period. All of the experiments were performed in accordance with the Animal Care and Guiding Principles for Experiments Using Animals, and the University of Kyung Hee Animal Care Committee approved this study (KHUASP (SE)-16-073). Date of approval: 30.08.2016

4.2. Reagents

AD was induced using American house dust-mite in the form of freeze-dried crude DFE (Greer Laboratories, Lenoir, NC, USA) and DNCB (Sigma-Aldrich, St. Louis, MO, USA) as a sensitizer. The DFE was dissolved in PBS containing 0.5% Tween 20 and the DNCB was dissolved in acetone and olive oil (AOO) in a 3:1 ratio. The bvPLA2 was obtained from Sigma-Aldrich and dissolved in phosphate buffered saline (PBS).

4.3. Experimental Protocol

All experimental protocols were performed in this study as previously described with some modification [10,22,31]. Briefly, the mice were divided into 3 groups (n = 3–5 per group). For the induction of atopic disease, the skin of inner ear lobe and outer ear lobe were removed using depilation lotion and surgical tape (Nichiban, Tokyo, Japan). A DNCB solution (20 μL 1%) dissolved in AOO was used in each ear, and 20 μL DFE (10 mg/mL) was then coated for the next 4 days. Over a period of 4 weeks, DNCB and DFE were repeatedly applied for AD induction in the ear. The mice were treated with bvPLA2 four times daily for 3 weeks. The ear thickness was measured twice a week using a dial gauge (Kori Seiki MFG, Co., Tokyo, Japan). On day 28, the mice were sacrificed, and the blood samples and the ear tissues were collected for further analysis. A schematic experimental protocol is depicted in below (Figure 6).

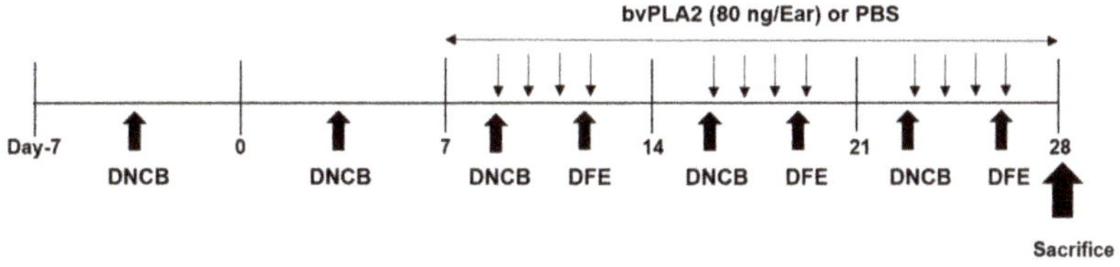

Figure 6. Schedule of experiment. Schematic of experimental design for the induction of atopic dermatitis (AD) in mice. One week after first boosting with 1% 2,4-dinitrochlorobenzene (DNCB), 1% DNCB (20 μL/ear) and *Dermatophagoides farinae* (DFE) (10 mg/mL, 20 μL/ear) was applied alternately to both ears once a week during 4 weeks. The bee venom phospholipase A2 (bvPLA2) (80 ng/ear) was topically applied four times a week for 3 weeks. On day 28, the mice were sacrificed for further analysis.

4.4. Measurement of Serum Immunoglobulin E (IgE)

For serum total IgE determination, the serum from the mice was analyzed. On day 28, the blood samples were collected from mice and centrifuged at 300 × g for 15 min to separate the serum. According to the manufacturer's instructions, the total serum IgE titers were determined using a mouse IgE ELISA Set (BD Pharmingen, San Diego, CA, USA, Sandwich enzyme-linked immunoassay kit). Briefly, a 96-well microtiter plate (Costar, NY, USA) was coated with 100 μL per well of capture antibody (anti-mouse IgE monoclonal antibody) diluted in coating buffer and incubated overnight at 4 °C. After washing three times with PBS containing 0.05% Tween 20 (Sigma-Aldrich), the plates were blocked with an assay diluent (PBS with 10% FBS) for 1 h at room temperature (RT). Next, the serum samples were diluted (1:250) with assay diluent (PBS with 10% FBS) and 100 μL of serum samples were incubated for 2 h at RT. The detection antibody with Streptavidin-Horseradish Peroxidase (SAv-HRP)reagent (secondary peroxidase-labeled biotinylated anti-mouse IgE monoclonal antibody) was incubated in assay diluent for 1 h at RT. Finally, the plates were treated with TMB substrate solution (BD Biosciences, San Jose, CA, USA) for 30 min, and the reaction was stopped via addition of 50 μL of stop solution. The optical density was measured at 450 nm with a microplate reader (SOFT max PRO, version 3.1 software, San Jose, CA, USA).

4.5. Assessment of Th1 and Th2 Cytokine Levels in Mouse Ear

According to the manufacturer's protocols, the levels of IL-4, IL-10, IFN-γ and IL-6 in the ear tissues were determined using a Mouse ELISA kit (BD Pharmingen, San Diego, CA, USA), which is a sandwich Enzyme-Linked Immunosorbent Assay kit). For protein extraction from the ear tissue, each tissue was homogenized in RIPA Buffer (50 mM Tris-HCl, pH 7.4, 150 mM NaCl, 0.25% deoxycholic acid, 1% Nonidet P-40, 1 mM EDTA) in the presence of protease inhbitor cocktail

(Roche Diagnostics, Mannheim, Germany). A 96-well microtiter plate (Costar, NY, USA) was coated with 100 μL per well of the capture antibody (anti-mouse IL-4, IL-10, IFN-γ and IL-6 Monoclonal antibody) diluted in coating buffer and incubated overnight at 4 °C. After washing, the plates were blocked with assay diluent (PBS with 10% FBS) for 1 h at RT. The ear tissue samples were diluted (1:5) with assay diluent (PBS with 10% FBS) and 100 μL of serum samples were incubated for 2 h at RT. The detection antibody containing the SAv-HRP reagent (secondary peroxidase-labeled biotinylated anti-mouse IL-4, IL-10, IFN-γ and IL-6 monoclonal antibody) was incubated in the assay diluent for 1 h at RT. Finally, the plates were treated with tetramethylbenzidine (TMB) substrate solution (BD Biosciences) for 30 min, and the reaction was stopped via the addition of 50 μL of stop solution. The optical density was measured at 450 nm with a microplate reader (SOFT max PRO, version 3.1, CA, USA). The total protein concentrations were determined using a BCA kit (Pierce Biotechnology Inc., Rockford, IL, USA). All results were expressed as relative units by normalization with total protein from each group.

4.6. Western Blot Assay

Proteins in the ear lysates were separated by SDS PAGE. After protein transfer onto membranes (DOGEN), membranes were blocked in 5% BSA solution for 1 h and treated with the 1st-antibodies (Foxp3, Santa Cruz, sc-31738; 1:1000 and β-actin Santa Cruz; 1:1000) for 12 h at 4 °C. The membranes were incubated with the 2nd antibodies for 2 h after washing in Tris Buffered Saline with Tween 20 (TBST) three times. After washing the 2nd antibodies, the blots were developed using Western blot analysis system (AbClon Inc., Seoul, Korea).

4.7. Histological Analysis

For the histopathological examinations, the ear samples were fixed with 4% paraformaldehyde solution overnight at 4 °C. The ear tissues were dehydrated, embedded in paraffin and cut into 4 μm sections using a rotary microtome. The sections were stained with hematoxylin and eosin (H&E; Sigma, St. Louis, MO, USA) to evaluate epidermal hyperplasia in the skin of mice as well as infiltration of immune cells into the dermis. IHC staining was performed to measure the degree of macrophage infiltration. Ear tissue sections were stained for CD11b. In brief, 0.3% H_2O-methanol incubated lung tissues were treated with CD11b antibody (abd-serotec, #mca74g) for 12 h at 4 °C. Subsequently, the tissues were treated following avidin-biotin peroxidase method then, the color was stained by DAB (Zymed Laboratories, South San Francisco, CA, USA).

Toluidine blue (TB) staining was performed to measure the degree of mast cell infiltration. Images of the lung tissue sections were acquired using an Olympus BX51 microscope (Olympus, Tokyo, Japan) and quantified using Image Pro-Plus 5.1 software (Media Cybemetics, Inc., Silver Spring, MD, USA).

4.8. Statistical Analyses

The statistical analyses of the data were conducted using Prism 5 software (GraphPad Software Inc., La Jolla, CA, USA). The data are presented as the means ± SD. The differences between the study groups were determined using one-way ANOVA followed by Newman–Keuls multiple comparisons tests. $p < 0.05$ was considered statistically significant.

Acknowledgments: This work was supported by the Basic Science Research Program of the National Research Foundation of Korea (NRF) funded by the Ministry of Science, ICT, and Future Planning (NRF-2017R1A2B3009574).

Author Contributions: H.B. conceived and designed the experiments; D.S. and W.C. performed the experiments; D.S. analyzed the data; H.B. contributed reagents/materials/analysis tools; D.S. wrote the paper.

Conflicts of Interest: The authors declare no conflict of interest.

References

1. Brown, S.; Reynolds, N.J. Atopic and non-atopic eczema. *BMJ* **2006**, *332*, 584–588. [CrossRef] [PubMed]
2. Berke, R.; Singh, A.; Guralnick, M. Atopic dermatitis: An overview. *Am. Fam. Phys.* **2012**, *86*, 35–42.
3. Silverberg, J.I.; Kleiman, E.; Lev-Tov, H.; Silverberg, N.B.; Durkin, H.G.; Joks, R.; Smith-Norowitz, T.A. Association between obesity and atopic dermatitis in childhood: A case-control study. *J. Allergy Clin. Immunol.* **2011**, *127*, 1180–1186, e1181. [CrossRef] [PubMed]
4. Son, J.H.; Chung, B.Y.; Kim, H.O.; Park, C.W. Clinical Features of Atopic Dermatitis in Adults Are Different according to Onset. *J. Korean Med. Sci.* **2017**, *32*, 1360–1366. [CrossRef] [PubMed]
5. Akdis, C.A.; Boguniewicz, M.; Leung, D.Y.M.; Novak, N.; Simons, F.E.R.; Weidinger, S.; Akdis, M.; Eigenman, P.; Lipozencic, J.K.; Platts-Mills, T.A.E.; et al. Diagnosis and treatment of atopic dermatitis in children and adults: European Academy of Allergology and Clinical Immunology American Academy of Allergy, Asthma and Immunology PRACTALL Consensus Report. *J. Allergy Clin. Immunol.* **2006**, *118*, 152–169. [CrossRef] [PubMed]
6. Novak, N. New insights into the mechanism and management of allergic diseases: Atopic dermatitis. *Allergy* **2009**, *64*, 265–275. [CrossRef] [PubMed]
7. Walling, H.W.; Swick, B.L. Update on the management of chronic eczema: New approaches and emerging treatment options. *Clin. Cosmet. Investig. Dermatol.* **2010**, *3*, 99–117. [CrossRef] [PubMed]
8. Chamlin, S.L.; Kao, J.; Frieden, I.J.; Sheu, M.Y.; Fowler, A.J.; Fluhr, J.W.; Williams, M.L.; Elias, P.M. Ceramide-dominant barrier repair lipids alleviate childhood atopic dermatitis: Changes in barrier function provide a sensitive indicator of disease activity. *J. Am. Acad. Dermatol.* **2002**, *47*, 198–208. [CrossRef] [PubMed]
9. Coondoo, A.; Phiske, M.; Verma, S.; Lahiri, K. Side-effects of topical steroids: A long overdue revisit. *Indian Dermatol. Online J.* **2014**, *5*, 416–425. [CrossRef] [PubMed]
10. Jung, K.H.; Baek, H.; Kang, M.; Kim, N.; Lee, S.Y.; Bae, H. Bee Venom Phospholipase A2 Ameliorates House Dust Mite Extract Induced Atopic Dermatitis Like Skin Lesions in Mice. *Toxins* **2017**, *9*. [CrossRef] [PubMed]
11. Castro, H.J.; Mendez-Lnocencio, J.I.; Omidvar, B.; Omidvar, J.; Santilli, J.; Nielsen, H.S., Jr.; Pavot, A.P.; Richert, J.R.; Bellanti, J.A. A phase I study of the safety of honeybee venom extract as a possible treatment for patients with progressive forms of multiple sclerosis. *Allergy Asthma Proc.* **2005**, *26*, 470–476. [PubMed]
12. Chen, J.; Lariviere, W.R. The nociceptive and anti-nociceptive effects of bee venom injection and therapy: A double-edged sword. *Prog. Neurobiol.* **2010**, *92*, 151–183. [CrossRef] [PubMed]
13. Mirshafiey, A. Venom therapy in multiple sclerosis. *Neuropharmacology* **2007**, *53*, 353–361. [CrossRef] [PubMed]
14. Park, M.H.; Choi, M.S.; Kwak, D.H.; Oh, K.W.; Yoon, D.Y.; Han, S.B.; Song, H.S.; Song, M.J.; Hong, J.T. Anti-cancer effect of bee venom in prostate cancer cells through activation of caspase pathway via inactivation of NF-kappaB. *Prostate* **2011**, *71*, 801–812. [CrossRef] [PubMed]
15. Habermann, E. Bee and wasp venoms. *Science* **1972**, *177*, 314–322. [CrossRef] [PubMed]
16. Raghuraman, H.; Chattopadhyay, A. Melittin: A membrane-active peptide with diverse functions. *Biosci. Rep.* **2007**, *27*, 189–223. [CrossRef] [PubMed]
17. Sobotka, A.K.; Franklin, R.M.; Adkinson, N.F., Jr.; Valentine, M.; Baer, H.; Lichtenstein, L.M. Allergy to insect stings. II. Phospholipase A: The major allergen in honeybee venom. *J. Allergy Clin. Immunol.* **1976**, *57*, 29–40. [CrossRef]
18. Dennis, E.A.; Rhee, S.G.; Billah, M.M.; Hannun, Y.A. Role of phospholipase in generating lipid second messengers in signal transduction. *FASEB J.* **1991**, *5*, 2068–2077. [CrossRef] [PubMed]
19. Granata, F.; Frattini, A.; Loffredo, S.; Del Prete, A.; Sozzani, S.; Marone, G.; Triggiani, M. Signaling events involved in cytokine and chemokine production induced by secretory phospholipase A2 in human lung macrophages. *Eur. J. Immunol.* **2006**, *36*, 1938–1950. [CrossRef] [PubMed]
20. Mukherjee, A.B.; Miele, L.; Pattabiraman, N. Phospholipase A2 enzymes: Regulation and physiological role. *Biochem. Pharmacol.* **1994**, *48*, 1–10. [CrossRef]
21. Kim, H.; Lee, H.; Lee, G.; Jang, H.; Kim, S.S.; Yoon, H.; Kang, G.H.; Hwang, D.S.; Kim, S.K.; Chung, H.S.; et al. Phospholipase A2 inhibits cisplatin-induced acute kidney injury by modulating regulatory T cells by the CD206 mannose receptor. *Kidney Int.* **2015**, *88*, 550–559. [CrossRef] [PubMed]

22. Park, S.; Baek, H.; Jung, K.H.; Lee, G.; Lee, H.; Kang, G.H.; Lee, G.; Bae, H. Bee venom phospholipase A2 suppresses allergic airway inflammation in an ovalbumin-induced asthma model through the induction of regulatory T cells. *Immun. Inflamm. Dis.* **2015**, *3*, 386–397. [CrossRef] [PubMed]
23. Brandt, E.B.; Sivaprasad, U. Th2 Cytokines and Atopic Dermatitis. *J. Clin. Cell. Immunol.* **2011**, 2. [CrossRef] [PubMed]
24. Leung, D.Y.; Boguniewicz, M.; Howell, M.D.; Nomura, I.; Hamid, Q.A. New insights into atopic dermatitis. *J. Clin. Investig.* **2004**, *113*, 651–657. [CrossRef] [PubMed]
25. Leung, D.Y. Infection in atopic dermatitis. *Curr. Opin. Pediatr.* **2003**, *15*, 399–404. [CrossRef] [PubMed]
26. Stone, K.D. Atopic diseases of childhood. *Curr. Opin. Pediatr.* **2002**, *14*, 634–646. [CrossRef] [PubMed]
27. Burke, J.E.; Dennis, E.A. Phospholipase A2 structure/function, mechanism, and signaling. *J. Lipid Res.* **2009**, *50*, S237–S242. [CrossRef] [PubMed]
28. Gittler, J.K.; Shemer, A.; Suarez-Farinas, M.; Fuentes-Duculan, J.; Gulewicz, K.J.; Wang, C.Q.; Mitsui, H.; Cardinale, I.; de Guzman Strong, C.; Krueger, J.G.; et al. Progressive activation of T(H)2/T(H)22 cytokines and selective epidermal proteins characterizes acute and chronic atopic dermatitis. *J. Allergy Clin. Immunol.* **2012**, *130*, 1344–1354. [CrossRef] [PubMed]
29. Spergel, J.M.; Mizoguchi, E.; Oettgen, H.; Bhan, A.K.; Geha, R.S. Roles of TH1 and TH2 cytokines in a murine model of allergic dermatitis. *J. Clin. Investig.* **1999**, *103*, 1103–1111. [CrossRef] [PubMed]
30. Hamid, Q.; Naseer, T.; Minshall, E.M.; Song, Y.L.; Boguniewicz, M.; Leung, D.Y. In vivo expression of IL-12 and IL-13 in atopic dermatitis. *J. Allergy Clin. Immunol.* **1996**, *98*, 225–231. [CrossRef]
31. Choi, J.K.; Kim, S.H. Rutin suppresses atopic dermatitis and allergic contact dermatitis. *Exp. Biol. Med.* **2013**, *238*, 410–417. [CrossRef] [PubMed]

toxins

MDPI

Article

Bee Venom Alleviates Atopic Dermatitis Symptoms through the Upregulation of Decay-Accelerating Factor (DAF/CD55)

Yenny Kim [1], Youn-Woo Lee [1], Hangeun Kim [2,*] and Dae Kyun Chung [1,2,*]

1 Graduate School of Biotechnology, Kyung Hee University, Yongin 17104, Korea; 5033743@naver.com (Y.K.); younwoo1093@naver.com (Y.-W.L.)

2 Skin Biotechnology Center, Kyung Hee University, Yongin 17104, Korea

* Correspondence: hkim93@khu.ac.kr (H.K.); dkchung@khu.ac.kr (D.K.C.); Tel.: +82-31-201-2465 (H.K.); +82-31-201-2465 (D.K.C.); Fax: +82-31-202-3461(D.K.C.)

Received: 20 March 2019; Accepted: 24 April 2019; Published: 26 April 2019

Abstract: Bee venom (BV)—a complex mixture of peptides and toxic proteins including phospholipase A2 and melittin—promotes blood clotting. In this study, we investigated the anti-atopic properties of BV and the mechanism associated with its regulation of the complement system. BV treatment upregulated the mRNA and protein levels of CD55 in THP-1 cells. Further experiments revealed that the phosphorylation of ERK was associated with upregulation of CD55. A complement-dependent cytotoxicity assay and a bacteria-killing assay showed that BV inactivated the complement system through the induction of CD55. The serum levels of C3 convertase (C3C) and Membrane attack complex (MAC) increased, while CD55 decreased in mice with AD-like lesions from DNCB treatment. However, the levels were inverted when the AD-like mice were treated with BV using subcutaneous injection, and we observed that the AD symptoms were alleviated. BV is often used to treat AD but its mechanism has not been elucidated. Here, we suggest that BV alleviates AD through the inactivation of the complement system, especially by the induction of CD55.

Keywords: Bee venom; complement system; decay accelerating factor; atopic dermatitis; complement dependent cytotoxicity; membrane attack complex

Key Contribution: This study highlights the potential mechanism of bee venom in the alleviation of atopic dermatitis, in which CD55-mediated inhibition of the complement system is involved.

1. Introduction

Bee venom (BV) is a secretion from the stinger of the worker bee; it is a complex mixture of proteins they use to protect themselves. Purified BV from honeybees has been used as a traditional medicine by the ancient Egyptians, Chinese, and Greeks [1]. BV contains pharmaceutically active peptides including melittin, apamin, adolapin, and the mast cell degranulating (MCD) peptide; enzymes (e.g., phospholipase A2, PLA2); biologically active amines (e.g., histamine and epinephrine); and nonpeptide components [2]. Melittin, the major component (50% of dry weight) of BV, has anti-inflammatory and anti-arthritis properties, which are driven by the inhibition of nuclear factor kappa B (NF-κB) [3]. Melittin has shown anticancer, antibacterial, and antiviral activities [4]. PLA2 from BV improved atopic dermatitis (AD)-like skin lesions induced by dust mite extract in mice. Topical application of PLA2 suppressed AD symptoms, including ear thickness, histological changes, inflammatory cytokines, and serum IgE concentration [5]. The anti-inflammatory effects of BV are expected to improve skin inflammatory diseases such as AD, but this has not been clearly demonstrated.

AD is the most common allergic skin disease, but its pathogenesis is complex and still not fully understood. Researchers have shown that the complex immune reactions associated with the Th2 response and IgE production affect AD, but Th22, Th17, and Th1 activation also occur in AD [6]. In addition, the complement system seems to affect AD. One study reported that complement components including C3, C4, and C3a are increased in AD patients compared to non-atopic controls [7,8]. Overactivation of the complement system has been shown to cause damage to the dermal-epidermal junction [9], which may aggravate AD. In addition, the anaphylatoxin C5a receptor is increased in AD mice, and treatment with a C5aR antagonist decreased IL-4 and IFN-γ levels in skin tissue, as well as the levels of IL-4, IFN-γ, histamine, and IgE in the serum, indicating that blocking C5aR can inhibit AD [10]. However, the role of complement inhibitory proteins (CIPs) in AD has not been elucidated.

In the current study, we investigated the role of BV in the regulation of the complement system. We examined the upregulation of CD55 in THP-1 cells and the levels and activities of C3 convertase (C3C) and membrane attack complex (MAC) from the serum of AD-like mice after treatment with BV or melittin.

2. Results

2.1. BV Increased CD55 Production in THP-1 Cells

Because complement cascades are regulated by membrane-bound complement regulators including membrane cofactor protein (MCP/CD46), decay-accelerating factor (DAF/CD55), and CD59 [11], we investigated whether THP-1 cells express cell membrane CIP in response to BV. CD46 mRNA was reduced by approximately 50% in the cells after stimulation with BV, while the mRNA levels of CD59 were reduced by a low level of BV (e.g., 0.001–0.1 µg/mL) and slightly upregulated by 1 µg/mL BV (Figure 1A). However, there were no statistically significant differences between the samples. These results suggested that BV-induced variations in CD46 and CD59 may not affect the activation of the complement system and AD symptoms. CD55, unlike other CIPs, showed a bell-shaped curve when cells were stimulated with BV. The highest induction of CD55 was caused by 0.01 µg /mL BV, and higher dosages of BV such as 0.1 and 1 µg/mL reduced CD55 expression compared with untreated cells (Figure 1B). When cells were treated with BV doses of up to 10 µg/mL, cell death was not observed, indicating that the induction and inhibition of CD55 by BV in cells were not associated with cell survivability (Figure 1C). Actually, the viability of THP-1 cells was significantly increased by BV (0.01–1 µg/mL), suggesting that BV may affect cell proliferation. When cells were treated with 0.01 µg/mL BV, CD55 mRNA peaked at 6 h and then declined (Figure 1D). The protein levels of CD55 also increased from 3 h after stimulation and peaked at 6 h; they then declined after 12 h but were still higher than in unstimulated cells (Figure 1E). Our data suggest that BV regulates CD55 expression in immune cells, which may affect complement cascades in the bloodstream.

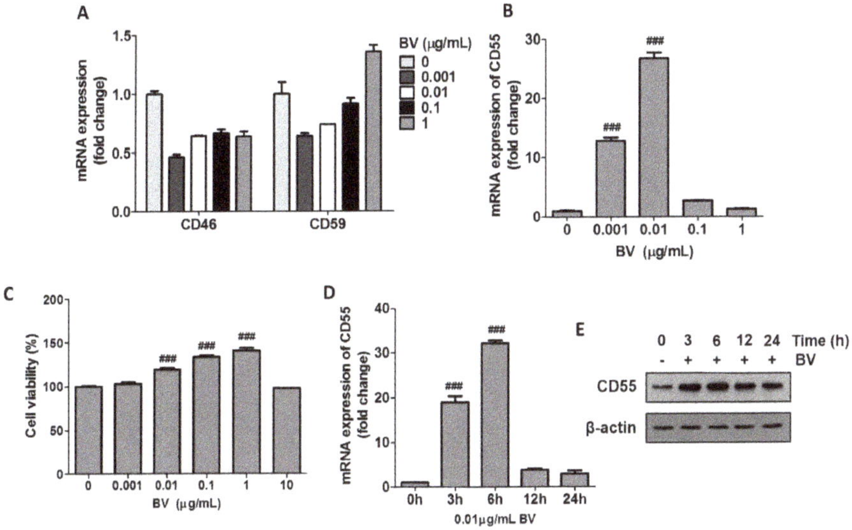

Figure 1. Bee venom (BV) increased CD55 production in THP-1 cells. (**A**) THP-1 cells were treated with the indicated doses of BV for 6 h. The mRNA levels of CD46 and CD59 were examined by qRT-PCR. (**B**) After BV treatment with the indicated doses for 6 h, CD55 mRNA was examined by qRT-PCR. (**C**) Cell viability was examined by WST-1 assay with THP-1 cells treated with the indicated dose of BV for 24 h. (**D**) THP-1 cells were treated with 0.01 μg/mL BV for the indicated time periods. Levels of mRNA were normalized with glyceraldehyde 3-phosphate dehydrogenase (GAPDH). The data are displayed as the mean ± SD of three independent experiments. Statistical analysis was conducted with one-way ANOVA Tukey statistical test. ### $p < 0.001$ compared to 0 μg/mL or 0 h. (**E**) THP-1 cells were treated with 0.01 μg/mL BV for the indicated times and CD55 protein was examined by Western blot. β-actin was used as the internal control. Data are representative of three independent experiments.

2.2. BV Induced CD55 Through the Activation of ERK

Next, we investigated the signaling pathway related to the BV-mediated induction of CD55 in THP-1 cells. After treatment with BV, activation of extracellular signal regulated kinases (ERKs) was observed, while other signaling molecules, including P38 mitogen-activated protein kinase (p38), protein kinase B (Akt), and c-Jun N-terminal kinases (JNK1/2), were not altered (Figure 2A). The densitometry analysis also indicates that the phosphorylation of ERK1/2 was activated by BV (Figure 2B). We examined the phosphorylation of NF-κB subunit p65, but it was not activated by BV treatment (data not shown). When cells were pretreated with the inhibitors for each signal, only the ERK inhibitor reduced CD55 expression in BV-treated cells, indicating that BV increases CD55 expression through the activation of ERKs in THP-1 cells (Figure 2C).

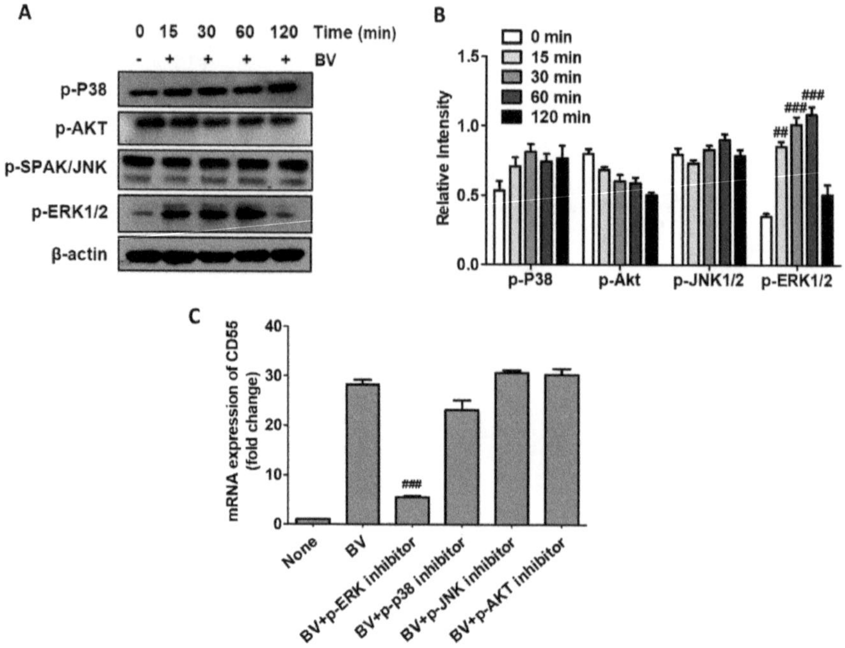

Figure 2. The extracellular signal regulated kinase (ERK) signaling pathway was associated with CD55 induction in THP-1 cells. (**A**) THP-1 cells were treated with 0.01 μg/mL BV for the indicated times and the phosphorylated signaling factors were examined by Western blot. β-actin was used as the internal control. The data are representative of two independent experiments. (**B**) Densitometry analysis for the phosphorylation of p38, Akt, JNK1/2, and ERK1/2. (**C**) THP-1 cells were treated with the signaling inhibitors for ERK, p38, JNK, and Akt for 30 min and then treated with 0.01 μg/mL BV for 6 h. The mRNA levels of CD55 were examined by qRT-PCR. Levels of mRNA were normalized with GAPDH. Data are displayed as the mean ± SD of three independent experiments. Statistical analysis was conducted with one-way ANOVA Tukey statistical test. ## $p < 0.01$; ### $p < 0.001$ compared to none or 0 min.

2.3. *BV Alleviated AD Symptoms*

Since BV has been used as traditional medicine [1], we examined whether BV has a treatment effects of AD. An AD-like condition was induced in mice (irritant contact dermatitis (ICD)) by treatment with 2.5% 2,4-Dinitrochlorobenzene (DNCB) for 14 days. The experimental group was subcutaneously injected with BV and treated with 0.2% DNCB, while the control group was treated only with 0.2% DNCB. Clinical assessment of the ICD mice was performed as described in the Materials and Methods section. The skin condition such as dryness, hemorrhage, excoriation, edema and redness, increased significantly in the DNCB-treated mice (ICD) at 14 days, but it was attenuated in the group that was injected with BV (ICD+BV) at 18 days and the skin almost completely recovered at 27 days (Figure 3A). The clinical skin score in the BV-injected group also significantly decreased at 12 days as compared to the control group (Figure 3B).

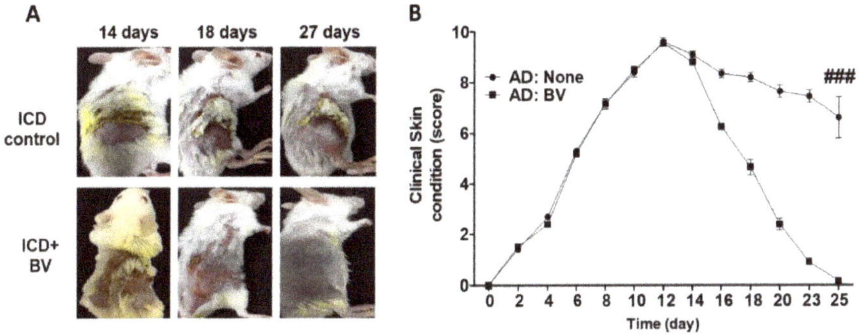

Figure 3. BV alleviated AD symptoms. An AD-like condition was induced in mice with 2.5% (*w*/*v*) DNCB for 3 days followed by 1% (*w*/*v*) DNCB at 3-day intervals. After 14 days of treatment with DNCB, one group ($n = 4$) was subcutaneously injected with 0.3 mg/kg BV and the other group ($n = 4$) was injected with PBS. Both groups were treated with 0.2% (*w*/*v*) DNCB to prevent spontaneous remission. (**A**) A photograph of the mice with the best effect among the experimental groups is shown. A representative mouse from each group is shown. (**B**) The severity of AD was quantified by individually scoring the symptoms (skin dryness, hemorrhage, edema, redness, and excoriation). Data are displayed as the mean ± SD of technical repeats of one representative experiment. We performed two independent experiments. Statistical analysis was conducted with two-way ANOVA Tukey statistical test. ### $p < 0.001$ compared to the none group.

2.4. BV Inactivated Complement System in AD-Like Mice

We examined whether BV affects complement in an AD-like mouse model, developed by treatment with 2.5% DNCB. First, we measured the levels of C3C and MAC in serum using sandwich ELISA kits. The serum levels of C3C and MAC in normal mice significantly decreased after BV treatment. In mice that developed ICD, both levels were significantly higher compared with the untreated mice ('none' in Figure 4A–D). When comparing the ICD mice groups, both C3C and MAC significantly decreased after BV injection compared with PBS injection (Figure 4A,B). On the other hand, the secreted CD55 serum levels decreased in ICD mice, but significantly increased after BV injection (Figure 4C). These data suggest that BV inhibits the complement system. The actual activity of complement was examined in mouse serum. The bactericidal activity decreased with BV in untreated mice, while it increased in ICD mice compared to the untreated mice. The bactericidal activity in ICD mice decreased in the BV-injected ICD mice (Figure 4D), indicating that MAC activity was increased in AD, which is consistent with previous studies [7,8], and it was decreased by BV injection. Since an abnormal increase of MAC damages skin tissues, it is necessary to maintain an appropriate level of MAC. Thus, BV may be a good candidate to maintain MAC homeostasis. Next, we examined the role of CD55 in complement-mediated tissue damage, using a complement-dependent cytotoxicity assay (Figure 4E). Normal human serum (NHS, 1:20 dilution) significantly decreased HaCaT cell viability, indicating that complement induced the cell death of keratinocytes. However, when NHS was added to the BV-treated cells, the viability increased in a BV dose-dependent manner. Increased viability, however, was not shown when cells were treated with anti-CD55 neutralization antibody prior to BV treatment, indicating that increased CD55 inhibits MAC activity, which may affect the alleviation of AD symptoms.

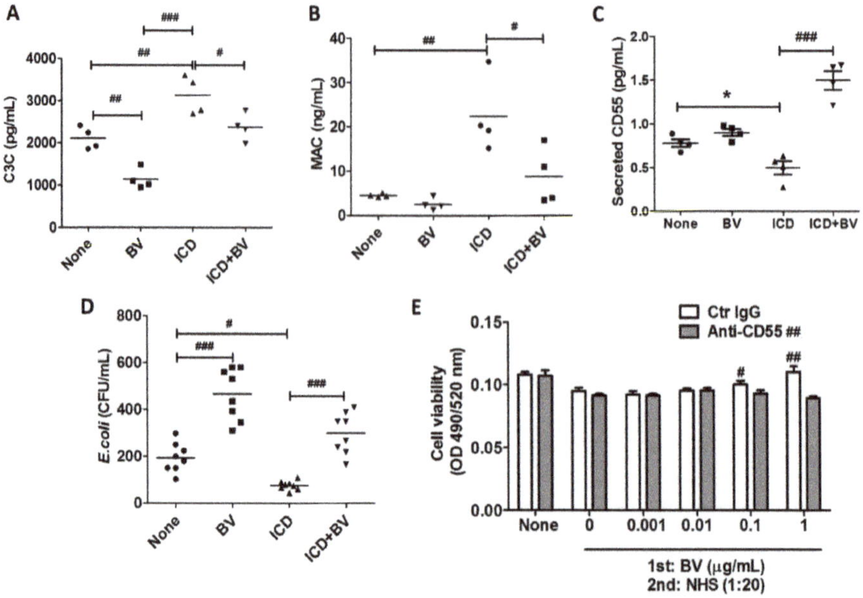

Figure 4. BV injection reduced the activities of C3C and membrane attack complex (MAC) through the induction of CD55. Blood samples were taken from AD-like mice ($n = 4$, each group) after 0.3 mg/kg BV injection, and serum was isolated. (**A** and **B**) Serum levels of C3C and MAC were examined by commercial ELISA kits. (**C**) The serum CD55 levels were examined by indirect ELISA using an anti-CD55 antibody. (**D**) The bactericidal activity of complement was examined with sera isolated from each group. (**E**) A complement-dependent cytotoxicity assay was performed with normal human serum (NHS) in HaCaT cells treated with 0.01 μg/mL BV for 24 h in the presence or absence of anti-CD55 antibody. Cell viability was examined by Calcein-AM (a fluorogenic, cell-permeant fluorescent probe) assay. NHS was isolated from blood supplied by the Blood Center of the Korean Red Cross. Data are displayed as the mean ± SD of four independent experiments. Statistical analysis was conducted with an unpaired two-tailed *t* test (**C**). * $p < 0.05$ compared to none, one-way ANOVA Tukey statistical test, # $p < 0.05$; ## $p < 0.01$; ### $p < 0.001$ compared between indicated groups (A to D) or compared to 0 μg/mL (E) or Two-way ANOVA, ## $p < 0.01$ compared between Control IgG and Anti-CD55.

2.5. Melittin Alleviated AD Symptoms Through the Regulation of Complement

When ICD mice were subcutaneously injected with melittin (0.15 mg/kg), AD symptoms were alleviated as seen in BV-injected mice. Melittin is a major component of BV and contributes anti-inflammatory and anticancer effects [2,4]. Skin condition was improved in melittin-injected ICD mice as well as in BV-injected ICD mice (Figure 5A). To examine whether melittin affects complement, C3C levels were examined in serum from ICD mice that were injected with melittin. As shown in Figure 5B, serum C3C levels were significantly lower compared with untreated ICD mice. The same reduction was shown in BV-injected mice. MAC levels in ICD mouse serum were also significantly decreased in melittin-injected mice compared to the ICD control mice, although they were slightly higher than in the BV-injected mice (Figure 5C). The secreted CD55 levels in melittin-injected ICD mice were significantly higher than in the control ICD mice, suggesting that the complement system was reduced (Figure 5D). These data suggest that BV, especially melittin, can alleviate AD via controlling complement.

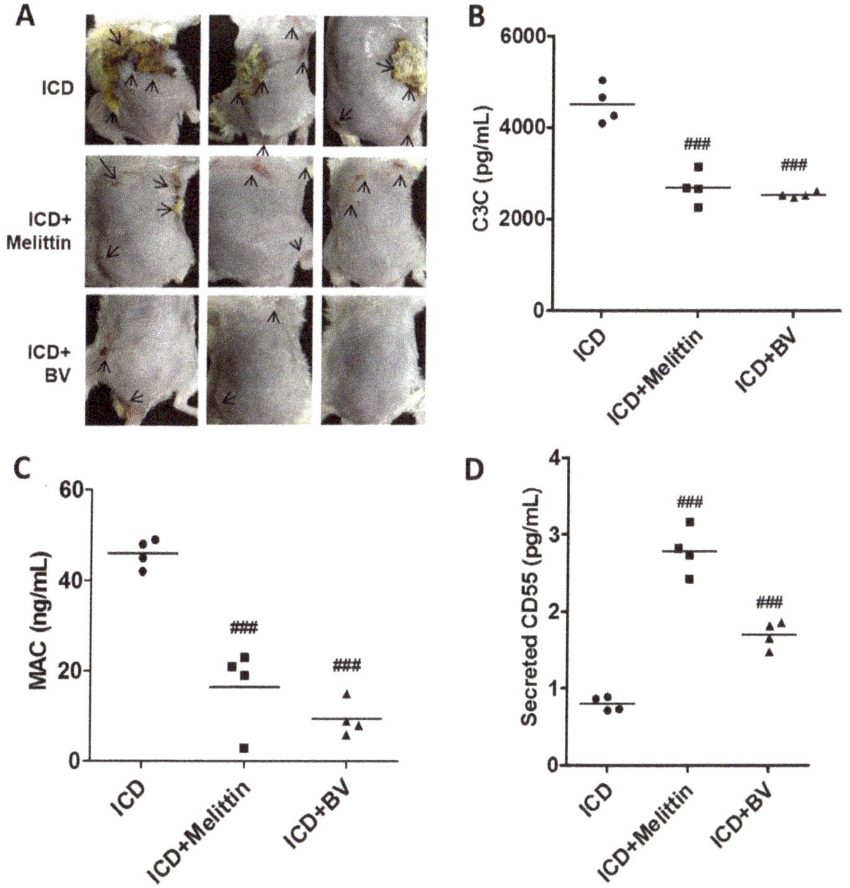

Figure 5. Melittin plays an important role in the BV-mediated alleviation of AD symptoms. An AD-like mouse model (n = 4, each group) was generated with 2.5% DNCB, and mice were subcutaneously injected with 0.3 mg/kg BV or 0.15 mg/kg melittin. (**A**) A photograph of the AD-like mice is shown after 14 daysof treatment with BV or melittin. Serum C3C (**B**) and MAC (**C**) levels were examined by commercial ELISA kits. (**D**) Secreted CD55 was examined by indirect ELISA with the sera isolated from BV- or melittin-treated mice and normal mice. Data are displayed as the mean ± SD of three independent experiments. Statistical analysis was conducted with one-way ANOVA Tukey statistical test. ### $p < 0.001$ compared to ICD. Arrows indicate AD-like skin condition (i.e., hemorrhage and excoriation, and edema and redness).

3. Discussion

Purified BV has anti-inflammatory and anticancer effects [12]. It also reduces AD symptoms, lowering serum IgE levels and dorsal skin thickness [13]. AD is a chronic skin inflammatory disease characterized by eczematous, dry, and chapped skin. AD is caused by the invasion of inflammatory immune cells including mast cells, eosinophils, monocytes/macrophages, and T lymphocytes into the skin barrier. The circulating eosinophils and serum IgE levels are increased in AD, which is associated with interleukin (IL)-4, IL-5, and IL-13 produced by the Th2 cells in most patients [14–18]. In addition, complement appears to aggravate AD. Patients with AD show increases in complement components including C3, C4, and C3a [7,8]. Because activation of complement generates a high level of MAC, MAC can worsen AD lesions, aggravating AD symptoms.

The complement system is an essential part of the immune system. It has long been described as belonging to the innate immune system, but recently a number of papers have demonstrated that it also contributes to adaptive immunity by regulating antigen-presenting cells [19]. As an inducer for innate immunity, MAC, a final product of complement cascades, kills invading bacteria, and anaphylatoxins, such as C3a and C5a, activate inflammatory responses. However, excessive activation of the complement system can induce tissue damage and inflammation, which aggravates symptoms in patients with AD [20,21]. Thus, it is necessary to reduce excessive activation of the complement system and restore homeostasis to defend against bacterial infection. Toxins isolated from insects and snakes seem to have beneficial effects against inflammatory diseases [22], but they are also toxic and potentially life-threatening for mammals [23].

BV has beneficial effects on idiopathic Parkinson's disease and oxaliplatin-induced neuropathic cold allodynia, and it is helpful in reducing glutamate-induced cell toxicity in neurodegenerative diseases [24–27]. BV components, especially melittin, inhibit complement cleavage and release bradykinin. These mechanisms are associated with coagulation, thrombolysis, hemolysis, and smooth muscle tone [28]. Recently, Shaldoum et al. have reported that BV may affect complement system. According to the results, in patients having various diseases, such as rheumatoid arthritis, back pain, diabetes mallets, arthritis, gastritis, sebaceous cyst, osteoarthritis, and hepatitis c virus, all abnormal levels of complement C3 returned to normal values, while abnormal C4 levels did not change when patients were exposure to natural BV [29]. In this study we have established the following mechanisms of BV in the alleviation of AD; (i) BV induced CD55 production through the activation of ERK1/2 pathways; (ii) increased CD55 downregulated formation of C3C and MAC; (iii) decreased MAC activity resulted in the alleviation of AD symptoms. BV may be a promising drug to treat AD, because it inhibits complement by inducing CD55. Among CIPs, only CD55 dramatically increased in BV-treated THP-1 cells and in serum from BV-injected mice. Although the mechanism for CD55-dependent inactivation of C3C is complex [30], the expected result is a reduction in MAC formation. The dose-dependent variation of CD55 in Figure 1B suggests that BV has different activities in mammalian cells. BV causes the activation of the immune system, which could increase the symptoms of atherosclerosis, diabetes-related endothelial damage, cancer, and autoimmune diseases [31]. However, it also has anti-inflammatory properties and is used in the treatment of liver fibrosis, atherosclerosis and other skin diseases [32].

We found that CD55 did not completely inhibit C3C in BV-injected mice. As shown in Figure 3, the increased C3C in BV-injected mice was restored to a normal state. The MAC level and activity also recovered to a normal state. These data suggest that BV does not inhibit the complement system completely, but can still protect against invasion of bacteria and support the repair of damaged tissues. However, the appropriate dosage should be considered when applied to mammals. The complement inhibitory effects of BV and melittin suggest that they can be used to treat complement-mediated diseases such as ischemia/reperfusion injury and autoimmune disorders, which are caused by excessive complement activation [21].

In conclusion, BV, especially melittin, appears to alleviate AD. This phenomenon appears to be mediated by ERK pathway activation leading to the induction of CD55 in BV-treated cells. CD55-mediated inhibition of complement alleviates AD symptoms, which can otherwise be aggravated by inflammation and MAC. Thus, BV can be considered as a therapeutic reagent to treat AD as well as inflammatory diseases.

4. Materials and Methods

4.1. Cell Culture

HaCaT and THP-1 cells were maintained with Dulbecco's modified Eagle's medium (DMEM, Welgene, Gyeongsangbuk-do, Korea) and RPMI 1640 (Welgene), respectively, supplemented with 10% (*v*/*v*) heat-inactivated fetal bovine serum (FBS, Welgene) and 1% (*v*/*v*) penicillin–streptomycin (P/S, Welgene). These cells were incubated at 37 °C in a CO_2 atmosphere.

4.2. Drugs

Bee venom was purchased from Guju Pharmaceutical Company, Ltd. (Gyeonggi-do, Korea). Bee venom was dissolved in Dulbecco's phosphate-buffered saline (DPBS).

4.3. Real-Time PCR

For real-time PCR, total RNA was extracted using TRIzol reagent (Takara Bio Inc., Shiga, Japan), following the manufacturer's instructions. cDNA was synthesized using an iScript cDNA synthesis kit (Bio-Rad, San Diego, CA, USA), following the manufacturer's instructions. CFX Connect™ Real-Time PCR Detection System (Bio-Rad) and SYBR Ex TaqTM^{II} (Takara) were used for real time PCR. The following forward and reverse primers were used; 5′-GTGAGGAGCCACCAACATTT-3′ and 5′-GCGG TCATCTGAGACAGGT-3′ for CD46; 5′-CAGCACCACCACAAATTGAC-3′ and 5′-CTGAACTGTTGG TGGGACCT-3′ for CD55; 5′-CCGCTTGAGGGAAAATGAG-3′ and 5′-CAGAAATGGAGTCACCAG CA-3′ for CD59; and 5′-AAGGTCGGAGTCAACGGATT-3′ and 5′-GCAGTGAGGGTCTCTC TCCT-3′ for GAPDH. The target gene expression was normalized with glyceraldehyde-3-phosphate dehydrogenase (GAPDH). The contamination of Mycoplasma was examined with EZ-PCR™ Mycoplasma Detection Kit (Catalog # SKU:20-700-20; Biological Industires, Cromwell, CT, USA) and we found no contamination.

4.4. Western Blot

THP-1 cells treated with BV (0.01 μg/mL) were lysed with Laemmli buffer and boiled for 5 min at 100 °C. Proteins were separated by 10% (*w*/*v*) or 12% (*w*/*v*) SDS-PAGE in a Glycine/Tris/SDS buffer and transferred onto polyvinylidene fluoride membranes for 2 h at 100 V. The membranes were blocked with 5% (*w*/*v*) bovine serum albumin in TBST (20 mM Tris-HCl, 150 mM NaCl, 0.05% (*v*/*v*) Tween 20) for 2 h at room temperature (RT) and washed three times with TBST. The membrane was incubated with the primary antibodies such as anti-phospho p38 (#9211), anti-phospho Akt (#9271), anti-phospho SPAK/JNK (#9251), anti-phospho ERK1/2 (#9101) (Those were purchased from Cell Signaling Technology Inc., Danvers, MA, USA), and β-actin (SC47778, Santa Cruz Biotechnology, Inc., Dallas, TX, USA) diluted in TBST (1:1000) for 2 h at RT and then washed with TBST three times. Next, the membrane was incubated with secondary HRP-conjugated anti-rabbit or anti-mouse antibody (diluted to 1:2000 in TBST) for 2 h at RT. After washing three times with TBST, the bands were detected by ECL reagent. β-actin was used as the internal loading control.

4.5. Animals

Male BALB/c mice (7 weeks old) were purchased from Nara Bio (Gyeonggi, Korea). They were kept in individual cages at 24 ± 2°C and 50 ± 10% moisture, and fed nutritionally balanced rodent food (Central Lab Animal Inc., Seoul, Korea) and sterilized water. The mice were cared for and used in accordance with the guidelines of the Animal Ethics Committee of Kyung Hee University (KHU14-021, the date of the approval: 26 October 2015).

4.6. Development of Irritant Contact Dermatitis (ICD) Mouse

Mice were shaved on the dorsal flank and back. They were left for 24 h to heal any abrasions that might have been caused by shaving. Olive oil and acetone were mixed at a ratio of 1:3, and then DNCB was added to make concentrations of 2.5% (2,4-dinitrochlorobenzene; Sigma-Aldrich Co., St. Louis, MO, USA), 1.0%, and 0.2%. Mice were topically treated with the 2.5% (*w*/*v*) DNCB mixture (200 μL). After 3 days exposure, mice were treated with 150 μL 1.0% (*w*/*v*) DNCB at 3-day intervals until 14 days, and then were treated with the 0.2% (*w*/*v*) DNCB mixture (100 μL).

4.7. Treatment of ICD Mice Using BV

The mice were randomly divided into four experimental groups of four animals each as follows (Figure 6): untreated normal group (None), BV-treated normal group (BV), untreated ICD group (ICD), and BV-treated ICD group (ICD+BV). The BV and ICD+BV groups were subcutaneously injected with 0.3 mg/kg BV at 2-day intervals, and None and ICD groups were subcutaneously injected with 60 μL PBS. For the analysis of serum complement components, blood samples were collected at day 25 before sacrifice. The blood samples were maintained at room temperature for 20 min. Then, the serum was separated by centrifugation at 12,000 rpm for 20 min. The serum samples were used to examine the MAC and C3C quantities and the bacterial killing assay.

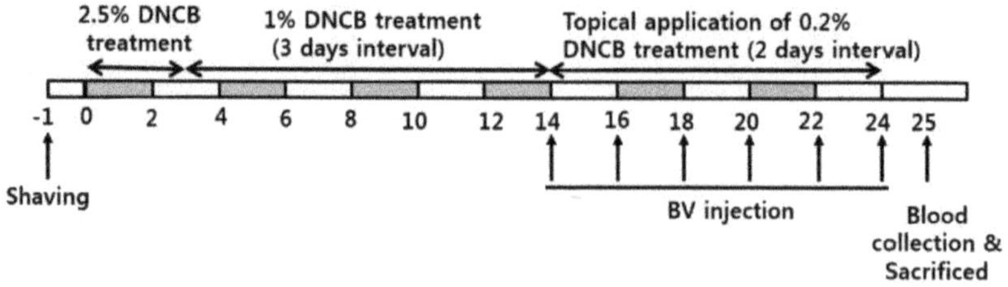

Figure 6. Time schedule for ICD generation, BV injection, and blood collection.

4.8. Clinical Skin Score.

Mice in each experimental group were photographed using a digital camera to analyze AD symptoms and the clinical appearance of the skin. AD symptoms were evaluated by scoring scaling and dryness, hemorrhage and excoriation, and edema and redness. The sum of the individual symptom scores was calculated (0 = normal, 1 = mild, 2 = moderate, 3 = severe). The total score for each animal ranged from 0 to 9 points.

4.9. Enzyme-Linked Immunosorbent Assay (ELISA)

The quantities of MAC and C3C were measured in mouse serum. The MAC and C3C ELISA kits were purchased from MyBioSource (San Diego, CA, USA). The assay was performed according to the manufacturer's instructions. The absorbance was measured at the wavelength of 450 nm using an ELISA reader. The concentrations of MAC and C3C were calculated using the standard included in the kits. For the detection of soluble CD55 from mouse serum, rabbit anti-CD55 (Santa Cruz, CA, USA) was coated on an ELISA plate (Corning Costar flat-bottom high-binding EIA/RIA 3690 plate) in PBS (pH 7.4; 0.4 μg/well) at 37 °C for 2 h. The plate was blocked with 2% BSA in PBS. Mouse serum was added in triplicate to experimental or control wells. After incubation at 4 °C overnight, wells were washed and bound CD55 was detected by serial addition of a biotin-labeled secondary antibody, avidin–horseradish peroxidase (HRP) conjugate (Pierce Immunopure streptavidin-HRP conjugate), and peroxidase substrate (Pierce Chemicals, IL). The absorbance at 450 nm was read and presented after subtraction of reagent-control values reacting against BSA-coated negative controls.

4.10. Complement-Dependent Cytotoxicity Assay

HaCaT cells were seeded on a 96-well plate with DMEM supplemented with 10% FBS and P/S. The cells were pretreated with BV together with a control IgG or an anti-CD55 neutralization antibody (Santa Cruz Biotechnology, SC51733, Dallas, TX, USA) and washed with DPBS. Normal human serum (NHS) was diluted 1:20 with DMEM without supplement, and the cells were incubated for 6 h. The cells were washed with DPBS and the viability of the HaCaT cells was measured with the Calcein AM cell viability assay system (EMD Millipore, #206700, Burlington, MA, USA). Briefly, 2 μM Calcein AM (final concentration) was added to each well and incubated for 15 min at 37 °C under CO_2. Fluorescence was examined at at 490 nm excitation and 520 nm emission wavelengths.

4.11. Bactericidal Assay

Escherichia coli was cultured in LB overnight. Then, the bacteria were washed and diluted with DPBS. *Escherichia coli* at 1×10^4 cells was cultured with mouse serum (1:50) at 37 °C for 60 min. The incubated bacteria were washed with DPBS and spread on an LB plate. After overnight culturing, CFU were counted.

4.12. Statistical Analysis

All the experiments were repeated at least three times. The data shown are representative results of the means ± SD of triplicate experiments. Statistical analyses were conducted with an unpaired two-tailed *t*-test, then one-way ANOVA, followed by Tukey's honestly significant difference (HSD) post hoc test, or two-way ANOVA. Prism 5 software was used for the analysis (Graphpad software Inc., Prism 5 (Version 5.01, San diego, CA, USA, 2007). $p < 0.05$ was considered significant. The * represents the *t*-tests while the # represents the ANOVA in the figures.

Author Contributions: H.K. conceived and designed the experiments; Y.K. and Y.-W.L. performed the experiments; D.K.C. analyzed the data; H.K. and Y.K. wrote the paper.

Funding: This research was supported by a grant from the regional innovation center program of the Ministry of Trade, Industry and Energy at the Skin Biotechnology Center of Kyung Hee University, Korea.

Acknowledgments: We thank members of Industrial Microbiology Laboratory (Kyung Hee University) for helpful discussions concerning this work.

Conflicts of Interest: The authors declare no conflicts of interest.

References

1. Hoyt, J. Literature Review of Bee Venom Therapy: Mechanisms of Action and Selected Therapeutic Uses. *Orient. Med. J.* **2015**, *23*, 6–9.
2. Lariviere, W.R.; Melzack, R. The bee venom test: A new tonic-pain test. *Pain* **1996**, *66*, 271–277. [CrossRef]
3. Lee, G.; Bae, H. Anti-Inflammatory Applications of Melittin, a Major Component of Bee Venom: Detailed Mechanism of Action and Adverse Effects. *Molecules* **2016**, *21*, E616. [CrossRef]
4. Raghuraman, H.; Chattopadhyay, A. Melittin: A membrane-active peptide with diverse functions. *Biosci. Rep.* **2007**, *27*, 189–223.
5. Jung, K.H.; Baek, H.; Kang, M.; Kim, N.; Lee, S.Y.; Bae, H. Bee Venom Phospholipase A2 Ameliorates House Dust Mite Extract Induced Atopic Dermatitis Like Skin Lesions in Mice. *Toxins* **2017**, *9*, 68. [CrossRef]
6. Gittler, J.K.; Shemer, A.; Suárez-Fariñas, M.; Fuentes-Duculan, J.; Gulewicz, K.J.; Wang, C.Q.F.; Mitsui, H.; Cardinale, I.; de Guzman Strong, C.; Krueger, J.G.; et al. Progressive activation of Th2/Th22 cytokines and selective epidermal proteins characterizes acute and chronic atopic dermatitis. *J. Allergy Clin. Immunol.* **2012**, *130*, 1344–1354. [CrossRef]
7. Kapp, A.; Wokalek, H.; schöpf, E. Involvement of complement in psoriasis and atopic dermatitis-measurement of C3a and C5a, C3, C4 and C1 inactivator. *Arch. Dermatol. Res.* **1985**, *277*, 359–361. [CrossRef]
8. Kapp, A.; Schöpf, E. Involvement of complement in atopic dermatitis. *Acta. Derm. Venereol. Suppl. (Stockh)* **1985**, *114*, 152–154.
9. Zhuang, Y.; Lyga, J. Inflammaging in skin and other tissues—the roles of complement system and macrophage. *Inflamm. Allergy Drug Targets* **2014**, *13*, 153–161. [CrossRef]
10. Dang, L.; He, L.; Wang, Y.; Xiong, J.; Bai, B.; Li, Y. Role of the complement anaphylatoxin C5a-receptor pathway in atopic dermatitis in mice. *Mol. Med. Rep.* **2015**, *11*, 4183–4189. [CrossRef]
11. Noris, M.; Remuzzi, G. Overview of Complement Activation and Regulation. *Semin. Nephrol.* **2013**, *33*, 479–492. [CrossRef] [PubMed]
12. Son, D.J.; Lee, J.W.; Lee, Y.H.; Song, H.S.; Lee, C.K.; Hong, J.T. Therapeutic application of anti-arthritis, pain-releasing, and anti-cancer effects of bee venom and its constituent compounds. *Parmacol. Ther.* **2007**, *115*, 246–270. [CrossRef] [PubMed]
13. Gu, H.; Kim, W.H.; An, H.J.; Kim, J.Y.; Gwon, M.G.; Han, S.M.; Leem, J.; Park, K.K. Therapeutic effects of bee venom on experimental atopic dermatitis. *Mol. Med. Rep.* **2018**, *18*, 3711–3718. [CrossRef] [PubMed]

14. Leung, D.Y.; Guttman-Yassky, E. Deciphering the complexities of atopic dermatitis: Shifting paradigms in treatment approaches. *J. Allergy Clin. Immunol.* **2014**, *134*, 769–779. [CrossRef]
15. Schlapbach, C.; Simon, D. Update on skin allergy. *Allergy* **2014**, *69*, 1571–1581. [CrossRef] [PubMed]
16. Lim, S.J.; Kim, M.; Randy, A.; Nam, E.J.; Nho, C.W. Effects of Hovenia dulcis Thunb. extract and methyl vanillate on atopic dermatitis-like skin lesions and TNF-α/IFN-γ-induced chemokines production in HaCaT cells. *J. Pharm. Pharmacol.* **2016**, *68*, 1465–1479. [CrossRef] [PubMed]
17. Galli, S.J.; Tsai, M.; Piliponsky, A.M. The development of allergic inflammation. *Nature* **2008**, *454*, 445–454. [CrossRef] [PubMed]
18. Owen, C.E. Immunoglobulin E: Role in asthma and allergic disease: Lessons from the clinic. *Pharmacol. Ther.* **2007**, *113*, 121–133. [CrossRef]
19. Killick, J.; Morisse, G.; Sieger, D.; Astier, A.L. Complement as a regulator of adaptive immunity. *Semin. Immunopathol.* **2018**, *40*, 37–48. [CrossRef]
20. Tsokos, G.C.; Fleming, S.D. Autoimmunity, complement activation, tissue injury and reciprocal effects. *Curr. Dir. Autoimmun.* **2004**, *7*, 149–164.
21. Markiewski, M.M.; Lambris, J.D. The Role of Complement in Inflammatory Diseases From Behind the Scenes into the Spotlight. *Am. J. Pathol.* **2007**, *171*, 715–727. [CrossRef]
22. Sales, T.A.; Marcussi, S.; da Cunha, E.F.F.; Kuca, K.; Ramalho, T.C. Can Inhibitors of Snake Venom Phospholipases A_2Lead to New Insights into Anti-Inflammatory Therapy in Humans? A Theoretical Study. *Toxins* **2017**, *9*, 341. [CrossRef]
23. Harris, J.B.; Scott-Davey, T. Secreted Phospholipases A2 of Snake Venoms: Effects on the Peripheral Neuromuscular System with Comments on the Role of Phospholipases A2 in Disorders of the CNS and Their Uses in Industry. *Toxins* **2013**, *5*, 2533–2571. [CrossRef]
24. Doo, K.H.; Lee, J.H.; Cho, S.Y.; Jung, W.S.; Moon, S.K.; Park, J.M.; Ko, C.N.; Kim, H.; Park, H.J.; Park, S.U. A Prospective Open-Label Study of Combined Treatment for Idiopathic Parkinson's Disease Using Acupuncture and Bee Venom Acupuncture as an Adjunctive Treatment. *J. Altern. Complement Med.* **2015**, *21*, 598–603. [CrossRef]
25. Lee, J.H.; Li, D.X.; Yoon, H.; Go, D.; Quan, F.S.; Min, B.I.; Kim, S.K. Serotonergic mechanism of the relieving effect of bee venom acupuncture on oxaliplatin-induced neuropathic cold allodynia in rats. *BMC Complement Altern. Med.* **2014**, *14*, 471. [CrossRef]
26. Lim, B.S.; Moon, H.J.; Li, D.X.; Gil, M.; Min, J.K.; Lee, G.; Bae, H.; Kim, S.K.; Min, B.I. Effect of bee venom acupuncture on oxaliplatin-induced cold allodynia in rats. *Evid. Based Complement Alternat. Med.* **2013**, *2013*, 369324. [CrossRef]
27. Lee, S.M.; Yang, E.J.; Choi, S.M.; Kim, S.H.; Baek, M.G.; Jiang, J.H. Effects of bee venom on glutamate-induced toxicity in neuronal and glial cells. *Evid. Based Complement Alternat. Med.* **2012**, *2012*, 368196. [CrossRef]
28. Mingomataj, E.C.; Bakiri, A.H. Episodic hemorrhage during honeybee venom anaphylaxis: Potential mechanisms. *J. Investig. Allergol. Clin. Immunol.* **2012**, *22*, 237–244.
29. Shaldoum, F.M.; Hassan, M.I.; Hassan, M.S. Natural Honey Bee venom Manipulates Human Immune Response. *Egypt J. Hosp. Med.* **2018**, *72*, 4252–4258.
30. Janeway, C.A., Jr.; Travers, P.; Walport, M.; Shlomchik, M.J. *Immunobiology: The Immune System in Health and Disease*, 5th ed.; Garland Science: New York, NY, USA, 2001; Available online: https://www.ncbi.nlm.nih.gov/books/NBK27100/ (accessed on 5 January 2019).
31. Park, J.H.; Yim, B.K.; Lee, J.H.; Lee, S.; Kim, T.H. Risk associated with bee venom therapy: A systematic review and meta-analysis. *PLoS ONE* **2015**, *10*, e0126971.
32. Lee, W.R.; Pak, S.C.; Park, K.K. The protective effect of bee venom on fibrosis causing inflammatory diseases. *Toxins* **2015**, *7*, 4758–4772. [CrossRef] [PubMed]

toxins

MDPI

Article

First Characterization of The Venom from *Apis mellifera syriaca*, A Honeybee from The Middle East Region

Jacinthe Frangieh [1], Yahya Salma [1,2], Katia Haddad [2], Cesar Mattei [3], Christian Legros [3], Ziad Fajloun [1,2,*] and Dany El Obeid [4,*]

1 Laboratory of Applied Biotechnology (LBA3B), Azm Center for Research in Biotechnology and its Applications, EDST, Lebanese University, Tripoli 1300, Lebanon; jacynthefrangieh@gmail.com (J.F.); yahyasalma@ul.edu.lb (Y.S.)
2 Faculty of Sciences 3, Lebanese University, Michel Slayman Tripoli Campus, Ras Maska 1352, Lebanon; khaddad@ul.edu.lb
3 Mitochondrial and Cardiovascular Pathophysiology – MITOVASC, Team 2, Cardiovascular Mechanotransduction, UMR CNRS 6015, INSERM U1083, Angers University, 49045 Angers, France; cesar.mattei@univ-angers.fr (C.M.); christian.legros@univ-angers.fr (C.L.)
4 Faculty of Agriculture & Veterinary Sciences, Lebanese University, Dekwaneh, Beirut 2832, Lebanon
* Correspondence: ziad.fajloun@ul.edu.lb (Z.F.); delobeid@gmail.com (D.E.O.); Tel. +961 03 31 51 74 (Z.F.); +961 70 19 51 34 (D.E.O.)

Received: 28 February 2019; Accepted: 27 March 2019; Published: 30 March 2019

Abstract: Bee venom is a mixture of several components with proven therapeutic benefits, among which are anti-inflammatory, analgesic, and various cardiovascular conditions. In this work, we analyzed for the first time the proteomic content and biological properties of the crude venom from *Apis mellifera syriaca*, a honeybee from the Middle East region. Using high-performance liquid chromatography-tandem mass spectrometry, we evidence the venom contains phospholipase A2, hyaluronidase, mast cell-degranulating peptide, adolapin, apamin, and melittin. The latter was purified by solid phase extraction method (SPE) and tested in parallel with crude venom for biological activities. Precisely, crude venom—but not melittin—exhibited antibacterial activity against *Staphylococcus aureus* and *Pseudomonas aeruginosa* strains. Alongside, hemolytic activity was observed in human blood subjected to the venom at high doses. *A. mellifera syriaca* venom displayed antioxidant activities, and not surprisingly, PLA2 catalytic activity. Eventually, the venom proved to exert antiproliferative effects against MCF-7 and 3T3 cancer cells lines. This first report of a new bee venom opens new avenues for therapeutic uses of bee venoms.

Keywords: *Apis mellifera syriaca*; bee venom; melittin; LC-ESI-MS; solid phase extraction; in vitro effects

Key Contribution: This study shows the separation of a new hymenoptera venom (*Apis mellifera syriaca*) using two different extraction techniques. This venom exhibits multiple in vitro biological activities.

1. Introduction

Bees use their venom as a defense tool against predators, intruders, and for colony defense [1]. Bee venom (BV) is a complex mixture of peptidyl toxins, enzymes, and other trace components, with a wide spectrum of biological activities such as anti-microbial, anti-cancerous, and antioxidant activities [2,3]. It has been used as a therapeutic tool in oriental medicine to treat several human inflammatory diseases such as rheumatism, arthritis, and to relieve back pain [4–6]. These medical claims have now found evidence in numerous studies showing that the use of BVs is not restricted to a single therapeutic area, but can be used for different conditions with various pathophysiological substrates, including for the nervous system, for immunity, or for the cardiovascular system [7]. The bioactive compounds of BV termed apitoxin can be divided into (i) proteins such as melittin, apamin, MCD-peptide (mast cell degranulating peptide) and adolapin, (ii) enzymes like phospholipase A2 (PLA2), hyaluronidase, α-glucosidase, acid phosphomonoesterase, and lysophopholipase, and (iii) also amino acids, phospholipids and volatile compounds [2]. Melittin, one of the major BV components, triggers the toxicity of the venom. It has pore-forming activity in the cell phospholipid bilayer, inducing membrane rupture [8]. Several studies have shown that melittin has a broad spectrum of biological, pharmacological, and toxicological activities including anti-bacterial, anti-viral, anti-inflammatory, and anti-tumor properties, together with hemolytic properties [8]. Apamin, another important peptide in BV, is the smallest venom-derived neurotoxin that blocks small-conductance Ca^{2+}-activated K^+ channels (SKCa) [9]. It exerts therapeutic benefits in mouse models of Parkinson disease [10]. Moreover, PLA2, the major enzyme present in BV, has the ability to cleave membrane phospholipids at the sn-2 position to release fatty acids, especially arachidonic acid and lysophospholipid. The arachidonic acid released is a precursor of eicosanoids such as prostaglandins and leukotrienes, which participate in the inflammatory reaction [11]. Then, BVs appear to harbor a large diversity of natural compounds which, as a mixture, contribute to the whole toxicity of the venom but, as single actors, could be used for their pharmaceutical properties [7]. The search for novel activities in BVs is then an attractive way of discovering future natural drugs for a variety of human pathologies [12].

Apis mellifera syriaca (Figure 1) is the endemic honey bee subspecies present in the Middle East (Lebanon, Syria, Jordan, Palestine, Iraq). Several ecological features make it an interesting venomous animal to study, and a valuable resource to use in agriculture. First, it has a good adaptation to the Mediterranean hot and dry climate [13]. Second, *A. mellifera syriaca*, when compared to the European species, exhibits a higher degree of pest and pathogen resistance, and efficient honey production [14]. But it is threatened by the arrival of commercial breeder lines which could alter its genetic advantages, which should lead to an import restriction or ban on foreign honey bee races [15]. To date, its venom has never been studied.

Figure 1. *A. mellifera syriaca* (copyright Dani El Obeid).

The aim of this work was to study and analyze for the first time the protein content and biological properties of the venom of *A. mellifera syriaca* (Figure 1). To separate the components of the venom, we performed two different extraction methods. First, the crude venom of *A. mellifera syriaca* was analyzed by LC-ESI-MS in order to identify and characterize the different components of this venom. Second, the SPE (solid phase extraction) method was used to separate the protein constituents from the venom. Eventually, the antibacterial, hemolytic, antioxidant, PLA2, and cytotoxic activities of the crude venom and its major component, melittin, extracted by SPE, were tested. Our work provides the first chemical and biological characterization of a new hymenoptera species.

2. Results

2.1. Separation and Analysis of Venom Compounds by LC-ESI-MS

To separate the venom components, we used two different strategies. First, the venom was eluted using liquid chromatography/electrospray ionization mass spectrometry (LC-ESI-MS). After precipitation in acetonitrile, the venom was further separated by a reverse-phase column (C18). Eight peaks corresponding to distinct proteic components of the *A. mellifera syriaca* venom extract at different retention times appear on the chromatogram (Figure 2A). Each peak corresponds to the elution of a single molecule of the venom, and its abundance is proportional to its intensity (absorbance at 220 nm). The venom eluent was subjected to online coupled ESI-MS analysis to measure the molecular weights and therefore identify all components eluted.

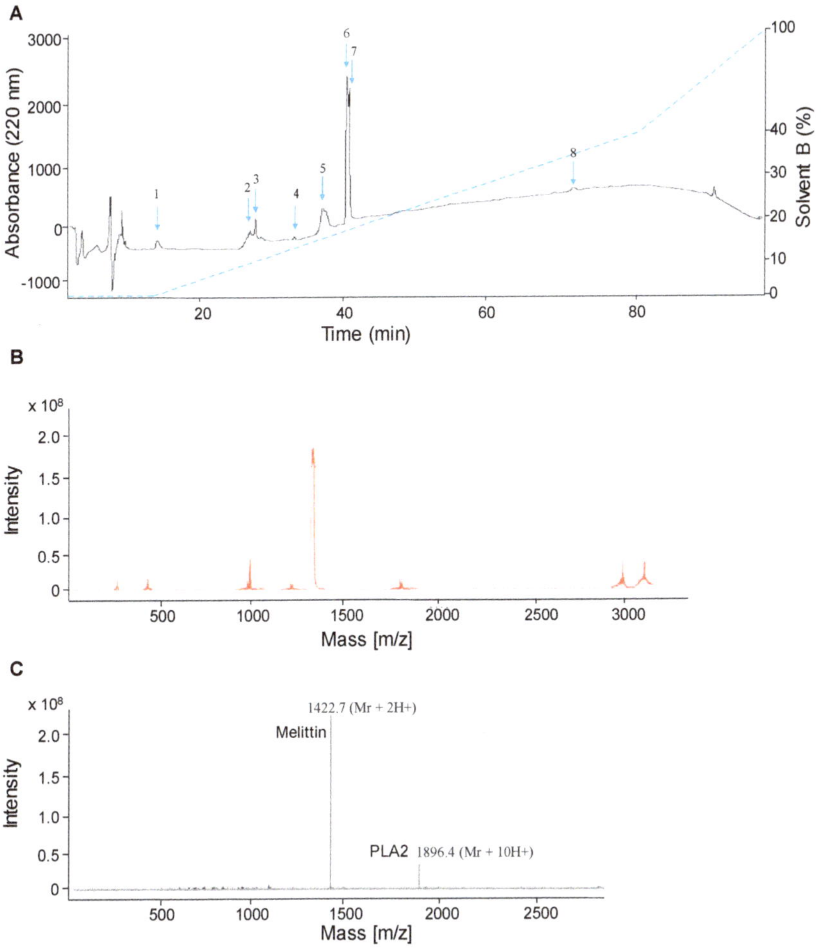

Figure 2. Fractionation of *A. mellifera syriaca* venom using LC-ESI-MS. (**A**) HPLC chromatogram showing reverse-phase C18 fractionation of the venom. (**B**) MS profile of the venom. (**C**) MS profile showing the peaks containing melittin and PLA2.

The mass spectrum (Figure 2B) from the venom chromatogram exhibited different peaks (Table S1), among which apamin (2 027 Da), melittin (2 846.4 Da), PLA2 (18 964 Da), MCD-peptide (2 599.8 Da), and hyaluronidase (53 875.6 Da), with two major peaks of the two most abundant molecules in the venom (melittin and PLA2) (Figure 2C). Separately, the compound spectra analysis of peak 1/HPLC showed the presence of a molecular mass of 2027.35 Da (1013.5 + 2 H^+) and that of peak 3/HPLC revealed a mass of 18 964 Da (1896.4 + 10 H^+), while the analysis of peak 6/HPLC revealed a molecular mass of 2 846.4 Da (1423.2 + 2 H^+) (see Figure S1).

2.2. Separation of Crude Venom Compounds by SPE

In order to separate and purify the different components of *A. mellifera syriaca* venom, we chose to use the SPE technique which allows the concentration of target compounds within the venom. It was operated with a C18 Cartridge by applying the same elution buffers of those used for HPLC. Five fractions were eluted, and each one could contain one or more molecules. These fractions were then analyzed with HPLC and the results obtained showed that the F4 fraction revealed a single peak in HPLC corresponding to the expected molecule—i.e melittin—according to the analysis with ESI-MS (Figure 3). This result suggests that SPE may be a relevant method for the separation of BV components and could be further used routinely in analytical toxicology.

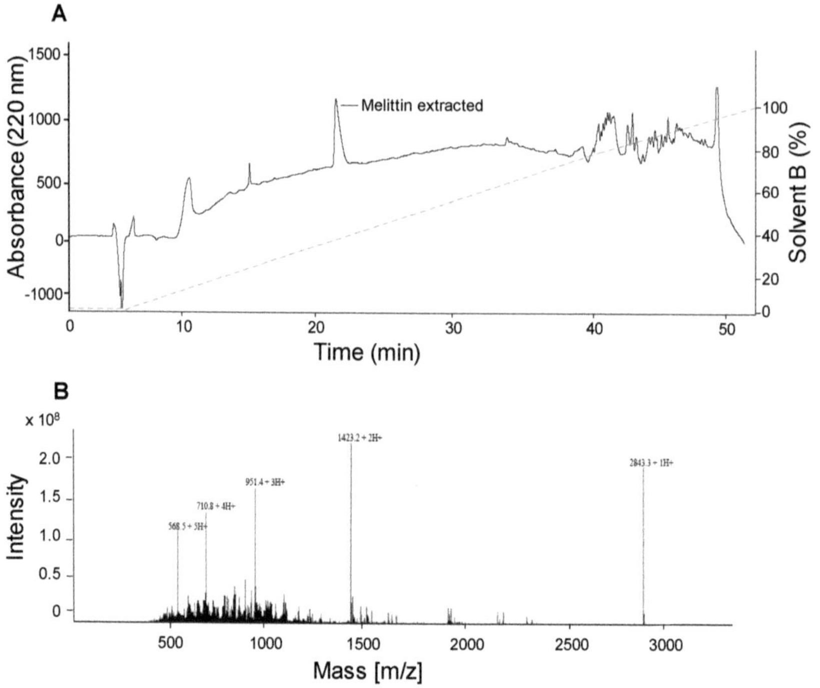

Figure 3. Isolation of melittin from the venom of *A. mellifera syriaca*. (**A**) HPLC profile of the melittin-containing fraction as a function of time and increasing % of acetonitrile solvent. (**B**) MS profile of the same fraction.

2.3. Antibacterial Activity

BVs are known to exert antibacterial effects [3]. We then decided to challenge the venom of *A. mellifera syriaca* against different bacterial strains, namely *Pseudomonas aeruginosa*, *Staphylococcus aureus*, *Bacillus subtilis*, *Proteus vulgaris*, *Enterococcus faecalis*, and *Escherichia coli*. At a concentration of 50 μg/mL, it exhibits antibacterial activity against *P. aeruginosa* and *S. aureus* strains only, with 38%

and 21.4% inhibitions respectively. No activity (very low or not significant) was observed against other strains studied, such as *Enterococcus faecalis*. To gain further insight into this activity, we decided to assess the putative antibacterial capacity of melittin (1 mg/mL). No or weak activity of melittin was observed on all strains (Figure 4).

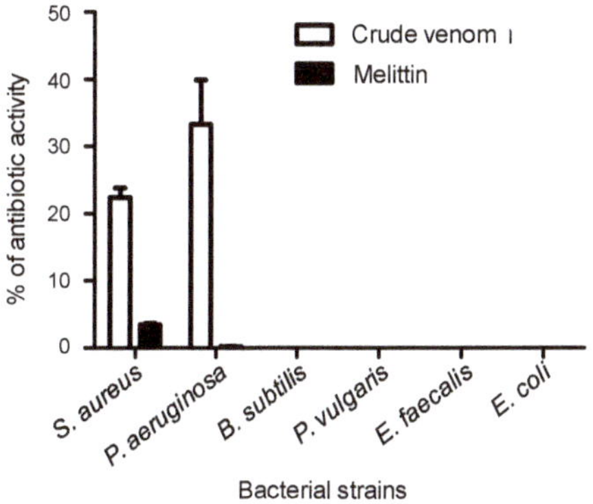

Figure 4. Antibacterial activity of *A. mellifera syriaca* venom. Growth inhibition of different bacterial strains by the venom as a percentage of the total inhibition by specific antibiotics (used as positive controls) is shown. H_2O was used as a negative control. Crude venom inhibits bacterial growth of *S. aureus* and *P. aeruginosa*. Data are expressed as mean ± SD.

2.4. Hemolytic Activity

In order to test the hemolytic activity of *A. mellifera syriaca* venom, a suspension of red blood cells (RBCs) was subjected to different concentrations of crude venom (2.5; 10; 20; 50; 100; 150; 200; 300 and 500 μg/mL). After incubation, the absorbance of the different supernatants was read at 540 nm. The values obtained determined the percentage of hemolysis in each supernatant as a function of the concentration of the venom used, by comparing them with both positive (H_2O) and negative controls (PBS). The venom of *A. mellifera syriaca* exhibits hemolytic activity: At a concentration of 2.5 μg/mL, 8.8% of hemolysis was observed, whereas, for a venom concentration of 10 μg/mL, 66.6% was recorded (Figure 5). A plateau was observed from a concentration of 20 μg/mL of the crude venom (100% hemolytic activity). These data suggest that *A. mellifera syriaca* venom harbors toxic components—namely melittin and PLA2—that act on the cell membrane of RBCs and induce their lysis.

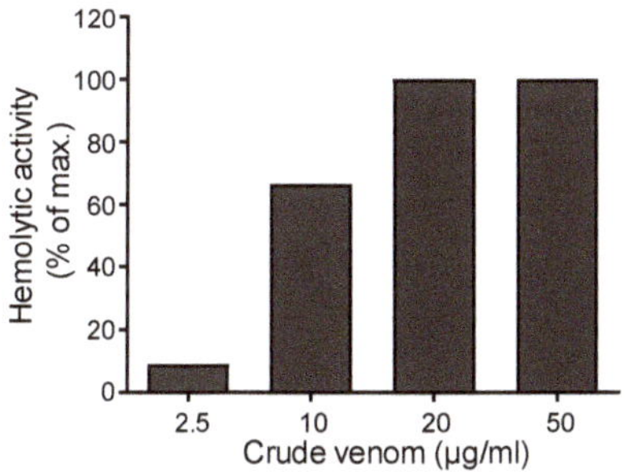

Figure 5. Hemolytic activity of *A. mellifera syriaca* venom. Different concentrations of venom (2.5–50 μg/mL) were used and hemolysis was quantified as a % of the maximal activity induced by H_2O. A dose-response effect was observed and a plateau was reached from 20 μg/mL.

2.5. Antioxidant Activity

1, 1-diphenyl-2-picrylhydrazyl (DPPH) assay was used to evaluate the antioxidant activity. *A. mellifera syriaca* crude venom showed dose-dependent antioxidant activity (Figure 6). All concentrations used could exert an antioxidant action at a level close or weaker than vitamin C, used as a positive control. In fact, from 2.5–200 µg/mL, the percentage of DPPH radicals scavenging activity varies in the range of 50–65%. This value is increased at 71.9% for a concentration of 300 µg/mL, and reaches a maximum of 86.6% for the highest concentration used (500 µg/mL). Melittin was also tested. At a concentration of 100 µg/mL melittin, 52.5% of the scavenging activity was recorded.

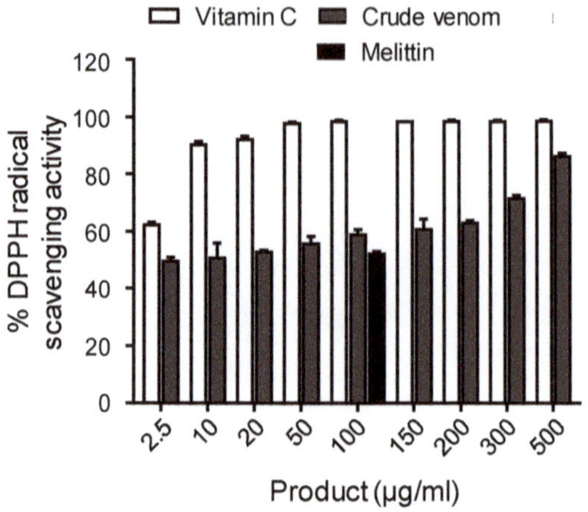

Figure 6. Antioxidant activity of *A. mellifera syriaca* venom. Different concentrations of venom (2.5–500 µg/mL) or melittin (100 µg/mL) were used, and absorbance was measured at 517 nm. Vitamin C was used as a positive control. The venom exhibits robust dose-dependent activity. Data are expressed as mean ± SD.

2.6. PLA2 Activity

BV produce inflammatory processes that lead to deep activation of nociceptors, thus inducing pain [16]. In snake venoms, it is well known that PLA2s are toxins which mediate pro-inflammatory activities. Indeed, they catalyze the hydrolysis of the sn-2 ester bond of membrane phospholipids, including phosphatidylcholine [17]. This phospholipid release favors the formation of arachidonic acid [18]. In the physiological context of inflammation, arachidonate is the molecular basis for the release of prostanglandins. As BV contain PLA2, we next challenged the crude venom and extracted melittin for their enzymatic activity to release free fatty acids, among which arachidonic acid is included. This test shows that crude venom has significant PLA2 activity (Figure 7). In fact, for a concentration of 5 µg/mL, 86.42% fatty acid release was observed. As expected, no effect was recorded for the melittin (100 µg/mL) extracted from the venom (not shown).

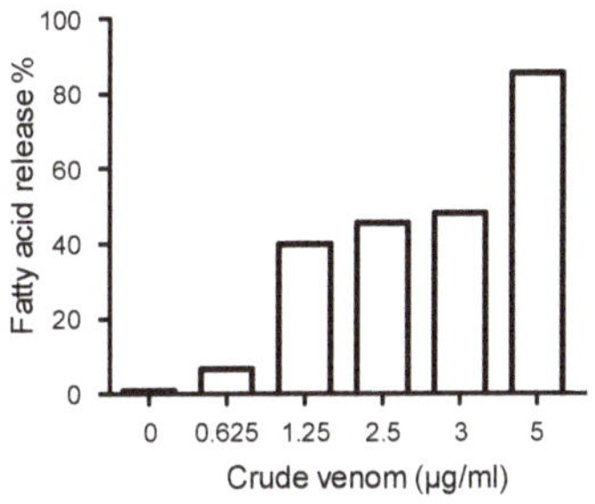

Figure 7. Effect of *A. mellifera syriaca* venom on fatty acid release. Phosphatidylcholine (PC) was subjected to increasing concentrations of the venom (0.625–5 µg/mL). Fatty acid release was measured as described in the methods. The venom exerts PLA2 activity.

2.7. Cytotoxic Activity on MCF-7 and 3T3 Cancer Cells

BV has long been studied for their anti-tumoral and cytotoxic effects, which can be explained by their capacities to promote necrosis and/or apoptosis. The cytotoxic activity of *A. mellifera syriaca* venom was evaluated on two types of cancer cell lines: MCF-7 and 3T3. Results obtained show that this venom has dose-dependent antiproliferative activity against both (Figure 8). Nevertheless, this toxicity was stronger against MCF-7 than 3T3 cancer cells.

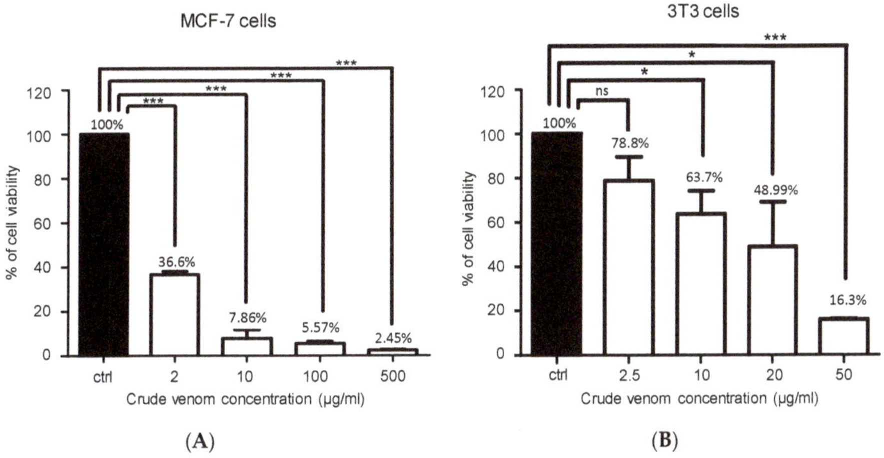

Figure 8. Cytotoxicity effect of *A. mellifera syriaca* venom on two different cancer cell lines. (**A**) Cytotoxicity activity of the venom on MCF-7 cells. (**B**) Cytotoxicity activity of the venom on 3T3 cells. Data are expressed as mean ± SD (n = 3–4). Unpaired t-test: ns (no significant), * $p < 0.05$; *** $p < 0.001$ when compared with the control. At 10 µg/mL, the crude venom showed a cytotoxic activity which is more significant against MCF-7 cancers cells.

3. Discussion

In this study, we intended to separate and characterize the components of *A. mellifera syriaca* venom, with the aim characterizing its chemical and biological properties. The identification of various toxins correlates with previous studies demonstrating their presence in BV [2,19,20]. HPLC with gradient elution is usually used for the separation of venom components like proteins and peptides. These molecules can then be identified and characterized using ESI-MS. In fact, ESI has been a workhorse for the MS analysis of proteins and peptides [21]. In our study, LC-ESI-MS analysis of crude venom showed that melittin was the most abundant compound, followed by PLA2, as it has been described in BVs [2]. SPE is an original approach that reduces solvent exposure and extraction time and is also used to gain high recoveries [22]. SPE allowed us to purify and obtain high quantities of the different components of *A. mellifera syriaca* venom, especially melittin.

The venom was then challenged in various assays to understand its biological activities. The antimicrobial effect against Gram+ and Gram- bacterial strains is in good accordance with previous studies describing the toxicity of *A. mellifera* venom [3]. It has been suggested that the toxicity of BV against bacteria is due in part to the presence of PLA2 and melittin [3,23]. In fact, our results showed that the antibacterial activity of *A. mellifera syriaca* crude venom was significant against some strains, mainly *S. aureus* and *P. aeruginosa*; however, at 1 mg/mL of purified melittin, activity was observed against *S. aureus* and not for the others bacterial strains. This confirms previous data showing that melittin was not effective against *P. aeruginosa* strains, but has a specific effect against *S. aureus* [24]. It has been reported that BV melittin is more active against Gram+ than Gram- bacteria, which suggests that PLA2 is the element responsible for the antibacterial activity observed on *P. aeruginosa* and *S. aureus* strains by causing a cleavage of membrane phospholipids and pore formation in the membrane followed by cell lysis. Alternatively, this activity may be produced through the synergetic action of melittin and PLA2 [3,23]. However, it would be necessary to challenge the antibacterial activity of *A. mellifera syriaca* PLA2 alone or in combination with melittin to confirm the origin of this toxicity against bacteria.

As shown in the past, melittin and PLA2 are together responsible for RBC lysis [25]. In fact, melittin, which is already known for its lytic activity, is considered to be the main cause of hemolysis. Melittin activates PLA2, in which catalytic activity causes the cleavage of the phospholipid bilayers, thus releasing lysophospholipids which become very active at the level of the membrane causing the destruction of RBCs [26]. Moreover, it has been reported that PLA2 purified from BV did not cause the lysis of RBCs, but when venom-purified melittin was added to the solution, hemolysis was observed [27]. As for the antioxidant effect of *A. mellifera syriaca* venom, our results are in accordance with previous studies on *Apis mellifera* venom reporting an antioxidant activity of BV extracts which inhibits the production of DPPH in a dose-dependent manner. Some data suggest that melittin alone exerts very poor antioxidant activity compared to BV extracts and this might be due to the influence of other venom components [2,28]. As for the pro-inflammatory effect of BV, it is due to PLA2, which has a phospholipid cleaving function: It releases arachidonic acid, a precursor of eicosanoids, which are robust inflammation mediators [29]. Moreover, melittin could play an indirect role by activating PLA2 which exerts its pro-inflammatory activity by increasing the secretion of chemical mediators such as pro-inflammatory cytokines [30].

Finally, the difference in BV cytotoxic activity against MCF-7 and 3T3 may be due to their specific membrane receptors. Previous studies demonstrated that BV components including melittin, apamin, and PLA2 exert anti-tumor activities against various types of cancer cell lines like mammary, renal, prostatic, and leukemic cells [31]. Moreover, previous data showed that BV acts as an anti-cancer agent through apoptosis, necrosis, and lysis induction of tumor cells via the activation of several signaling pathways involving a Bcl-2 protein, caspase 3, in synovial fibroblasts [32]. For instance, the antitumor effect of melittin is caused by the suppression of the production of matrix metalloproteinase: MMP-9 inhibition is correlated to the invasion inhibition of MCF-7 cells and the inhibition of caspase activity [33]. So, melittin derived from *A. mellifera syriaca* venom and/or even the crude venom may inhibit the proliferation of cancer cells and favor their apoptosis.

4. Conclusions

We characterized for the first time the *A. mellifera syriaca* venom. Its chemical composition reveals the presence of molecules already known in BVs such as apamin, melittin, PLA2, MCD-peptide, and hyaluronidase. MS analysis discloses unidentified experimental molecular masses, which may correspond to novel molecules with potential therapeutic interests. Several biological activities of *A. mellifera syriaca* crude venom were evaluated in vitro and the results obtained showed that this venom was able to inhibit the growth of certain bacterial strains that develop antibiotic resistance. The most significant result obtained in this work is its anti-tumoral effects, which revealed an antiproliferative action against MCF-7 and 3T3 cancer cells, making this BV a good natural precursor for the design of novel anticancer drugs. Paradoxical biological activities have emerged from this study, but this is a common feature in animal venoms. The next step will identify unknown components which could exhibit novel pharmacological properties.

5. Materials and Methods

5.1. Materials

5.1.1. Bees

Healthy hives of local strains (*Apis mellifera syriaca*) were selected. The apiary was located in Ramlieh, Aley (Lebanon). The forage there is mainly from wild plantations and the flowers were fully blooming.

5.1.2. Venom

The venom was collected from healthy colonies of local *A. mellifera syriaca* strains. There was sufficient pollen in nature and in the hives (two frames of pollen in each colony). The collection was locally made following the standard electroshock method [34] and was installed at the top of the hive. When the wires were electrified and a mild shock was applied to the bees, they covered the surface of the wired glass plate and stung the surface of the glass plate in response to the electrical stimulation. Secreted venom from bee sting dried rapidly when exposed to the air. Dried venom was scraped off with a sharp scalpel and transferred to the laboratory and was stored at a temperature of −20 °C until further analysis. Extraction was made for 15–20 min on each colony and was repeated twice every 2 weeks.

5.1.3. Reagents

Acetonitrile (Acn), trifluoroacetic acid (TFA), phosphatidylcholine (PC), triton, dibasic sodium phosphate (Na2HPO4), monopotassium phosphate (KH2PO4), dimethylsulfoxide (DMSO), "Dulbecco's Modified Eagle's Medium" culture medium (DMEM, which contain 4500 mg/L glucose, L-glutamine, and sodium bicarbonate, without sodium pyruvate), the MTT kit, and vitamin C were purchased from Sigma Aldrich (Ibra Hadad, Beirut, Lebanon). The bacterial strains were provided by the microbiology laboratory of the Faculty of Public Health 3 of the Lebanese University in Tripoli.

5.2. Methods

5.2.1. Chemical Characterization of The Crude Venom by LC-ESI-MS

Chromatographic separation was carried out using a Discovery® HS C18 25 cm × 4.6 mm, 5 µm column. 5.5 mg of freeze-dried crude venom was dissolved in 1 mL of ultrapure water, then this amount was filtered using a syringe filter. 100 µL of the solution was injected into the HPLC. The collection process requires an elution gradient of 0–40% acetonitrile for 80 min at a flow rate of 1 mL/min, and a UV detector at 220 nm to separate the different components of the venom. The elution gradient used is composed of two eluents: Eluent A (0.1% TFA in water), and eluent B (0.1% TFA in acetonitrile). The fractions obtained and collected by HPLC were subjected to an ESI-MS analysis in order to identify and characterize the components of these fractions. This analysis was carried out in a scanning mode between 100 and 3000 m/z, following the same elution gradient conditions used for HPLC analysis, and similarly, the absorbance was measured at 220 nm. Data acquisition was recorded with the HyStar ™ and Esquire data system.

5.2.2. Separation of Crude Venom Compounds using Solid Phase Extraction (SPE)

SPE of crude venom was performed using a C18 Cartridge. Also, 10 mg of *A. mellifera syriaca* venom was dissolved in 5 mL of ultrapure water, then this solution was filtered in a syringe filter. Two eluents were used: Buffer A (0.1% TFA in H_2O) and buffer B (0.1% TFA in acetonitrile). Different % of elution were used to extract the compounds of the venom and these different elution gradients were chosen on the basis of HPLC data relating to the elution of each compound from *A. mellifera syriaca* venom. This indicates at what percentage of elution (eluent A: H_2O and eluent B: Acetonitrile) each compound appears as. The first step of the SPE technique is the conditioning of the cartridge made by 100% acetonitrile to hydrate the silica. After this step, 100% H_2O was added through the column, and the column was loaded with the venom sample (dissolved in H_2O). After loading the sample, the various eluents were applied to the column, and the different fractions corresponding to the different eluents were collected and analyzed using HPLC in order to observe the quality of their contents and to evaluate their purity.

5.2.3. Antibacterial Activity

Ten mg of freeze-dried venom of *A. mellifera syriaca* were dissolved in 200 µL of ultrapure water. Similarly, a purified melittin solution from *A. mellifera syriaca* venom with an initial concentration of 1 mg/mL was tested as a standard. Crude venom and melittin were tested against six bacterial strains using sensitivity tests onto diffusion discs [35,36]. The first step of the procedure consists in the enrichment of the bacteria. Thus, the peptone water was prepared in tubes and then autoclaved at 121 °C for 15 min, then using a sterile loop, the bacteria were put in peptone water and mixed. The tubes containing bacteria were incubated at 37 °C for 24 h. Bacterial strains were then seeded on Petri dishes using a sterile loop and incubated at 37 °C for 24 h. Finally, six tubes were prepared (each one containing 3 mL of sterile water); their contents were poured into dishes containing Mueller Hinton medium, then the suspension was spread over the entire surface of the agar, and the excess was removed. After drying, the sterile filter paper discs were placed on the Petri dishes with sterile forceps. A volume of 10 µL/disc of the corresponding solution was added. Finally, the Petri dishes were incubated for 24 h at 37 °C. The area of inhibition was measured using a caliper. The antibacterial test was performed in duplicate. A specific antibiotic was used for each strain as a positive control (maximal activity) and H_2O as a negative control. Data show the % of maximal effect.

5.2.4. Hemolytic Activity

Hemolytic effect of *A. mellifera syriaca* crude venom was performed using human red blood cells (RBCs) [37]. Fresh blood was collected from healthy volunteers in EDTA tubes and centrifuged at 3000 rpm for 5 min. The supernatant containing serum and white blood cells were removed. The RBC pellet was washed three times with PBS and centrifuged each time at 3000 rpm for 5 min. A suspension of pure RBC was obtained. From this suspension, a volume of 100 µL was taken in each tube and treated with the venom at different concentrations from an initial stock solution (5 mg/mL in PBS). Two control tubes were prepared: One is considered as a positive control, which contains RBCs and distilled H_2O, while the other corresponds to the negative control containing RBCs and PBS. All tubes were incubated at 4 °C for 30 min and then centrifuged at 3000 rpm for 5 min. Then, the absorbance of the supernatant was measured at 540 nm, and the absorbance values obtained determined the percentage of hemolysis in each tube.

The hemolysis was calculated according to the following formula (where A designates Absorbance) [37]:

$$\text{Hemolysis (\%)} = [(A_{\text{Tube}} - A_{\text{Negative Control}})/A_{\text{Positive Control}}] \times 100$$

5.2.5. Antioxidant Activity Assay

The antioxidant activity was evaluated by applying the free radical scavenging method using DPPH (2,2-diphenyl-1-picrylhydrazyl) [2]. Five mg of lyophilized crude venom was dissolved in 1 mL of ultrapure water, and several samples (5 mg/mL) were prepared to test the antioxidant activity. Melittin (1 mg/mL) purified from the venom was also tested. Vitamin C was used as a positive control. A blank tube was performed for each dose of the venom as a negative control. Absorbance was measured at 517 nm before incubation using a spectrophotometer. The tubes were then incubated in the dark for 30 min, after which the optical density was measured at the same wavelength (517 nm). The antioxidant activity was performed in duplicate.

The % of DPPH radical scavenging activity was determined using the following formula (where A designates Absorbance):

$$\text{DPPH radical scavenging activity (\%)} = [(A_{\text{DPPH}} - A_{\text{Echantillon}})/A_{\text{DPPH}}] \times 100$$

5.2.6. Measurement of PLA2 Activity

To study the specific PLA2 activity of *A. mellifera syriaca* venom, phosphatidylcholine was used as a substrate [38]. The effect was measured by using a spectrophotometer based on pH change due to the release of free fatty acids from L-α-phosphatidylcholine [38]. First, a solution of the reaction medium was prepared as followed: 3.5 mM L-α-phosphatidylcholine (PC) ws placed in an Erlenmeyer flask into which 7 mM Triton X-100 (408 μL) was added. Also, distilled water was added to complete the volume to 50 ml. This solution was subjected to magnetic agitation for 1 h to promote the solubilization of the PC. Then, 10 mM CaCl2, 2H2O, 100 mM NaCl and 0.055 mM phenol red are added (pH 7.6). Finally, distilled water was added to reach a final volume of 100 mL. Different concentrations of crude venom were prepared from an initial solution of 5 mg/mL (in distilled water). Each tube contained 1 mL of the PC solution and the volume of the specified concentrations of venom. The tube representing the negative control consisted only of the PC solution. Absorbance was measured at 558 nm before incubation, and after incubation of the tubes at 37 °C for 5 min using a spectrophotometer.

The percentage of in vitro PLA2 catalytic activity was calculated using the following formula (where A designates absorbance):

$$\text{Fatty acid release (\%)} = [A_{\text{negative control}} - (A_{\text{sample}}/A_{\text{negative control}})] \times 100$$

5.2.7. Cytotoxic Activity Assay on MCF-7 and 3T3 Cancer Cells

The venom cytotoxicity was investigated using the MTT viability test [39]. An initial solution of 5 mg/mL of crude venom was prepared and this solution was filtered using a syringe filter. Cells were cultured in DMEM culture medium, until confluence. A plate of 24 wells was used in each well 1 mL of each prepared solution at different concentration was deposited. For each solution, triplicate copies were made for the MCF-7 cells, and four copies were made for 3T3 cells. This plate was incubated at 37 °C for 24 h. After incubation, a volume of 10 μL MTT was added in each well. This step was performed in dark, as MTT is photosensitive. The plate was stirred and then incubated at 37 °C for 1h. The medium was then removed and 1 mL DMSO was added to each well to solubilize Formazan crystals. Absorbance quantification was read at 560 nm.

5.2.8. Statistical Analysis

Results were expressed as the mean ± standard deviation (SD). Statistical significance between different samples was analyzed using a two-tailed unpaired t-test. Statistical significance was defined as * $p < 0.05$, ** $p < 0.01$ and *** $p < 0.001$. This analysis was carried out using GraphPad Prism 7.02 (GraphPad Software, San Diego, USA).

Supplementary Materials: The following are available online at http://www.mdpi.com/2072-6651/11/4/191/s1, Table S1: Experimental molecular weights (Da) obtained by ESI-MS analysis of the *A. mellifera syriaca* venom eluted from HPLC, Figure S1: High resolution mass spectra of melittin, apamin and PLA2 obtained from the venom of *A. mellifera syriaca*.

Author Contributions: Z.F. and D.E.O. conceived and designed the experiments; J.F. performed the experiments; Y.S. and K.H. contributed to cell culture and hemolytic tests; J.F., C.M., C.L., Z.F. and D.E.O. interpreted the results; J.F., C.M., C.L., Z.F. and D.E.O. wrote the manuscript.

Funding: This research was funded by the Lebanese University.

Acknowledgments: We would like to thank M. Zeenny Kheir, President of the Federation of Zgharta Caza municipalities, and Marc Karam, Asma Chbani for helpful discussions.

Conflicts of Interest: The authors declare no conflict of interest.

References

1. Li, R.; Zhang, L.; Fang, Y.; Han, B.; Lu, X.; Zhou, T.; Feng, M.; Li, J. Proteome and phosphoproteome analysis of honeybee (*Apis mellifera*) venom collected from electrical stimulation and manual extraction of the venom gland. *BMC Genom.* **2013**, *14*, 766. [CrossRef] [PubMed]
2. Sobral, F.; Sampaio, A.; Falcão, S.; Queiroz, M.J.R.; Calhelha, R.C.; Vilas-Boas, M.; Ferreira, I.C. Chemical characterization, antioxidant, anti-inflammatory and cytotoxic properties of bee venom collected in Northeast Portugal. *Food Chem. Toxicol.* **2016**, *94*, 172–177. [CrossRef] [PubMed]
3. Zolfagharian, H.; Mohajeri, M.; Babaie, M. Bee Venom (*Apis Mellifera*) an Effective Potential Alternative to Gentamicin for Specific Bacteria Strains. *J. Pharm.* **2016**, *19*, 225–230. [CrossRef]
4. Matysiak, J.; Schmelzer, C.E.; Neubert, R.H.; Kokot, Z.J. Characterization of honeybee venom by MALDI-TOF and nanoESI-QqTOF mass spectrometry. *J. Pharm. Biomed. Anal.* **2011**, *54*, 273–278. [CrossRef]
5. Han, S.; Lee, K.; Yeo, J.; Kim, W.; Park, K. Biological effects of treatment of an animal skin wound with honeybee (*Apis melifera.* L) venom. *J. Plast. Reconstruct. Aesthet. Surg.* **2011**, *64*, e67–e72. [CrossRef] [PubMed]
6. Cherniack, E.P.; Govorushko, S. To bee or not to bee: The potential efficacy and safety of bee venom acupuncture in humans. *Toxicon* **2018**, *154*, 74–78. [CrossRef] [PubMed]
7. Zhang, S.; Liu, Y.; Ye, Y.; Wang, X.R.; Lin, L.T.; Xiao, L.Y.; Zhou, P.; Shi, G.X.; Liu, C.Z. Bee venom therapy: Potential mechanisms and therapeutic applications. *Toxicon* **2018**, *148*, 64–73. [CrossRef] [PubMed]
8. Chen, J.; Guan, S.M.; Sun, W.; Fu, H. Melittin, the Major Pain-Producing Substance of Bee Venom. *Neuroscience* **2016**, *32*, 265–272. [CrossRef] [PubMed]
9. Sah, P.; Faber, E.L. Channels underlying neuronal calcium-activated potassium currents. *Prog. Neurobiol.* **2002**, *66*, 345–353. [CrossRef]
10. Alvarez-Fischer, D.; Noelker, C.; Vulinović, F.; Grünewald, A.; Chevarin, C.; Klein, C.; Oertel, W.H.; Hirsch, E.C.; Michel, P.P.; Hartmann, A. Bee Venom and Its Component Apamin as Neuroprotective Agents in a Parkinson Disease Mouse Model. *PLoS ONE* **2013**, *8*, e61700. [CrossRef]
11. Bae, G.L.H. Bee Venom Phospholipase A2: Yesterday's Enemy Becomes Today's Friend. *Toxins* **2016**, *8*, 48.
12. Pak, S.C. An Introduction to the Toxins Special Issue on "Bee and Wasp Venoms: Biological Characteristics and Therapeutic Application". *Toxins* **2016**, *8*, 315. [CrossRef]
13. Zaitoun, S.T.; Al-Ghzawi, A.M.; Shannag, H.K. Population dynamics of the Syrian Honeybee, Apis mellifera syriaca, under semi-arid Mediterranean conditions. *Zool. Middle East* **2000**, *21*, 129–132. [CrossRef]
14. Haddad, N.; Mahmud Batainh, A.; Suleiman Migdadi, O.; Saini, D.; Krishnamurthy, V.; Parameswaran, S.; Alhamuri, Z. Next generation sequencing of *Apis mellifera* syriacaidentifies genes for *Varroaresistance* and beneficial beekeeping traits. *Insect Sci.* **2016**, *23*, 579–590. [CrossRef]
15. Zakour, M.K.; Ehrhardt, K.; Bienefeld, K. First estimate of genetic parameters for the *Syrian honeybeeApis mellifera syriaca*. *Apidologie* **2012**, *43*, 600–607. [CrossRef]
16. Chen, J.; Luo, C.; Li, H.L.; Chen, H.S. Primary hyperalgesia to mechanical and heat stimuli following subcutaneous bee venom injection into the plantar surface of hindpaw in the conscious rat: A comparative study with the formalin test. *Pain* **1999**, *83*, 67–76. [CrossRef]
17. Teixeira, C.F.P.; Landucci, E.C.T.; Antunes, E.; Chacur, M.; Cury, Y. Inflammatory effects of snake venom myotoxic phospholipases A2. *Toxicon* **2003**, *42*, 947–962. [CrossRef] [PubMed]

18. Smith, W.L.; DeWitt, D.L.; Garavito, R.M. Cyclooxygenases: Structural, cellular, and molecular biology. *Annu. Rev. Biochem.* **2000**, *69*, 145–182. [CrossRef]
19. Rybak-Chmielewska, H.; Szczêsna, T. HPLC study of chemical composition of honeybee (*Apis mellifera* L.) venom. *J. Apicult. Sci.* **2004**, *48*, 103–109.
20. Giralt, M.M.A.E. Three valuable peptides from bee and wasp venoms for therapeutic and biotechnological use: Melittin, apamin and mastoparan. *Toxins* **2015**, *7*, 1126–1150.
21. Loo, J.A. Studying non covalent protein complexes by electrospray ionization mass spectrometry. *Mass Spectrom. Rev.* **1997**, *16*, 1–23. [CrossRef]
22. Camel, V.R. Solid phase extraction of trace elements. *Spectrochim. Acta Part B* **2003**, *58*, 1177–1233. [CrossRef]
23. Leandro, L.F.; Mendes, C.A.; Casemiro, L.A.; Vinholis, A.H.; Cunha, W.R.; Almeida, R.D.; Martins, C.H. Antimicrobial activity of apitoxin, melittin and phospholipase A_2 of honey bee (*Apis mellifera*) venom against oral pathogens. *Anais Academia Brasileira Ciencias* **2015**, *87*, 147–155. [CrossRef] [PubMed]
24. Fratini, F.; Cilia, G.; Turchi, B.; Felicioli, A. Insects, arachnids and centipedes venom: A powerful weapon against bacteria. *Toxicon* **2017**, *130*, 91–103. [CrossRef]
25. Tosteson, M.T.; Holmes, S.J.; Razin, M.; Tosteson, D.C. Melittin Lysis of Red Cells. *J. Membr. Biol.* **1985**, *87*, 35–44. [CrossRef]
26. Vetter, R.S.; Visscher, P.K.; Camazine, S. Mass Envenomations by Honey Bees and Wasps. *West J. Med.* **1999**, *170*, 223.
27. CezaryWatala, J.K.K. Hemolytic potency and phospholipase activity of some bee and wasp venoms. *Comp. Biochem. Physiol. Part C Comp. Pharmacol.* **1990**, *97*, 187–194.
28. Somwongin, S.; Chantawannakul, P.; Chaiyana, W. Antioxidant activity and irritation property of venoms from *Apis* species. *Toxicon* **2018**, *145*, 32–39. [CrossRef]
29. Murakami, M.; Kudo, I. Phospholipase A2. *J. Biochem.* **2002**, *131*, 285–292. [CrossRef]
30. Tusiimire, J.; Wallace, J.; Woods, N.; Dufton, M.; Parkinson, J.; Abbott, G.; Clements, C.; Young, L.; Park, J.; Jeon, J. Effect of bee venom and its fractions on the release of pro-inflammatory cytokines in PMA-differentiated U937 cells co-stimulated with LPS. *Vaccines* **2016**, *4*, 11. [CrossRef]
31. Oršolić, N. Bee venom in cancer therapy. *Cancer Metastasis Rev.* **2012**, *31*, 173–194. [CrossRef] [PubMed]
32. Premratanachai, P.; Chanchao, C. Review of the anticancer activities of bee products. *Afr. J. Microbiol. Res.* **2014**, *4*, 337–344. [CrossRef]
33. Wang, J.; Li, F.; Tan, J.; Peng, X.; Sun, L.; Wang, P.; Jia, S.; Yu, Q.; Huo, H.; Zhao, H. Melittin inhibits the invasion of MCF-7 cells by downregulating CD147 and MMP-9 expression. *Oncol. Lett.* **2017**, *13*, 599–604. [CrossRef] [PubMed]
34. Pence, R.J. Methods for producing and bio-assaying intact honeybee venom for medical use. *Am. Bee J.* **1981**, *121*, 726–731.
35. Bauer, A.W.; Kirby, W.M.M.; Sherris, J.C.; Turck, M. Antibiotic susceptibility testing by a standardized single disk method. *Am. J. Clin. Pathol.* **1966**, *45*, 493–496. [CrossRef] [PubMed]
36. Surendra, N.S.; Jayaram, G.N.; Reddy, M.S. Antimicrobial activity of crude venom extracts in honeybees (*Apis cerana, Apis dorsata, Apis florea*) tested against selected pathogens. *Afr. J. Microbiol. Res.* **2011**, *5*, 2765–2772.
37. Accary, C.; Rima, M.; Kouzayha, A.; Hleihel, W.; Sadek, R.; Desfontis, J.C.; Fajloun, Z.; Hraoui-Bloquet, S. Effect of the Montivipera bornmuelleri snake venom on human blood: Coagulation disorders and hemolytic activities. *Open J. Hematol.* **2014**, 5. [CrossRef]
38. Accary, C.; Hraoui-Bloquet, S.; Hamze, M.; Sadek, R.; Hleihel, W.; Desfontis, J.-C.; Fajloun, Z. Preliminary proteomic analysis and biological characterization of the crude venom of Montivipera bornmuelleri; a viper from Lebanon. *Recent Adv. Biomed. Chem. Eng. Mater. Sci.* **2014**, *1*, 167–173.
39. Amini, E.; Baharara, J.; Nikdel, N.; Salek Abdollahi, F. Cytotoxic and Pro-Apoptotic Effects of Honey Bee Venom and Chrysin on Human Ovarian Cancer Cells. *Asia Pac. J. Med. Toxicol.* **2015**, *4*, 68–73.

toxins

MDPI

Article

Pharmacokinetic Properties of the Nephrotoxin Orellanine in Rats

Deman Najar [1], Börje Haraldsson [1], Annika Thorsell [2], Carina Sihlbom [2], Jenny Nyström [1] and Kerstin Ebefors [1,*]

[1] Department of Physiology, Institute of Neuroscience and Physiology, Sahlgrenska Academy, University of Gothenburg, 41390 Gothenburg, Sweden; deman.najar@neuro.gu.se (D.N.); borje.haraldsson@gu.se (B.H.); jenny.nystrom@gu.se (J.N.)

[2] Proteomics Core Facility, Sahlgrenska Academy, University of Gothenburg, 41390 Gothenburg, Sweden; annika.thorsell@gu.se (A.T.); carina.sihlbom@gu.se (C.S.)

* Correspondence: kerstin.ebefors@gu.se; Tel.: +46-722-032-9103

Received: 6 July 2018; Accepted: 15 August 2018; Published: 17 August 2018

Abstract: Orellanine is a nephrotoxin found in mushrooms of the Cortinarius family. Accidental intake of this substance may cause renal failure. Orellanine is specific for proximal tubular cells and could, therefore, potentially be used as treatment for metastatic renal cancer, which originates from these cells. However, more information is needed about the distribution and elimination of orellanine from the body to understand its potential use for therapy. In this study, 5 mg/kg orellanine (unlabeled and ^{3}H-labeled) was injected intravenously in rats (Wistar and Sprague Dawley). Distribution was measured (Wistar rats, $n = 10$, $n = 12$) using radioluminography and the highest amount of orellanine was found in the kidney cortex and bladder at all time-points investigated. The pharmacokinetic properties of orellanine were investigated using LC-MS/MS and β-scintillation to measure the amount of orellanine in plasma. Three groups of rats were investigated: control rats with intact kidneys ($n = 10$) and two groups with bilateral renal artery ligation ($n = 7$) where animals in one of these groups were treated with peritoneal dialysis ($n = 8$). Using LC-MS/MS, the half-life of orellanine was found to be 109 ± 6 min in the controls. In the groups with ligated renal arteries, orellanine had a half-life of 756 ± 98 min without and 238 ± 28 min with dialysis. Thus, orellanine was almost exclusively eliminated by glomerular filtration as well as by peritoneal dialysis.

Keywords: orellanine; clearance; fungal toxin; half-life

Key Contribution: Orellanine is a potent nephrotoxin that may be used as treatment against metastatic renal cancer. No previous data on the pharmacokinetic properties of this toxin has been published. Therefore, this study contributes to a better understanding of the properties of orellanine.

1. Introduction

Orellanine is a natural toxin found in the *Cortinarius* family of mushrooms found in North America and in Europe. Its selective renal toxicity was recognized already in the 1950s [1]. Each year, several people suffer from renal damage ranging from reduced to complete loss of renal function by accidentally ingesting the mushroom and there is still no specific antidote to orellanine poisoning. It is well known that orellanine specifically targets the tubular epithelium [2,3], but the toxicological properties and mechanisms are still not fully known. Several potential mechanisms have been described and all of them point towards oxidative stress [3–6]. Data from our group suggest that orellanine nephrotoxicity in vivo is mediated by a combination of increased oxidative radical formation and orellanine-induced down-regulation of several intracellular anti-oxidative enzymes [3]. Since orellanine specifically targets the tubular epithelial cells, our group has suggested that it could be used for treatment of metastasizing renal cancer originating from the tubular epithelium [7,8].

Orellanine is a bipyridine *N*-oxide (3,3′,4,4′-tetrahydroxy-2,2′-bipyridine-*N*,*N*′-dioxide) with a molecular weight of 252.19 g/mol. When purified, it is a colorless fine crystalline substance. The structure of orellanine was first described in 1979 [9]. In 1985, the photodecomposition of the compound was described by the same group [10] and it was synthesized in the same year [11]. Orellanine decomposes when heated above 150 °C, when exposed to UV, or by reacting with hydrogen in the presence of platinum as a catalyst. If orellanine is reduced, it yields through the toxic form orellinine, the nontoxic substance orelline with the structure of 3,3′,4,4′-tetrahydroxy-2,2′-bipyridyl and molecular weight of 220.18 g/mol. Orellanine has four pKa values at approximately 0.5, 1.0, 7.0, and 7.4. The net charge at physiological pH is close to −4. The structure of orellanine is shown in Figure 1. In mushrooms, orellanine mainly exists in its di-glucosylated form known as orellanine-4,4′-diglucopyranoside [7,12,13]. Small amounts of orellinine and orelline compared to the amount of orellanine are also detected in the mushrooms [14]. Intoxication with orellanine varies in severity depending on the dose ingested but two to three mushrooms have been estimated as enough to develop dialysis dependent kidney failure [15,16]. Studies of intoxication with orellanine in rats show no signs of acute toxicity apart from renal failure and no sign of damage to organs other than the kidney [17]. In a retrospective case control study, orellanine-intoxicated patients where compared to patients with renal failure due to other causes. No differences between the groups were seen in damage to other organs or in cause of death [18]. It is known that orellanine is excreted from the body into the urine during the first 24 h after intake [17] and that plasma levels are undetectable a few days after intoxication [19]. In contrast, there is a case report of high orellanine concentration (6.12 mg/L) measured 10 days after the suspected intake of mushrooms [20]. No study of the pharmacokinetic properties of the substance has been published as of yet.

The primary aim of this study was to perform a pharmacokinetic and distribution study of orellanine in rats to further explore if the toxin could have a future as a therapeutic option for treating renal cancer. Unlabeled and ^{3}H-labeled (tritium) orellanine were intravenously administrated to anesthetized healthy rats with or without renal function (ligated and un-ligated kidneys), including one group without renal function, but on dialysis, and the elimination from plasma was measured. The ^{3}H-labeled orellanine was used to capture elimination of orellanine and any metabolites formed in the rats. ^{3}H-labeled orellanine was also used for quantitative whole body radioluminography to measure the distribution of orellanine in different organs after administration. The results are relevant for our understanding of the dynamics of orellanine intoxication and for future potential therapeutic clinical use of orellanine in treating patients with metastatic renal cancer.

2. Results

2.1. Radioluminography

In order to investigate the distribution of orellanine in the body, two setups of radiolumionography experiments were conducted. In the first experiment, rats were injected with a single dose of ^{3}H-labeled orellanine (see Figure 1 for structure of orellanine and ^{3}H-labeled orellanine). Rats were then sacrificed at 0.5, 1, 6, 12 and 24 h. At all time-points, the highest concentration of orellanine was found in the urinary bladder (at 0.5 h with 560 nmol-eq/g tissue) and the kidney cortex (at 0.5 h with 76 nmol-eq/g tissue, see Table 1). The radioactivity in the blood declined from the highest concentration of 8.6 nmol-eq/g tissue at 30 min after administration to 1.3 nmol-eq/g at the last time point. High concentrations of radioactivity compared to blood was also found at the first time point in the liver (35 nmol-eq/g tissue), the bone marrow (15 nmol-eq/g tissue), and the connective tissues (9.7 nmol-eq/g tissue). Note that the radioactivity signal cannot discriminate between orellanine and any metabolites formed. The other organs investigated had lower radioactivity than the blood at all-time points studied (see Table 1 and Figure 2a). In order to investigate if reduced renal function and repeated exposure to the toxin affected the distribution pattern, a second set of experiments was performed. In this set, the rats were pre-treated with a dose of unlabeled orellanine before administration of the radioactive labeled substance. The results obtained were similar to the single dose experiment with the highest levels of radioactivity seen in the kidney cortex and in the urinary bladder (see Table 2 and Figure 2b).

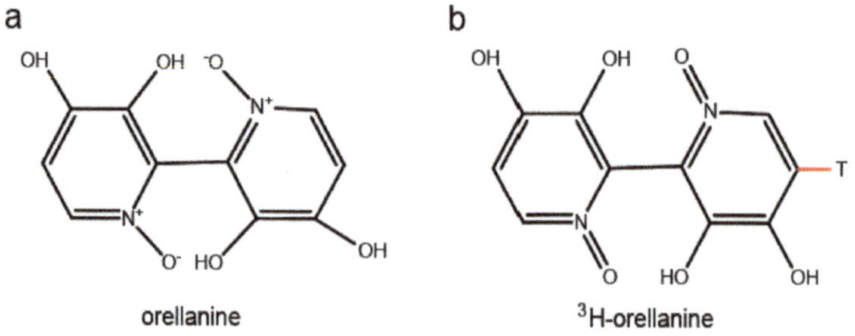

Figure 1. The structural formula of orellanine (**a**) and structure of ^{3}H-labeled orellanine (**b**).

Table 1. Individual tissue concentrations (nmol-eq/g tissue) after a single dose of ^{3}H-labeled orellanine.

Time Point (h)	0.5	1	6	12	24
Adrenal gland	5.2	2.6	1.1	1.6	<LOQ
Bone marrow	15.0	4.7	3.9	4.0	3.0
Brain	<LOQ	<LOQ	<LOQ	<LOQ	<LOQ
Brown fat	<LOQ	<LOQ	<LOQ	<LOQ	<LOQ
Connective Tissue	9.7	3.2	1.6	2.0	<LOQ
Dermis	3.2	1.4	<LOQ	<LOQ	<LOQ
Epidermis	4.0	1.9	<LOQ	<LOQ	<LOQ
Gastric mucosa	3.2	1.8	1.1	<LOQ	<LOQ
Heart blood	8.6	4.9	2.0	1.7	1.3
Intestinal mucosa	5.3	4.2	0.8	0.8	1.0
Kidney cortex	76.0	73.0	19.0	18.0	15.0
Lens (eye)	<LOQ	<LOQ	<LOQ	<LOQ	<LOQ
Liver	35.0	25.0	7.8	6.1	4.0
Lung	8.0	4.0	1.6	1.4	0.9
Lymph node	3.9	2.1	0.8	14.0	<LOQ
Myocardium	6.0	2.6	1.1	0.8	0.6
Pancreas	3.9	1.9	1.6	1.0	<LOQ
Salivary gland	4.5	2.6	1.2	1.0	0.7
Skeletal muscle	1.8	1.6	<LOQ	<LOQ	<LOQ
Spleen	6.5	5.6	4.3	3.8	3.3
Testicle	1.9	1.3	<LOQ	<LOQ	<LOQ
Thymus	4.9	2.5	1.2	1.1	1.1
Thyroid gland	5.2	2.5	1.7	0.9	0.6
Urinary bladder	560.0	2600.0	200.0	37.0	18.0
Limit of Quantification (LOQ)	0.6	0.6	0.7	0.52–0.67	0.6

Measurements are the mean value of three measurements for each tissue. The CV is 7.5–14.3%.

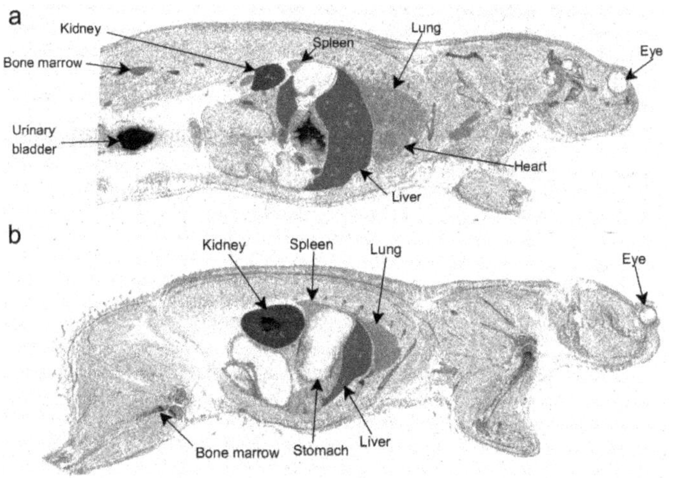

Figure 2. Distribution of orellanine after injection of ^{3}H-labeled orellanine in rats shown by radioluminography. Rats injected with a single dose of ^{3}H-labeled orellanine 30 min after injection. (**a**) shows the highest radioactivity in the kidney and the bladder. Rats that first received a single dose of orellanine 72 h before injection with ^{3}H-labeled orellanine showed the same pattern with the highest radioactivity in the kidney and bladder after 30 min (**b**).

Table 2. Individual tissue concentrations (nmol-eq/g tissue) after a single dose of ^{3}H-labeled orellanine 72 h after a single dose of unlabeled orellanine.

Time Point (h)	0.5	1	6	12
Adrenal gland	5.1	8.0	2.1	0.7
Bone marrow	10.0	5.8	6.1	4.4
Brain	<LOQ	<LOQ	<LOQ	<LOQ
Brown fat	<LOQ	1.3	<LOQ	<LOQ
Connective tissue (skin)	10.0	12.0	1.4	0.8
Dermis	2.0	2.0	<LOQ	<LOQ
Epidermis	4.5	6.9	1.9	<LOQ
Gastric mucosa	3.8	4.8	1.9	1.0
Heart blood	9.0	12.0	3.5	1.8
Intestinal mucosa	3.3	9.2	3.2	2.1
Kidney cortex	63.0	90.0	23.0	22.0
Lens (eye)	<LOQ	<LOQ	<LOQ	<LOQ
Liver	29.0	37.0	18.0	7.2
Lung	8.8	10.5	3.0	1.3
Lymph node	2.7	7.0	3.7	2.6
Myocardium	4.7	6.3	2.6	1.8
Pancreas	2.9	5.1	2.5	1.7
Pituitary	3.1	5.6	2.2	2.1
Retina	8.3	11.0	2.4	1.0
Salivary gland	3.8	6.2	3.3	2.3
Skeletal muscle	1.9	2.1	2.0	0.9
Spleen	4.6	8.8	6.8	3.5
Testicle	2.4	2.7	1.0	<LOQ
Thymus	4.0	6.2	4.3	2.2
Thyroid gland	4.5	6.6	2.4	1.2
Urinary bladder	590.0	600.0	93.0	15.0
LOQ	0.6	0.6	0.43–0.65	0.54–0.70

Measurements are a mean value of three measurements for each tissue. The CV is 8.5–10.7%.

2.2. Pharmacokinetics Study: General Condition of Animals

For the pharmacokinetic study of orellanine in rats, three groups of animals were used. One group of rats has an intact kidney function and two groups had ligated renal arteries to remove kidney function. One of the groups with ligated kidneys underwent dialysis as renal replacement therapy and dialysis was initiated immediately after administration of orellanine. Plasma was collected at 10, 30, 45, 60, 90, 180, and 360 min after administration of orellanine. The body weights of the animals were similar in the three groups, which are 363 ± 54 g, 333 ± 43 g, and 325 ± 61 g for the controls including rats with ligated renal arteries and rats with ligated renal arteries and dialysis respectively (n = 10, 7, and 8). The mean arterial blood pressure (MAP) started at similar levels. Thus, MAP was 94 ± 8 mmHg in the control group, 94 ± 4 mmHg in the ligated group, and 92 ± 7 mmHg in the group with ligated kidneys undergoing dialysis. After more than 6 h of anesthesia, the animals had lower MAP values, which were 81 ± 18 mmHg, 79 ± 13 mmHg, and 65 ± 14 mmHg, respectively. For the rats undergoing dialysis, the dialysis resulted in a slight net removal of fluid (ultrafiltration), which likely explains the lower MAP in that group. However, the animals were all in good condition and the MAP levels were acceptable for all animals, which is shown by the highly efficient dialysis (see below).

2.3. Parallel Reaction Monitoring (PRM) of Orellanine

Targeted tandem mass spectrometry (LC-MS/MS) analyses were used to determine the profile of orellanine elimination over time in all groups. The linearity for the method was determined and the response of orellanine in the standard curve showed a linearity in the examined concentration range of 0.039 µg/mL to 15 µg/mL.

In the analysis of the study samples, a reference plasma sample was analyzed between the time series from the different animals. The reference sample contained a known amount of orellanine and was used to roughly estimate the orellanine concentrations in study samples and was used to enable a comparison between animals and groups. The lowest level of orellanine was observed in the control group at all time points while higher levels were observed in the animals without renal perfusion and without urine production (data not shown).

The peak in the extracted chromatogram corresponding to the orellanine peak in the different groups and in the reference eluted at a retention time (RT) of 5.2 min (Figure 3). Unexpectedly, peaks at a RT around 4 min were also detected in the samples from a later time point. During the PRM analysis, full fragment ion spectrum of orellanine (m/z 253.04 Da) was monitored continuously throughout the entire LC separation and the most intense fragment (m/z 236.2 Da) was extracted for the quantitation. Inspection of the full fragment spectra corresponding to orellanine and the additional peaks revealed identical fragment ion spectra (data not shown). These findings suggest a time-dependent formation of orellanine metabolites that are in-source fragmented during the analysis and, therefore, detected as orellanine. To confirm that the early eluting peaks are due to ion source fragmentation, parameters in the MS-method were set to minimize ion source fragmentation. As a result, the intensity of these peaks was significantly decreased and it was verified that the metabolites are unstable during the ionization in the analysis. Rats without kidney function (ligated kidneys and ligated + dialysis) had different profiles of metabolites than the control group, which indicates elevated amounts of metabolites remaining in the blood in these groups. Figure 4 shows how the formation of the metabolites in rats correlate with the decrease of orellanine levels over time in the ligated animals.

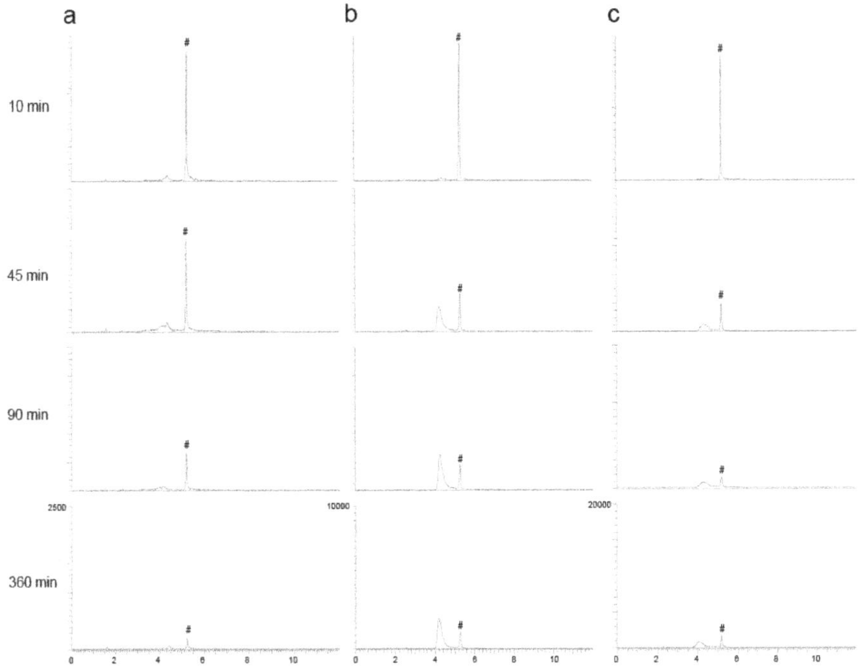

Figure 3. Representative extracted LC-MS/MS chromatograms from the analysis of plasma samples. Orellanine (#) has a retention time at 5.2 min. The orellanine peak is well separated from the metabolite peak seen with retention times around 4 min. Study samples from one representative animal with intact renal function (**a**) ligated kidneys (**b**) and ligated kidneys and dialysis (**c**) at four different time points after iv administration of orellanine. Metabolites are formed with time and rats with ligated kidneys and no urine production form most metabolites.

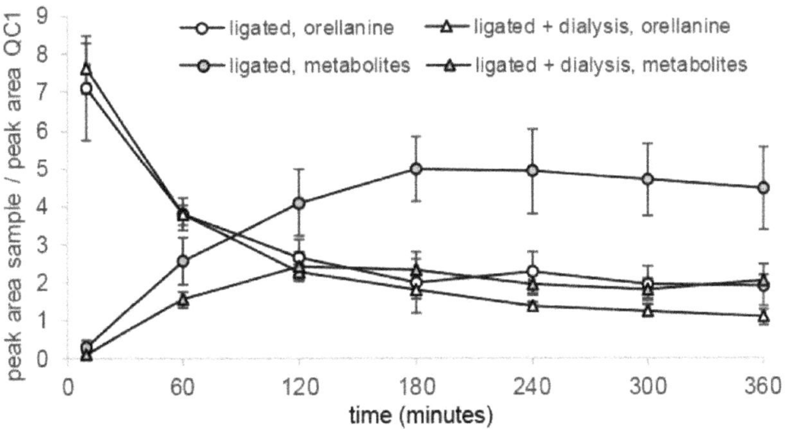

Figure 4. Formation of metabolites correlates with the elimination of orellanine. Rats with ligated kidneys and therefore no urine production with and without undergoing dialysis showed formation of the orellanine metabolites with time. This correlated with the profile of orellanine elimination. Rats that did not undergo dialysis showed higher levels of metabolites than the rats undergoing dialysis.

2.4. Plasma Concentration of Orellanine Versus Time and Half-Life

In order to investigate the level of orellanine in the plasma of the animals, two different methods were used: LC-MS/MS and β-scintillation. One difference between the methods is that LC-MS/MS detects orellanine while β-scintillation detects all molecules containing the ^{3}H-label and therefore cannot differentiate orellanine from any metabolites or break down products.

Orellanine levels measured by LC-MS/MS corresponds to the elimination of orellanine from plasma. After 60 min, the orellanine levels had decreased more than 50% in all groups due to the orellanine being taken up by renal cells or ending up in the urine, which is shown in the distribution experiments. Dialysis was initiated for one group of rats with ligated kidneys at time point 0 and the first change of dialysis fluid took place at +45 min. This renders a larger distribution volume in these rats and, therefore, the concentration of orellanine in the blood is the lowest in this group at time points 30 min and 45 min. After 360 min, rats with intact renal function and the rats with ligated kidneys receiving dialysis had eliminated most of their orellanine from the blood. The rats with ligated kidneys still had over 20% of the initial concentration left after 360 min (see Figure 5a).

Measurements of ^{3}H-labeled orellanine levels in the plasma showed a slower elimination during the distribution phase (0–45 min) than the LC-MS/MS measurements of orellanine. After 360 min and after measurement with beta scintillation of ^{3}H-labeled orellanine, all three groups of rats had higher amounts of the initial dose of orellanine left than when measuring orellanine with LC-MS/MS. The difference is due to measurements of the radioactive substance reflecting both orellanine and metabolites being formed. The half-life of orellanine was determined using data obtained from the time points between 45 min to 360 min. The measurements using LC-MS/MS resulted in half times of orellanine being 109 ± 6 min for rats with intact kidneys, 756 ± 98 min for rats with ligated kidneys, and 238 ± 28 min for rats with ligated kidneys undergoing dialysis. Measurements with beta scintillation using ^{3}H rendered half times for orellanine and its metabolites of 225 ± 10 min, 1033 ± 183 min and 583 ± 30 min, respectively (see Table 3).

Table 3. Pharmacokinetics of orellanine calculated from time-point 45 min to 360 min.

Parameter	LC-MS/MS, Orellanine			Beta Scintillation, ^{3}H-Orellanine		
	Intact Renal Function	**Ligated Kidneys**	**Ligated Kidney Dialysis**	**Intact Renal Function**	**Ligated Kidneys**	**Ligated Kidney Dialysis**
Intercept (nmol/L)	2115 ± 342	3028 ± 610	2521 ± 149	1471 ± 71	2538 ± 115	2773 ± 135
Rate of elimination (nmol/mL/min)	0.00651 ± 0.00032	0.000101 ± 0.00013	0.00338 ± 0.00063	0.0031 ± 0.00012	0.00085 ± 0.00019	0.00122 ± 0.00006
Half-life ($t^{1/2}$, min)	109 ± 60	756 ± 98	238 ± 28	225 ± 10	1033 ± 183	583 ± 30
Dose injected (nmol)	87.3 ± 1.5	85.4 ± 1.3	86.9 ± 1.2	87.6 ± 1.2	84.9 ± 1.3	87.2 ± 1.1
Area under the curve (AUC, nmol × min/L)	503,870 ± 70,670	3,494,674 ± 783,934	645,606 ± 99,420	821,601 ± 52,808	3,223,824 ± 230,783	2,823,938 ± 123,739
Volume of Distribution (V_D, mL)	31.7 ± 4.8	35.1 ± 8.6	49.1 ± 4.7	35.3 ± 1.7	37.7 ± 4.9	26.3 ± 1.7

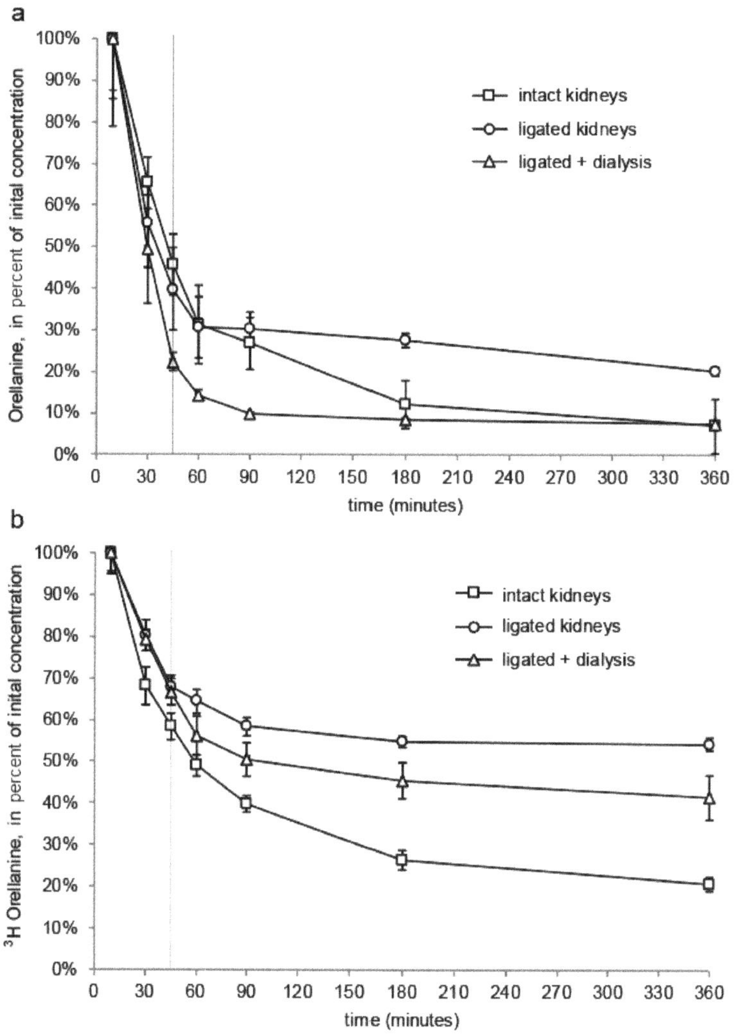

Figure 5. Concentration of orellanine over time presented as a percentage of the concentration measured at 10 min from LC-MS/MS analysis (**a**) and ^{3}H-labeled orellanine and its corresponding metabolites by β-scintillation (**b**). The graphs can be divided into two parts with the first part showing the distribution phase and the second part showing the elimination phase. Measurements of orellanine using LC-MS/MS show that rats with intact kidney function have the most efficient elimination under the elimination phase when compared to rats without renal function (ligated kidneys). For rats without renal function, rats not undergoing dialysis have the highest amount of orellanine left after 360 min when compared to rats without renal function but undergoing dialysis. ^{3}H-labeled orellanine determined the concentration of both orellanine and its corresponding metabolites using beta scintillation. After 360 min, the rats with intact kidneys have the lowest levels of orellanine and its metabolites left.

2.5. Clearance of Orellanine

The clearance of orellanine was calculated from Equation 3 (see Methods section). The average clearances were 483 ± 24 μL/min for the controls, 75 ± 10 μL/min for rats with ligated renal arteries, and 251 ± 47 μL/min for ligated animals on dialysis. The differences between the groups are all statistically significant ($p < 0.05$).

The protein binding of orellanine was determined for plasma from different species and found to be 33.5% for rats (unpublished data). Correcting for this degree of protein binding, the clearance for ´free, unbound´ orellanine was calculated. The renal clearance of orellanine was estimated as the difference between the clearance of the controls and the rats without renal function (i.e., including the effect of ligating renal arteries and protein binding), which resulted in an average value of 613 ± 52 μL/min. Similarly, the clearance under acute peritoneal dialysis was determined as the difference between the rats with ligated renal arteries with and without dialysis and was 263 ± 86 μL/min.

The clearance of orellanine and its metabolites was determined from the elimination of ^{3}H-labeled orellanine using similar calculations as the calculations of orellanine in the LC-MS/MS experiments. The average clearance was 232 ± 9 μL/min for the controls, 63 ± 14 μL/min for rats with ligated kidneys, and 90 ± 5 μL/min for ligated kidneys on dialysis.

Note that the clearance of ^{3}H-orellanine in rats with normal renal function was much lower than for orellanine-treated rats ($p < 0.001$) but higher than for ^{3}H-orellanine of rats with ligated kidneys without dialysis and rats with ligated kidneys on dialysis ($p < 0.05$), which suggests a higher degree of protein binding for the metabolites.

3. Discussion

Although research so far has explored the effects of orellanine toxicity and effects on kidney function [13,21], there is a need for a better understanding of the pharmacokinetic properties of this nephrotoxin. This might help in understanding future deadly webcap poisoning cases in improving clinical management [18] as well as opening the way for orellanine as a potential cure for metastatic end stage renal carcinoma [8]. New and curative therapy options for this type of cancer is needed since the outcome for patients with metastatic disease still is poor even though several therapeutic options have been suggested [22–25]. We have shown in an earlier paper that orellanine toxicity extends to the renal carcinoma cells in vitro and to human renal cell carcinomas tumors on rats [8], which indicates that orellanine could have a future as an anti-renal cancer treatment.

In this study, we have determined the pharmacokinetic properties of orellanine and its metabolites. Our work shows that orellanine is eliminated rapidly from plasma with a half-life of 109 min in anesthetized rats and mainly ends up in the kidney cortex and urine. Reduced renal function in the rats obtained by repeated dosing of orellanine did not affect the distribution pattern. The half-life of orellanine and its metabolites was twice as long (222 min), which most likely reflects a higher degree of protein binding of the metabolites. The renal clearance of orellanine not bound to protein was 613 μL/min, which is roughly 50% of the glomerular filtration rate (GFR) of awake rats reported to be 1310 μL/min [26]. Orellanine was easily removed by acute peritoneal dialysis with a clearance of 263 μL/min. Metabolites of orellanine were eliminated by peritoneal dialysis and by the kidneys but removal is far slower than orellanine.

There are three potential explanations for the clearance of orellanine being lower than the expected GFR level: First, the discrepancy suggests that orellanine is freely filtered across the glomerular barrier to urine where the compound is reabsorbed by proximal tubular cells and returned to plasma. Thus, with a 95% interval of confidence of renal clearance for orellanine of 491–735 μL/min, between 37% to 56% of the filtered orellanine is likely to have been reabsorbed. Second, anesthesia and abdominal surgery may reduce GFR even though the effect is expected to be small with isoflurane as anesthetic and minimal surgical procedures [27,28]. Third, orellanine may acutely reduce GFR by 50% due to a direct toxic effect on nephrons [3,5,29]. However, there are no reports of direct effects of orellanine on the renal vasculature, the glomerular capillaries, or the mesangial cells that could explain such an immediate reduction of GFR. Therefore, the first alternative seems most plausible albeit speculative since independent measurements of GFR are lacking.

The LC-MS/MS analysis suggests a time-dependent formation of orellanine metabolites eluting at a shorter RT compared to orellanine. Both the fragmentation pattern of the metabolites and evidence that their detection is effected by in-source fragmentation suggests that the metabolites are orellanine conjugated with a charged group at the hydroxyl group/s that falls off during the ionization. We were not able to determine the m/z of the intact metabolites due to the lower ionization efficiency obtained when changing the MS-parameters to also reduce the in-score fragmentation. In a previous study of mushrooms extracts, mono-glucopyranoside and diglucopyranoside were demonstrated to be naturally occurring glucosides of orellanine [7]. Furthermore, these mushroom glucosides were eluting before orellanine as well as had a similar fragmentation pattern as orellanine compared to the metabolites detected in the present study. Hydroxyl groups in aromatic systems are easily conjugated with glucuronic acid, which results in a more polar metabolite compared to the parent compound. These conjugates can be unstable in the ion source and are, therefore, detected as the parent compound in the MS-analysis. Moreover, the glucosides in the mushroom extract were shown to hydrolyze to orellanine in an acidic environment [7]. Re-analysis of the plasma samples after storage in an acidic environment also indicated that these metabolites are hydrolyzed over time. Therefore, we speculate that the metabolites formed in the present study could be glucosides of orellanine formed in the circulation. The two most well-known metabolites of orellanine are orellinine and orelline, but they are not the metabolites found in this experiment.

The slower elimination of orellanine measured with β-scintillation was most likely due to the formation of the metabolites during the experiment, which is shown in Figure 4. In the LC-MS/MS analysis, the elimination of orellanine was monitored while, in the β-scintillation measurement, orellanine and its metabolites cannot be differentiated and both are measured together.

The distribution of orellanine in the rats after injection with the toxin supports a rapid elimination from the blood with most orellanine ending up in the kidney cortex and urine. Other organs displayed higher levels of ^{3}H than the blood after 24 h except for the kidney cortex and urine, the liver, spleen, and bone marrow even though these levels were much lower than for the kidney and urinary system. There are no reports of patients with orellanine intoxication having any damage to any other organs except the kidneys [18].

These results seem to suggest the presence of specific renal transporters responsible for uptake of orellanine from urine into tubular cells and possibly back in intact form to blood. Hereby, the half-life may be longer than expected due to a small solute being freely filtered across the glomerular wall. It was not within the scope of this paper to identify the proteins responsible for such uptake of orellanine. However, revealing these transporters will be key for the understanding of the toxicological and potentially therapeutic effects of orellanine.

4. Conclusions

In conclusion, orellanine is mainly eliminated by renal excretion involving free glomerular filtration and tubular reabsorption. The compound also forms metabolites and they appear to have stronger protein binding properties when compared to the intact orellanine. Therefore, they remained longer in the system. The metabolites formed are likely to be glucosides of orellanine, which are naturally occurring in orellanine-containing mushrooms. In light of this, we conclude that this nephrotoxic compound may be eliminated rapidly through the urine or by dialysis.

5. Materials and Methods

5.1. Test Solution of Orellanine

Pure orellanine was synthesized by Ramidus AB, IDEON Lund, Sweden, as a 99% pure freeze-dried powder without detectable contaminations of its metabolites. The substance was kept dry and protected from light at room temperature. The orellanine was dissolved in 3 M HCl (Sigma-Aldrich, Steinheim, Germany) and the pH was carefully raised by adding portions of small amounts of 10 M NaOH (Sigma-Aldrich, Steinheim, Germany). In this process, the orellanine precipitates before becoming a clear solution. After a clear solution was obtained, the pH was normalized with 3 M HCl to a pH of 7.4 to 7.5. The solution was further diluted in PBS without magnesium and calcium (Lonza, Verviers, Belgium) to obtain an orellanine stock solution of 7.6 mg/mL containing 1 M NaCl. The solution was sterile filtered, aliquoted, and stored in −80 °C until use. The whole process took place in a dark room.

Radiolabeling of orellanine with tritium (^3H) was done by the Red Glead Discovery AB, Lund, Sweden. The procedure resulted in > 95% bound ^3H-labeled orellanine with a specific activity of 35.5 Ci per mmol orellanine. The structure of ^3H-labeled orellanine is shown in Figure 1b.

5.2. Rat Experiments

For the radioluminography experiments, male Wistar rats (Taconic, Ebjy, Denmark) were used which weighed, on average, 200 g on arrival. For the pharmacokinetic experiments, male Sprague Dawley rats (Charles River, Wilmington, MA, USA) with, on average, 200 g body weight on arrival were used.

5.3. Radioluminography

Two setups of radioluminography experiments were conducted. One setup with one single dose of orellanine (5 mg/kg) was conducted and one setup with two doses was conducted. For the single dose experiment, 10 male Wistar rats were administered ^3H-labeled orellanine formulated to 0.4 mCi/mL and 1.25 mg/mL in physiological saline intravenously in the tail vein. Two rats were sacrificed at each 0.5 h, 1 h, 6 h, 12 h, and 24 h after administration and one animal per time point was embedded in a gel of aqueous carboxymethyl cellulose and frozen in ethanol at −70 °C. For the second set up with 2 doses of orellanine, 12 male Wistar rats were pre-treated with unlabeled orellanine (1.25 mg/mL), intraperitoneally. After 72 h, the animals were administered ^3H-labeled orellanine (0.471 mCi/mL, 1.28 mg/mL in physiological saline) intravenously in the tail vein. Three rats were sacrificed at each 0.5 h, 1 h, 6 h, and 12 h after administration of the ^3H-labeled formulation. For both setups, 30 μm sections were cut at different levels from each embedded animal. The obtained sections were freeze-dried at −20 °C sections and put on ^3H-imaging plates. Together with the ^3H calibration standards (^3H-radioactivity mixed with whole blood), the images were exposed 70 h to 96 h. After exposure on imaging plates, the plates were scanned at a pixel size of 50 μm using BAS 2500 (Fuji Film Sverige AB, Stockholm, Sweden) and quantified using AIDA, version 4.19 (Raytest, Straubenhardt, Germany). For each time point, the radioactivity was determined as the mean value of measurements of three separate sections for each tissue.

5.4. Pharmacokinetic Studies

Sprague Dawley rats were randomly divided into three groups including sham operated control rats (n = 10), rats with bilaterally ligated renal arteries and hence no urine production (n = 7), and rats with ligated renal arteries treated with peritoneal dialysis (PD) (n = 8). Rats undergoing PD got a PD-catheter (PE-50, Solomon Scientific, Skokie, IL, USA) inserted into the abdominal cavity. After stabilization of the rats, 5 mg/kg body weight of orellanine containing trace amounts of ^3H-orellanine was injected into the jugular vein. After a flush of 1 mL glucose-bicarbonate-NaCl solution, a continuous slow infusion of the same solution was started (infusion rate of 17 μL/min, i.e., 1 mL/h). Approximately 400 µL of blood was drawn at the time points: 10, 30, 45, 60, 90, 180, and 360 min. For the group of rats undergoing PD, dialysis was immediately initiated by filling the abdominal cavity with 15 mL of 1.5% glucose solution (Gambro AB, Lund, Sweden). Every 45 min (± 10 min) the PD-fluid was drained and collected. After the experiment, the animals were euthanized with an anesthetic overdose of isoflurane and cardiac excision.

5.5. Beta Scintillation of ^3H

To each of the beta scintillation tubes containing plasma, 3 mL of quenching solution was added. Radioactivity was measured using a Beta Scintillator (Liquid Beckman LS 6500 Scintillation Counter, *Beckman* Coulter Inc., Brea, CA, USA) and the appropriate protocol for tritiated material (^{3}H) was used according to the manufacturer.

5.6. Bioanalysis of Orellanine

Plasma samples were filtered through a molecular weight cut-off filter (Nanosep 30k Omega filters, Pall Life Sciences, Port Washington, NY, USA) in order to remove plasma proteins and higher molecular weight biomolecules. Orellanine was extracted from the plasma (20 µL) with the addition of formic acid (final concentration 4% (*v/v*), final volume 84 µL), mixed for 10 min followed by centrifugation (30 min at 1200 rpm). Flow-through samples were collected and transferred into vials. A standard curve within the concentration range 0.039 µg/mL to 15 µg/mL (0.039, 0.78, 0.16, 0.31, 0.63, 1.25, 2.5, 5, 10, and 15 µg/mL was prepared with the addition of orellanine to human plasma. Reference plasma (quality control) samples were prepared with the addition of orellanine to human plasma and final concentrations of 1.25 µg/mL and 5 µg/mL. The linearity was determined to be 0.99 (r^2) within this concentration range. Reference plasma samples and standard curves were filtered in parallel, which is described in Section 5.5. The precision between multiple injections was less than 10% deviation and the accuracy was high with less than 5% between the theoretical and experimental amounts for the reference sample. In the analysis of the study samples, a reference plasma sample was analyzed between the time series from the different animals in each LC-MS/MS run. The reference sample was used to estimate the orellanine concentrations in study samples as well as enable a comparison between animals, groups, and LC-MS/MS experiments. The plasma protein binding of orellanine was determined after 4 h of dialysis (data not shown and found to be quite low, i.e., 33.5% for rat and 42.5% for humans).

The method for LC-MS/MS used was modified from Herrmann et al. [7]. The samples were analyzed using parallel reaction monitoring (PRM) on an LTQ Orbitrap Velos mass spectrometer interfaced to an UltiMate 3000 system (Thermo Fisher Scientific, Waltham, MA, USA). Samples (7.5 µL injection volume) were separated on an Acquity UPLC Peptide CSH C18 column (100 × 2.1 mm, 1.7 µm, Waters, Milford, MA, USA) using a gradient starting with 2.5 min isocratic separation with 5% B followed by a rise from 5% to 60% of B within 1 min and finally made isocratic with 60% B for 2.5 min at 45 °C with a flow rate of 200 µL/min. Mobile phase A was 0.2% formic acid in 2 mM ammonium formate and mobile phase B was acetonitrile in 0.2% formic acid. Orellanine parent ion mass (*m/z* 253.05 Da) was isolated in the ion trap with a 2 Da isolation window. The collision energy was set to 25 with a scan range *m/z* 200.00-260.00. The most intense fragment (*m/z* 236.2 Da, corresponding to loss of 17 Da and OH) was selected for quantification. The peak areas were determined by the extraction and integration of this fragment in the fragmentation spectra using XCalibur (Thermo Fisher Scientific, Waltham, MA, USA) and were used for the determination of the profiles for half-life calculation. Each study sample was injected twice and the average peak area was calculated. The reference samples at 1.25 µg/mL and 5 µg/mL were analyzed before and after each time point from an animal to compensate for day-to-day variation in the analyses and to be able to compare between UPLC-MS runs. The average peak area of orellanine in the study samples was divided by the average peak area for the reference sample at 1.25 µg/mL in the analysis sequence. To roughly estimate the concentrations in study samples, the ratios were multiplied by 1.25 µg/mL.

5.7. Pharmacokinetic Analysis

For the pharmacokinetic analysis, a first order kinetic model was used. According to the model, the elimination of a solute after an i.v. injection is determined by an elimination rate constant, k_{el}. Thus, the concentration at time t can be estimated using the following expression where C_0 is the concentration at the time of injection.

$$C(t) = C_0 \times e^{-k_{el} \times t} \tag{1}$$

The constant k_{el} is dependent on the clearance (K) and the distribution volume (V) and it has been shown that $k_{el} = K/V$. Rearranging Equation 1 and inserting K/V gives the relationship below.

$$\frac{Kt}{V} = Ln\left(\frac{C(t)}{C_0}\right) \tag{2}$$

Clearance of orellanine can be determined with the equation below.

$$K = k_{el} \times V \tag{3}$$

For clearance calculations, V was assumed to equal the extracellular fluid volume (ECV) estimated independently of the kinetic modeling from data in the literature [26].

The exact dose of ^3H-labeled orellanine given was estimated by determining the activity in the injected solution (cpm/mg) and the exact weight of the solution injected. The latter was determined by weighing the syringe before and after injection with 0.1 mg precision. Similarly, the amount of intact orellanine was determined from the concentration of the test solution and the precise injected volume. The half-time is given by Equation (4).

$$t_{1/2} = ln2/k_{el} \tag{4}$$

The area under the curve is calculated by the equation below.

$$AUC = \int_{t=0}^{t=\infty} Cp(t)dt \tag{5}$$

The volume of distribution can be derived from Equation (6).

$$V_D = \frac{Dose}{AUC \cdot k_{el}} \tag{6}$$

The accuracy of Equation (6) depends on how many h of sampling is done and becomes most accurate for sampling periods of 24 h, which was not technically feasible.

5.8. Statistics

Results are presented as the mean ± standard error of the mean. Differences between groups were determined using parametrical tests such as the Student t-test. Statistical significance was defined as $p < 0.05$.

5.9. Ethical Approval

All applicable national and institutional guidelines for the care and use of animals were followed. All procedures performed in studies involving animals were in accordance with the ethical standards of the institution at which the studies were conducted. The Regional Ethical Board in Gothenburg, Sweden approved the experiments on 12 June 2012, no. 144–12.

Author Contributions: Conceptualization, J.N., B.H., and K.E. Methodology, B.H., D.N., C.S., and A.T. Writing-Original Draft, D.N., K.E., B.H., and J.N. Writing-Review & Editing, D.N., B.H., A.T., C.S., J.N., and K.E. Funding Acquisition, J.N., B.H., and D.N.

Funding: This research was funded by the Swedish Research Council grant no. 09898 (B.H.) and 14764 (J.N.), the Swedish Cancer foundation (B.H.), the Inga-Britt and Arne Lundberg Research Foundation (B.H.), and the John and Brit Wennerström Research Foundation (D.N.).

Acknowledgments: The Proteomics Core Facility at the Sahlgrenska Academy, Gothenburg University performed the LC-MS/MS analysis. The radioluminograph study was partly performed by Active Biotech, Lund, Sweden.

Conflicts of Interest: The following authors are shareholders in a company, Oncorena AB, formed to commercially explore the potential of orellanine: B.H. & J.N. The other authors declare no conflict of interest.

References

1. Grzymala, S. Massenvergiftungen durch den orangefuchsigen hautkopf. *Z. Pilzkd* **1957**, *23*, 139–142.
2. Mottonen, M.; Nieminen, L.; Heikkila, H. Damage caused by two finnish mushrooms, cortinarius speciosissimus and cortinarius gentilis on the rat kidney. *Z. Naturforsch C* **1975**, *30*, 668–671. [CrossRef] [PubMed]
3. Nilsson, U.A.; Nystrom, J.; Buvall, L.; Ebefors, K.; Bjornson-Granqvist, A.; Holmdahl, J.; Haraldsson, B. The fungal nephrotoxin orellanine simultaneously increases oxidative stress and down-regulates cellular defenses. *Free Radic. Biol. Med.* **2008**, *44*, 1562–1569. [CrossRef] [PubMed]
4. Schumacher, T.; Hoiland, K. Mushroom poisoning caused by species of the genus cortinarius fries. *Arch. Toxicol.* **1983**, *53*, 87–106. [PubMed]
5. Cantin-Esnault, D.; Richard, J.M.; Jeunet, A. Generation of oxygen radicals from iron complex of orellanine, a mushroom nephrotoxin; preliminary esr and spin-trapping studies. *Free Radic. Res.* **1998**, *28*, 45–58. [CrossRef] [PubMed]
6. Richard, J.M.; Cantin-Esnault, D.; Jeunet, A. First electron spin resonance evidence for the production of semiquinone and oxygen free radicals from orellanine, a mushroom nephrotoxin. *Free Radic. Biol. Med.* **1995**, *19*, 417–429. [CrossRef]
7. Herrmann, A.; Hedman, H.; Rosen, J.; Jansson, D.; Haraldsson, B.; Hellenas, K.E. Analysis of the mushroom nephrotoxin orellanine and its glucosides. *J. Nat. Prod.* **2012**, *75*, 1690–1696. [CrossRef] [PubMed]
8. Buvall, L.; Hedman, H.; Khramova, A.; Najar, D.; Bergwall, L.; Ebefors, K.; Sihlbom, C.; Lundstam, S.; Herrmann, A.; Wallentin, H.; et al. Orellanine specifically targets renal clear cell carcinoma. *Oncotarget* **2017**, *8*, 91085–91098. [CrossRef] [PubMed]
9. Antkowiak, W.Z.; Gessner, W.P. The structures of orellanine and orelline. *Tetrahedron Lett.* **1979**, *20*, 1931–1934. [CrossRef]
10. Antkowiak, W.Z.; Gessner, W.P. Photodecomposition of orellanine and orellinine, the fungal toxins ofcortinarius orellanus fries andcortinarius speciossimus. *Cell Mol. Life Sci.* **1985**, *41*, 769–771. [CrossRef]
11. Dehmlow, E.V.; Schulz, H.-J. Synthesis of orellanine the lethal poison of a toadstool. *Tetrahedron Lett.* **1985**, *26*, 4903–4906. [CrossRef]
12. Spiteller, P.; Spiteller, M.; Steglich, W. Occurrence of the fungal toxin orellanine as a diglucoside and investigation of its biosynthesis. *Angew. Chem. Int. Ed. Engl.* **2003**, *42*, 2864–2867. [CrossRef] [PubMed]
13. Dinis-Oliveira, R.J.; Soares, M.; Rocha-Pereira, C.; Carvalho, F. Human and experimental toxicology of orellanine. *Hum. Exp. Toxicol.* **2016**, *35*, 1016–1029. [CrossRef] [PubMed]
14. Oubrahim, H.; Richard, J.M.; Cantin-Esnault, D.; Seigle-Murandi, F.; Trecourt, F. Novel methods for identification and quantification of the mushroom nephrotoxin orellanine. Thin-layer chromatography and electrophoresis screening of mushrooms with electron spin resonance determination of the toxin. *J. Chromatogr. A* **1997**, *758*, 145–157. [CrossRef]
15. Calvino, J.; Romero, R.; Pintos, E.; Novoa, D.; Guimil, D.; Cordal, T.; Mardaras, J.; Arcocha, V.; Lens, X.M.; Sanchez-Guisande, D. Voluntary ingestion of cortinarius mushrooms leading to chronic interstitial nephritis. *Am. J. Nephrol.* **1998**, *18*, 565–569. [CrossRef] [PubMed]
16. Delpech, N.; Rapior, S.; Cozette, A.P.; Ortiz, J.P.; Donnadieu, P.; Andary, C.; Huchard, G. Outcome of acute renal failure caused by voluntary ingestion of cortinarius orellanus. *Presse Med.* **1990**, *19*, 122–124. [PubMed]
17. Prast, H.; Pfaller, W. Toxic properties of the mushroom cortinarius orellanus (fries). II. Impairment of renal function in rats. *Arch. Toxicol.* **1988**, *62*, 89–96. [CrossRef] [PubMed]
18. Hedman, H.; Holmdahl, J.; Mölne, J.; Ebefors, K.; Haraldsson, B.; Nyström, J. Long-term clinical outcome for patients poisoned by the fungal nephrotoxin orellanine. *BMC Nephrol.* **2017**, *18*, 121. [CrossRef] [PubMed]

19. Rohrmoser, M.; Kirchmair, M.; Feifel, E.; Valli, A.; Corradini, R.; Pohanka, E.; Rosenkranz, A.; Poder, R. Orellanine poisoning: Rapid detection of the fungal toxin in renal biopsy material. *J. Toxicol. Clin. Toxicol.* **1997**, *35*, 63–66. [CrossRef] [PubMed]
20. Rapior, S.; Delpech, N.; Andary, C.; Huchard, G. Intoxication by cortinarius orellanus: Detection and assay of orellanine in biological fluids and renal biopsies. *Mycopathologia* **1989**, *108*, 155–161. [CrossRef] [PubMed]
21. Danel, V.C.; Saviuc, P.F.; Garon, D. Main features of cortinarius spp. poisoning: A literature review. *Toxicon* **2001**, *39*, 1053–1060. [CrossRef]
22. Capitanio, U.; Montorsi, F. Renal cancer. *Lancet* **2016**, *387*, 894–906. [CrossRef]
23. Thorstenson, A.; Bergman, M.; Scherman-Plogell, A.H.; Hosseinnia, S.; Ljungberg, B.; Adolfsson, J.; Lundstam, S. Tumour characteristics and surgical treatment of renal cell carcinoma in sweden 2005–2010: A population-based study from the national swedish kidney cancer register. *Scand. J. Urol.* **2014**, *48*, 231–238. [CrossRef] [PubMed]
24. Hu, S.L.; Chang, A.; Perazella, M.A.; Okusa, M.D.; Jaimes, E.A.; Weiss, R.H.; American Society of Nephrology Onco-Nephrology Forum. The nephrologist's tumor: Basic biology and management of renal cell carcinoma. *J. Am. Soc. Nephrol.* **2016**, *27*, 2227–2237. [CrossRef] [PubMed]
25. Albiges, L.; Choueiri, T.; Escudier, B.; Galsky, M.; George, D.; Hofmann, F.; Lam, T.; Motzer, R.; Mulders, P.; Porta, C.; et al. A systematic review of sequencing and combinations of systemic therapy in metastatic renal cancer. *Eur. Urol.* **2015**, *67*, 100–110. [CrossRef] [PubMed]
26. Davies, B.; Morris, T. Physiological parameters in humans and laboratory animals. *Pharm. Res.* **1993**, *10*, 1093–1095. [CrossRef] [PubMed]
27. Mercatello, A. Changes in renal function induced by anesthesia. *Ann. Fr. Anesth. Reanim.* **1990**, *9*, 507–524. [CrossRef]
28. Burchardi, H.; Kaczmarczyk, G. The effect of anaesthesia on renal function. *Eur. J. Anaesthesiol.* **1994**, *11*, 163–168. [PubMed]
29. Pfaller, W.; Gstraunthaler, G.; Prast, H.; Rupp, L.; Ruedl, C.; Michelitsch, S.; Moser, M. *Effects of the Fungal Toxin Orellanine on Renal Epithelium*; Marcel Dekker: New York, NY, USA, 1991; pp. 63–69.

toxins

MDPI

Article

Biological Activities of Cationicity-Enhanced and Hydrophobicity-Optimized Analogues of an Antimicrobial Peptide, Dermaseptin-PS3, from the Skin Secretion of *Phyllomedusa sauvagii*

Yining Tan [1], Xiaoling Chen [1,*], Chengbang Ma [1], Xinping Xi [1,*], Lei Wang [1], Mei Zhou [1], James F. Burrows [1], Hang Fai Kwok [2] and Tianbao Chen [1,*]

1 Natural Drug Discovery Group, School of Pharmacy, Queen's University Belfast, Belfast BT9 7BL, Northern Ireland, UK; ytan07@qub.ac.uk (Y.T.); c.ma@qub.ac.uk (C.M.); l.wang@qub.ac.uk (L.W.); m.zhou@qub.ac.uk (M.Z.); j.burrows@qub.ac.uk (J.F.B.)

2 Faculty of Health Sciences, University of Macau, Avenida de Universidade, Taipa, Macau, China; hfkwok@umac.mo

* Correspondence: xchen19@qub.ac.uk (X.C.); x.xi@qub.ac.uk (X.X.); t.chen@qub.ac.uk (T.C.); Tel.: +44-28-9097-2200 (X.C.); Fax: +44-28-9024-7794 (X.C.)

Received: 4 July 2018; Accepted: 3 August 2018; Published: 7 August 2018

Abstract: The skin secretions of the subfamily Phyllomedusinae have long been known to contain a number of compounds with antimicrobial potential. Herein, a biosynthetic dermaseptin-precursor cDNA was obtained from a *Phyllomedusa sauvagii* skin secretion-derived cDNA library, and thereafter, the presence of the mature peptide, namely dermaseptin-PS3 (DPS3), was confirmed by LC–MS/MS. Moreover, this naturally occurring peptide was utilized to design two analogues, $K^{5,17}$-DPS3 (introducing two lysine residues at positions 5 and 17 to replace acidic amino acids) and $L^{10,11}$-DPS3 (replacing two neutral amino acids with the hydrophobic amino acid, leucine), improving its cationicity on the polar/unipolar face and hydrophobicity in a highly conserved sequence motif, respectively. The results in regard to the two analogues show that either increasing cationicity, or hydrophobicity, enhance the antimicrobial activity. Also, the latter analogue had an enhanced anticancer activity, with pretreatment of H157 cells with 1 μM $L^{10,11}$-DPS3 decreasing viability by approximately 78%, even though this concentration of peptide exhibited no haemolytic effect. However, it must be noted that in comparison to the initial peptide, both analogues demonstrate higher membrane-rupturing capacity towards mammalian red blood cells.

Keywords: antimicrobial peptide (AMP); dermaseptin; anuran skin secretion; drug design; antimicrobial activity; anticancer activity

Key Contribution: This study reports the identification of a naturally occurring AMP, dermaseptin, from skin secretions and examines its potential biological activity. It also aims to improve the biological activity (e.g., antimicrobial and anticancer effects) via optimization of the peptide's cationic or hydrophobic properties through residue substitutions.

1. Introduction

Of the many anuran skin-derived peptides that are known, dermaseptin and dermaseptin-like peptides are the most remarkable candidates for developing new antibiotics in Hylidae frogs [1–3]. Although there is much heterogeneity in either the peptide sequence, or length, among dermaseptins, the family nevertheless shares several common structural characteristics, including Trp at position 3 and a conserved sequence of AA(G)KAALG(N)A in the mid-region [4]. In addition, dermaseptins commonly possess a high propensity to adopt an α-helical conformation in hydrophobic media, since the first dermaseptin peptide with 80% of α-helical conformation was isolated from Hylidae frogs [3,5,6]. Pharmacologically, apart from broad-spectrum antimicrobial activity (e.g., dermaseptin S4 and B), haemolytic activity and anticancer activity have been reported (dermaseptin-PH and B2) [3,7,8].

Numerous studies indicate that the net charge is a key factor influencing the binding of AMPs to membranes, as AMPs bind to the membrane by electrostatic interaction and competitively replace the divalent cation [9–11]. Therefore, it is believed that changing the number of positive charges present in an AMP can likely change its membrane binding ability, resulting in a change in antimicrobial activity. Also, in general, approximately half of AMP amino acid residues are hydrophobic, and their hydrophobicity and their activity can be altered by changing the number of Leu, Ile and Val residues in these peptides. However, both antibacterial and haemolytic activities of these AMPs tend to get increased simultaneously by increasing the hydrophobicity, due to the fact that the hydrophobic groups play a key role in their insertion into the cell membrane [10].

Herein, we describe the discovery of a biosynthetic precursor, preprodermaseptin, encoding an antimicrobial peptide, DPS3, from the skin secretion of *Phyllomedusa sauvagii* using a combination of shotgun cloning and mass spectrometry. The corresponding chemically synthesised replicate exerted weak antibacterial activity towards pathogenic microorganisms and weak cytotoxic activity towards tumour cells. Therefore, we designed two analogues of this naturally occurring peptide, $K^{5,17}$-DPS3 and $L^{10,11}$-DPS3, to potentially optimize its cationicity on the polar/unipolar face and hydrophobicity in conserved sequence motif of dermaseptin, respectively.

2. Results

2.1. Molecular Cloning of a DPS1 Precursor cDNA from a Skin Secretion-Derived cDNA Library

Using the shotgun cloning strategy, the nucleotide sequence of a full-length biosynthetic precursor-encoding cDNA was consistently cloned among the artificially reconstructed cutaneous secretion-derived cDNA library from *Phyllomedusa sauvagii*. More specifically, the domain architecture of this preprodermaseptin transcript (Figure 1) compromises 70 amino acid residues, encoding a single copy of a peptide termed DPS3, where the C-terminus was subjected to post-translational modification with carboxyl-terminal amide formation. From the translated open reading frame, the KR is a typical convertase processing site in vivo and the resulting mature peptide consisted of 23 amino acid residues (ALWKDILKNAGKAALNEINQIVQ-amide). The cDNA precursor was deposited in GenBank database under an accession no. of MH536746.

```
    M  A  F   L  K  K  S   L  F  L   V  L  F   L  G  L  V
  1ATGGCGTTCC TAAAGAAATC TCTTTTCCTT GTACTATTCC TTGGATTGGT
   TACCGCAAGG ATTTCTTTAG AGAAAAGGAA CATGATAAGG AACCTAACCA
    S  L  T   I  C  E  E   E  K  R   E  N  E   D  E  M  E
 51CTCTCTTACT ATCTGTGAAG AAGAGAAAAG AGAAAATGAA GATGAAATGG
   GAGAGAATGA TAGACACTTC TTCTCTTTTC TCTTTTACTT CTACTTTACC
     Q  D  D   D  E  Q   S  E  M  K   R  A  L   W  K  D
101AACAAGATGA TGATGAGCAA AGTGAAATGA AGAGAGCTCT GTGGAAAGAT
   TTGTTCTACT ACTACTCGTT TCACTTTACT TCTCTCGAGA CACCTTTCTA
    I  L  K   N  A  G  K   A  A  L   N  E  I  N   Q  I  V
151ATATTAAAAA ATGCAGGAAA GGCTGCTTTA AATGAAATTA ATCAAATAGT
   TATAATTTTT TACGTCCTTT CCGACGAAAT TTACTTTAAT TAGTTTATCA
      Q  G  G   *  *
201ACAAGGAGGA TAATAAAGTA AGGAAGATAT AAAATGTAAT TAAATCAATT
   TGTTCCTCCT ATTATTTCAT TCCTTCTATA TTTTACATTA ATTTAGTTAA
251TGTTCCTCCT ATTATTTCAT TCCTTCTATA TTTTACATTA ATTTAGTTAA
   ACAAGGAGGA TAATAAAGTA AGGAAGATAT AAAATGTAAT TAAATCAATT
301ATCAATAATT GTGCCAACCC TATATTAAAG CATGCTGAAC CGAAAAAAAA
   TAGTTATTAA CACGGTTGGG ATATAATTTC GTACGACTTG GCTTTTTTTT
351AAGAAAAAAA AAAAAAAAAA AAAAAAA
   TTCTTTTTTT TTTTTTTTTT TTTTTTT
```

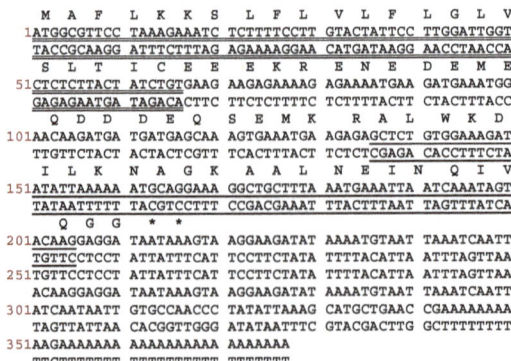

Figure 1. Nucleotide and translated open reading frame amino acid sequence of the cDNA encoding the biosynthetic precursor of a novel peptide from the skin secretion of *Phyllomedusa sauvagii*. The putative N-terminal signal peptide sequence is double-underscored, putative mature peptide sequence is single-underscored and an asterisk indicates the stop codon.

2.2. Isolation and Structural Characterisation of DPS3

The predicted amino acid sequence identified via cDNA cloning suggested the existence of a peptide in the skin secretion of *Phyllomedusa sauvagii*, so the lyophilized skin secretion was directly analysed to determine if this peptide was present. The presence of the mature DPS3 peptide was confirmed by RP–HPLC isolation, with the retention time at approximately 108 min, and MS/MS fragmentation sequencing (Figure S1, Figure 2 and Table 1, respectively).

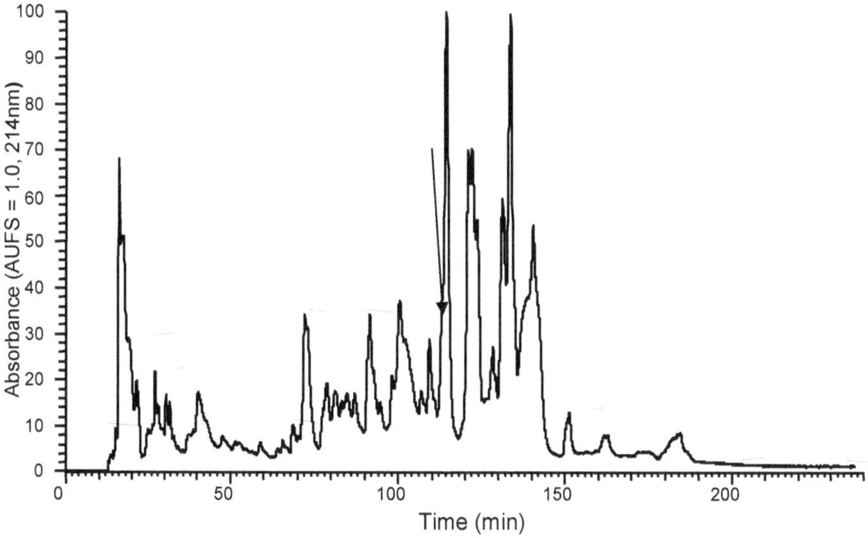

Figure 2. Region of RP–HPLC chromatogram of *Phyllomedusa sauvagii* skin secretion indicating the absorbance peak by an arrow that corresponds to DPS3.

Table 1. Predicted b-ion and y-ion MS/MS fragment ion series (singly and doubly charged) of DPS3. The observed ions are indicated by single-underlined. The unit for MS data is m/z.

#1	b (1+)	b (2+)	Seq.	y (1+)	y (2+)	#2
1	72.04440	36.52584	A			23
2	185.12847	93.06787	L	2478.41923	1239.71325	22
3	371.20779	186.10753	W	2365.33516	1183.17122	21
4	499.30276	250.15502	K	2179.25584	1090.13156	20
5	614.32971	307.66849	D	2051.16087	1026.08407	19
6	727.41378	364.21053	I	1936.13392	968.57060	18
7	840.49785	420.75256	L	1823.04985	912.02856	17
8	968.59282	484.80005	K	1709.96578	855.48653	16
9	1082.63575	541.82151	N	1581.87081	791.43904	15
10	1153.67287	577.34007	A	1467.82788	734.41758	14
11	1210.69434	605.85081	G	1396.79076	698.89902	13
12	1338.78931	669.89829	K	1339.76929	670.38828	12
13	1409.82643	705.41685	A	1211.67432	606.34080	11
14	1480.86355	740.93541	A	1140.63720	570.82224	10
15	1593.94762	797.47745	L	1069.60008	535.30368	9
16	1707.99055	854.49891	N	956.51601	478.76164	8
17	1837.03315	919.02021	E	842.47308	421.74018	7
18	1950.11722	975.56225	I	713.43048	357.21888	6
19	2064.16015	1032.58371	N	600.34641	300.67684	5
20	2192.21873	1096.61300	Q	486.30348	243.65538	4
21	2305.30280	1153.15504	I	358.24490	179.62609	3
22	2404.37122	1202.68925	V	245.16083	123.08405	2
23			Q-Amidated	146.09241	73.54984	1

2.3. Physicochemical Properties and Secondary Structures of DPS3 and Its Analogues

Both DPS3 and $L^{10,11}$-DPS3 possessed the same net positive charge of +2, which increased to +6 in the case of the cationicity-enhanced analogue (Table 2). Additionally, DPS3 and $K^{5,17}$-DPS3 had a similar degree of hydrophobicity, which was increased in $L^{10,11}$-DPS3. The helical wheel projects showed that DPS3 and its analogues have the same direction of summed vectors of hydrophobicity (Figure 3). Meanwhile, $K^{5,17}$-DPS3 had one more positive charge on both hydrophilic and hydrophobic faces than the other two analogues, and $L^{10,11}$-DPS3 showed an enlarged hydrophobic face. Also, although these three peptides existed in random coils in aqueous solution, they all adopted α-helical conformations in membrane-mimicking solution, presenting obviously negative peaks at 222 nm and 208 nm, with the natural peptide presenting the largest proportion of α-helical domain (44.9% of its secondary structure) (Figure 3 and Table 2).

Table 2. Physicochemical properties of DPS3 and its two analogues.

Peptide	Hydrophobicity (H)	Hydrophobic Moment (μH)	% Helix [1]	Net Charge
ALWKDILKNAGKAALNEIN QIVQ-NH_2	0.373	0.437	44.9	+2
ALWKKILKNAGKAALNKIN QIVQ-NH_2	0.349	0.437	39	+6
ALWKDILKNLLKAALNEIN QIVQ-NH_2	0.508	0.517	28.8	+2

[1] In 50% 2,2,2-trifluoroethanol (TFE)/10 mM ammonium acetate (NH_4Ac) solution.

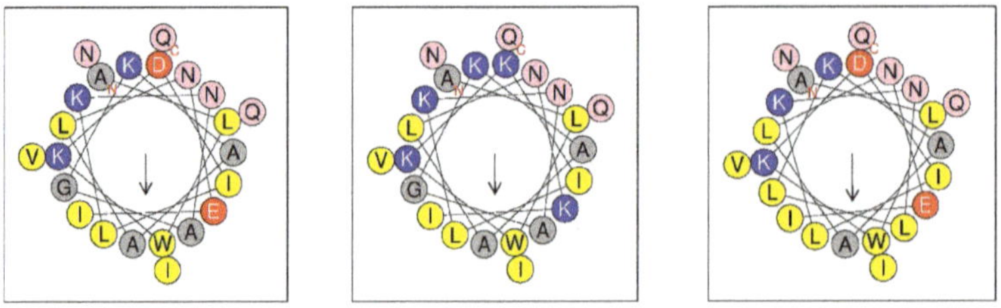

(a) (b) (c)

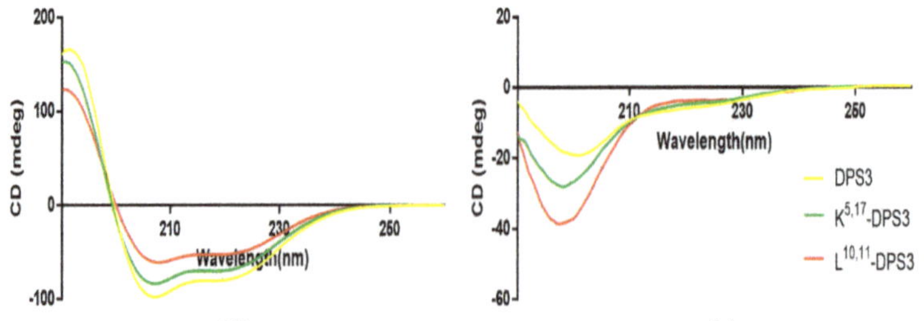

(d) (e)

Figure 3. Predicted helical wheel projections of the three peptides, (**a**) DPS3, (**b**) $K^{5,17}$-DPS3 and (**c**) $L^{10,11}$-DPS3; CDspectra recorded for 100 μM of DPS3 (yellow), $K^{5,17}$-DPS3 (green) and $L^{10,11}$-DPS3 (red) peptides in (**d**) 10 mM NH_4Ac/water solution and in (**e**) 50% TFE/10 mM NH_4Ac/water solution.

2.4. Antimicrobial Activity

The parent peptide, DPS3, generally showed weak antimicrobial activity, although it did exhibit better activity against Gram-negative bacteria. As expected, when compared to the parent peptide, both artificial analogues displayed enhanced antimicrobial activity against all the microorganisms examined (Table 3). In particular, $K^{5,17}$-DPS3 displayed MIC values of 8 μM or less against Gram-positive (*Staphylococcus aureus*) and Gram-negative (*Escherichia coli*) bacteria, as well as yeast (*Candida albicans*).

Table 3. Antimicrobial activity of the parent DPS3 peptide and its two analogues against various microorganisms.

Microorganisms	DPS3	$K^{5,17}$-DPS3	$L^{10,11}$-DPS3
		MIC (μM)	
S. aureus (NCTC 10788)	256	8	8
E. coli (NCTC 10418)	32	8	16
C. albicans (NCYC 1467)	64	4	16

2.5. Cytotoxicity of Peptides on Human Cancer and Normal Cells

DPS3 and its two artificial analogues all exhibited inhibitory effects on the proliferation of the two tested human cancer cell lines and normal cell line (Figure 4). Increasing the cationicity of the parent peptide had no significant influence on its antiproliferative activity, whereas altering its hydrophobicity markedly enhanced its antiproliferative activity, with this peptide exhibiting an IC_{50} value more than 10-fold lower than either of the other peptides (Table 4).

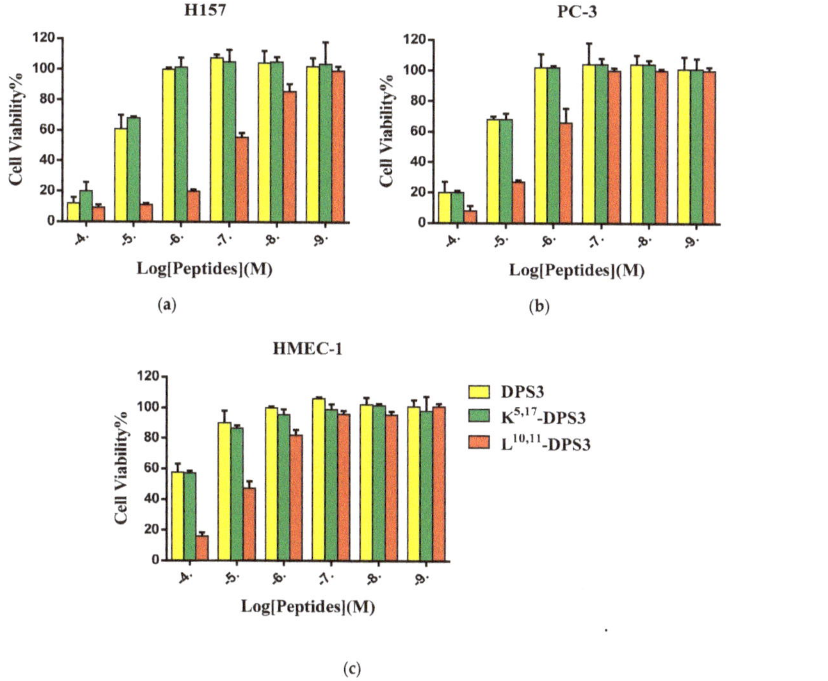

Figure 4. The cytotoxic effect of DPS3 (yellow), $K^{5,17}$-DPS3 (green) and $L^{10,11}$-DPS3 (red) on the human cancer cell lines (**a**) H157 and (**b**) PC-3, and normal cell line HMEC-1 (**c**).

Table 4. Induced cytotoxicity of DPS3 and analogues on the human cancer cells. IC_{50}s were calculated from the normalized curves in Figure 4 using GraphPad Prism 6 (GraphPad Software, USA).

Peptide	IC_{50} for H157 (μM)	IC_{50} for PC3 (μM)	IC_{50} for HMEC-1 (μM)
DPS3	15.67	18.20	132.10
$K^{5,17}$-DPS3	18.20	18.20	123.00
$L^{10,11}$-DPS3	0.12	1.85	8.76

2.6. Haemolysis Activity

All three peptides exhibited some haemolytic activity against healthy red blood cells (Figure 5). However, both artificial analogues exhibited a greater effect than the parent peptide, with the $L^{10,11}$-DPS3 analogue showing marked haemolysis even at lower concentrations. The HC_{50s} of DPS3, $K^{5,17}$-DPS3 and $L^{10,11}$-DPS3 are 138.1, 14.98 and 3.44, respectively.

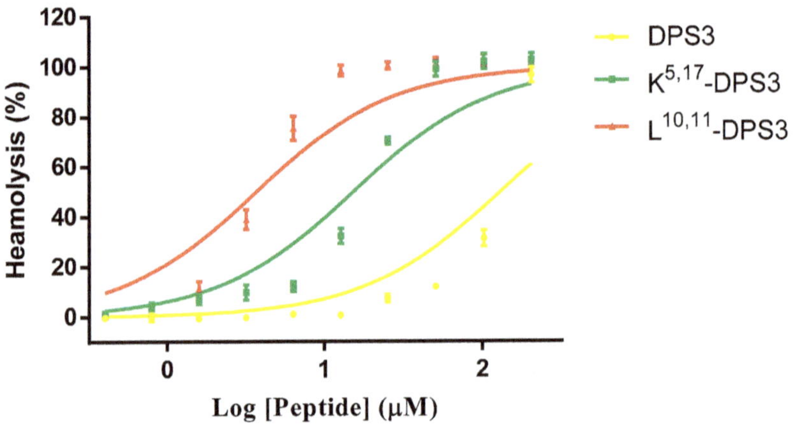

Figure 5. Haemolytic activity of DPS3 (yellow), $K^{5,17}$-DPS3 (green) and $L^{10,11}$-DPS3 (red) against horse red blood cells. The HC_{50s} of DPS3, $K^{5,17}$-DPS3 and $L^{10,11}$-DPS3 are 138.1, 14.98 and 3.44, respectively.

3. Discussion

Typically, most naturally occurring dermaseptins are 28 to 34 amino acid residues in length [12–14], therefore, the 23-mer DPS3 reported here is relatively short for this family. Huang and colleagues have previously isolated a similar-length dermaseptin, dermaseptin-PH [3]. Compared with other similar dermaseptins, the antimicrobial activity of DPS3 and dermaseptin-PH indicate that these two native truncated dermaseptins are less potent as AMPs than other longer dermaseptin peptides, possibly suggesting increasing peptide length in this family is potentially related to a higher antimicrobial activity. Besides, the physicochemical properties, including charge and hydrophobicity, are the main factors affecting antimicrobial activity, and therefore they are considered as one of the design parameters to optimize in AMPs [7,15–17]. In terms of the antibacterial mechanism of action of AMPs, it is mostly thought to concern the electrostatic interaction and hydrophobic engagement between AMPs and bacterial cell membranes [18]. Compared with other members of the dermaseptin family (Figure S2), DPS3 shares sequence similarities and they are canonical cationic and α-helical amphipathic peptides [1,3,5,8,19,20]. Besides, the conformational transition feature from coil to helix upon binding to lipid bilayers, in general, concerns a membrane-damaging mode of action in the dermaseptin family. In particular, the polycationic properties, as well as a large number of hydrophobic amino acids in the primary structure, and conformational alternation from random coil to helical frame among dermaseptins, along with their membrane-lytic activity, suggest their mechanism of action is likely to involve membrane disruption. Indeed, previous studies have found leakage and morphological alterations in the artificial bacterial membrane after treatment with fluorescent-labelled dermaseptins [21,22]. More recently, using the electron microscopy, a dermaseptin peptide, DS1, was found to distort the cell wall surface, proposing that the cytolysis or cell membrane disruption of *C. albicans* eventually cause cell death [23].

Early research has revealed a strong correlation between α-helical domain and antimicrobial activity, involving the local fusion of the membrane leaflets, pore formation, cracks, as well as the depth of membrane insertion [24–26]. Also, the α-helical conformation of dermaseptins is normally considered as one of the main factors in the hydrophobic interaction of AMPs and lipid layer [24,26]. In our investigation, the CD spectra of the three peptides showed that the amount of helical conformation is similar across all three. Our results indicate that an increase in antimicrobial activity can be achieved via optimization of cationic or hydrophobic properties of the dermaseptin peptides through residue substitution. However, both artificial analogues induced more haemolysis than the parent peptide, though $K^{5,17}$-DPS3 showed minimal haemolysis at the concentrations that exhibited antimicrobial activity, indicating it could still represent an interesting AMP. It is often supposed that the electrostatic interaction between anionic molecules (such as LPS, teichoic acids or acidic phospholipids) and positively charged AMPs is an essential step allowing the cationic peptides to selectively aggregate on the bacterial membrane; therefore, it is also believed that changing the number of positive charges present in an AMP can likely change its membrane binding ability [9–11]. Also, the selective interactions between lysine/arginine residues and anionic lipid membranes and reduced selectivity upon increasing hydrophobic properties of antimicrobials have been supported by X-ray scattering data [27,28]. We speculate that the cationicity-enhanced analogue shows more sensitive membrane binding ability towards bacterial cells than mammalian cells, therefore resulting in less haemolysis. The electrical attraction of these peptides to the membrane is important for their antimicrobial activity, and this could explain why the cationicity-enhanced analogue shows more potent antimicrobial activity than $L^{10,11}$-DPS3, but less haemolysis. Similarly, K_4K_{20}-S4, a dual lysine-substituted dermaseptin-S4, also shown increased antimicrobial activity, along with two-fold haemolytic potency and higher lipophilic affinity [29]. Previously, 2D-NMR has shown that the consensus motif AA(G)KAALG(N) among dermaseptins is adopting a well-defined α-helical structure [1,7]. Herein, we enhanced the hydrophobicity in this highly conserved motif with amphipathic α-helix and found more potent cytolytic action towards mammalian erythrocytes. On the other hand, as the hydrophobicity is important for membrane disruption, increasing the hydrophobicity could improve the peptides' ability to disrupt membranes, but not in a way that favours microorganism selectivity. In this regard, we proposed that it possibly results from stronger hydrophobic interaction within the core of the bacterial membrane. Taken together, the data presented indicate that it is possible to improve the membrane-lytic activity of these AMPs without increasing the peptide length.

DPS3, and its analogues, all exhibit antiproliferative effects on the tested cancer cells, although $L^{10,11}$-DPS3 exhibited an enhanced antiproliferative impact. Recent studies have shown that cancer cell membranes, similar to the bacterial membrane, carry a negative charge due to overexpression of anionic molecules, such as phosphatidylserine, sialic acid and glycosaminoglycans (GAGs) [7,17,18]. Therefore, although it is unclear how the dermaseptins mediate their anticancer activity, it could result from their interaction with and disruption of the cell membrane, similar to their antimicrobial action. However, dermaseptins have been found to interact with, and aggregate on, the surface of cancer cells, as well as being able to penetrate into cancer cells, without compromising the cell membrane [8,14,19,22]. Also, Dos Santos and colleagues found that an Alexa- and biotin-labelled version of the dermaseptin peptide, DRS-B2, could internalise into cancer cells, and implicated its non-protein binding partner GAGs, suggesting GAGs are possibly involved in dermaseptin internalization [8]. Notably, the viability of H157 cells pretreated with $L^{10,11}$-DPS3 (1 μM) decreased by approximately 78%, whilst no membrane lysis was observed in mammalian erythrocytes exposed to the same concentration of peptide, possibly suggesting that a nonlytic mechanism may be involved, at least at lower concentrations. At higher peptide concentrations (>1 μM), the higher anticancer cell impact is consistent with increasing haemolysis activity, suggesting that this is possibly due to cell membrane disruption. However, further investigations will be needed to confirm this further.

4. Conclusions

The 23-mer peptide DPS3 reported here is relatively short in comparison to other naturally occurring dermaseptins, which are commonly 28 to 34 amino acids. Although DPS3 exhibited relatively weak bioactivity for dermaseptin family members, we rationally designed two DPS3 analogues with the aim of improving its bioactivity. In particular, $K^{5,17}$-DPS3 and $L^{10,11}$-DPS3 both exhibited more potent antimicrobial activity, whilst $L^{10,11}$-DPS3 also exhibited enhanced anticancer activity, even at concentrations that had minimal impact upon healthy mammalian cells. This would suggest that $K^{5,17}$-DPS3 and $L^{10,11}$-DPS3 may have promise as antimicrobial agents, and $L^{10,11}$-DPS3 could have potential as an anticancer agent. However, further investigations will be required to determine the exact mode of action that these peptides utilise against the microbial and cancer cells.

5. Materials and Methods

5.1. Acquisition of *Phyllomedusa sauvagii* Dermal Secretions

Three specimens of *Phyllomedusa sauvagii* (4–6 cm snout-to-vent length) obtained from a commercial source in the United States were exposed to 12 h of light at 20–25 °C daily, and multivitamin -loaded crickets were provided as the fodder three-times/week. Following four-month breeding, dermal secretions were collected via surface electrical stimulation [30]. In summary, the skin surface was moistened with deionized water, followed by mild transdermal electric stimulation (5 V, 100 Hz, 140 ms pulse width). Finally, the secretions were collected by gently flushing the frog skin with deionized water, and they were lyophilized and stored at −20 °C until mRNA extraction. Animal procedure was performed according to the guidelines in the UK Animal (Scientific, Procedures) Act 1986, project license PPL 2694, issued by the Department of Health, Social Services and Public Safety, Northern Ireland. Procedures had been vetted by the IACUC of Queen's University Belfast, and approved on 1 March 2011.

5.2. Shotgun Cloning of a cDNA Encoding DPS3 Peptide Biosynthetic Precursor

The shotgun cloning employed the rapid amplification of cDNA ends (RACE) technique with a degenerate primer, which has been described previously [31]. Briefly, the mRNA from the skin secretion of *Phyllomedusa camba* was reverse-transcript to the cDNA library. The 3′-RACE was conducted using the cDNA library and the primer (5′-ACTTTCYGAWTTRYAAGMCCAAABATG-3′) that was designed to a segment of the 5′-untranslated region of cDNAs from *Phyllomedusa* species (accession nos. AJ251876, AJ005443). The RACE products were subjected to purification and cloned, and finally sequenced by the ABI 3100 Automatic Capillary Sequencer.

5.3. Identification and Analysis of Amino Acid Sequence

The identification of DPS3 from the skin secretion was performed according to our previous study [31]. Briefly, 5 mg of lyophilized skin secretion was separated using the RP–HPLC column (Jupiter C-18 250 mm × 10 mm, Phenomenex, Macclesfield, UK). The fraction was collected every minute and then analyzed using MALDI–TOF MS (Perseptive Biosystems, Voyager DE, Perseptive Biosystems, Framingham, MA, USA) and LCQ-Fleet ion-trap MS for sequencing (Thermo Fisher Scientific, San Francisco, CA, USA).

5.4. Design and Synthesis of DPS3 and Its Two Analogues

To investigate the effect of the increasing cationicity and hydrophobicity, the peptide DPS3 was used as the framework to design two analogues where the two acidic amino acids (at positions 5 and 17) and two neutral amino acid residues (at positions 10 and 11) were substituted with lysine and leucine residues, respectively. Accordingly, theses analogues are named $K^{5,17}$-DPS3 (ALWKKILKNAGKAALNKINQIVQ-NH_2) and $L^{10,11}$-DPS3 (ALWKDILKNLLKAALNEINQIVQ-NH_2). The mean hydrophobicity, hydrophobic moment and helical wheel projections of peptides were predicted by Heliquest (http://heliquest.ipmc.cnrs.fr/). Peptide synthesis was carried out using Tribute peptide synthesizer (Protein Technologies, Tucson, AZ, USA) along with Rink amide resin and standard Fmoc chemistry, which was described in a previous study [32].

5.5. CD Analysis of Synthetic Peptides

The secondary structure was determined by a CD spectrometer (Jasco J851, Tokyo, Japan). The analysis method was performed as previous study [32]. The obtained spectra were analysed by the BeStSel CD online analysis program (http://bestsel.elte.hu [33]) to calculate the proportion of α-helical conformation of each peptide.

5.6. Antimicrobial Assays

The antimicrobial activity of three peptides was evaluated via the minimal inhibitory concentrations (MICs) against *S. aureus* (NCTC 10788), *E. coli* (NCTC 10418) and *C. albicans* (NCYC 1467) [32]. Microorganisms were inoculated with peptide (1 to 512 μM) in a 96-well plate and determined at 550 nm using a Synergy HT plate reader (Biotech, Winooski, VT, USA).

5.7. Cell Viability of Human Cancer and Normal Cells

Cell viabilities were achieved using a typical MTT assay [32]. Briefly, non-small cell lung cancer cell line, H157, human prostate carcinoma cell line, PC-3, and dermal microvascular endothelium cell line, HMEC-1, were treated with synthesized peptides from 10^{-4} to 10^{-9} M. After adding MTT, the formazan crystals were dissolved and read by the plate reader at 570 nm.

5.8. Haemolysis Assay

The haemolytic activity of each peptide was determined using a 2% suspension of horse blood cells (supplied by TCS Biosciences Ltd., Buckingham, UK) as previous study [32]. The red blood cell suspension was treated by the peptide (512−1 μM) at 37 °C for 2 h. The release of haemoglobin detected by the plate reader at λ550 nm. 1% Triton X-100 and PBS were applied as the positive and negative controls, respectively.

Supplementary Materials: The following are available online at http://www.mdpi.com/2072-6651/10/8/320/s1, Figure S1: Annotated fragment ion spectrum of DPS3, Figure S2: Alignment of amino acid sequences of DPS3 and other dermaseptins.

Author Contributions: Conceived of and designed the experiments: T.C., M.Z. and L.W.; Performed the experiments: Y.T., X.C. and C.M.; Analyzed the data: H.F.K., X.C. and X.X.; Wrote the paper: Y.T., X.C. and X.X.; Edited the paper: J.F.B., X.X. and X.C.

Funding: This research received no external funding.

Conflicts of Interest: The authors declare no conflict of interest.

References

1. Nicolas, P.; El Amri, C. The dermaseptin superfamily: A gene-based combinatorial library of antimicrobial peptides. *Biochim. Biophys. Acta (BBA)-Biomembr.* **2009**, *1788*, 1537–1550. [CrossRef] [PubMed]
2. Nicolas, P.; Ladram, A. Dermaseptins. In *Handbook of Biologically Active Peptides*, 2nd ed.; Kastin, A.J., Ed.; Academic Press: Boston, MA, USA, 2013; Volume 3, pp. 350–363. ISBN 9780123850959.

3. Huang, L.; Chen, D.; Wang, L.; Lin, C.; Ma, C.; Xi, X.; Chen, T.; Shaw, C.; Zhou, M. Dermaseptin-PH: A Novel Peptide with Antimicrobial and Anticancer Activities from the Skin Secretion of the South American Orange-Legged Leaf Frog, *Pithecopus (Phyllomedusa) hypochondrialis*. *Molecules* **2017**, *22*, 1805. [CrossRef] [PubMed]
4. Amiche, M.; Ladram, A.; Nicolas, P. A consistent nomenclature of antimicrobial peptides isolated from frogs of the subfamily Phyllomedusinae. *Peptides* **2008**, *29*, 2074–2082. [CrossRef] [PubMed]
5. Mor, A.; Van Huong, N.; Delfour, A.; Migliore-Samour, D.; Nicolas, P. Isolation, amino acid sequence and synthesis of dermaseptin, a novel antimicrobial peptide of amphibian skin. *Biochemistry* **1991**, *30*, 8824–8830. [CrossRef] [PubMed]
6. Strahilevitz, J.; Mor, A.; Nicolas, P.; Shai, Y. Spectrum of antimicrobial activity and assembly of dermaseptin-b and its precursor form in phospholipid membranes. *Biochemistry* **1994**, *33*, 10951–10960. [CrossRef] [PubMed]
7. Kustanovich, I.; Shalev, D.E.; Mikhlin, M.; Gaidukov, L.; Mor, A. Structural requirements for potent versus selective cytotoxicity for antimicrobial dermaseptin S4 derivatives. *J. Biol. Chem.* **2002**, *277*, 16941–16951. [CrossRef] [PubMed]
8. Dos Santos, C.; Hamadat, S.; Le Saux, K.; Newton, C.; Mazouni, M.; Zargarian, L.; Miro-Padovani, M.; Zadigue, P.; Delbé, J.; Hamma-Kourbali, Y.; et al. Studies of the antitumor mechanism of action of dermaseptin B2, a multifunctional cationic antimicrobial peptide, reveal a partial implication of cell surface glycosaminoglycans. *PLoS ONE* **2017**, *12*, e0182926. [CrossRef] [PubMed]
9. Brown, K.L.; Hancock, R.E. Cationic host defense (antimicrobial) peptides. *Curr. Opin. Immunol.* **2006**, *18*, 24–30. [CrossRef] [PubMed]
10. Schmidtchen, A.; Pasupuleti, M.; Malmsten, M. Effect of hydrophobic modifications in antimicrobial peptides. *Adv. Colloid Interface Sci.* **2014**, *205*, 265–274. [CrossRef] [PubMed]
11. Zhang, L.; Chen, X.; Zhang, Y.; Ma, C.; Xi, X.; Wang, L.; Zhou, M.; Burrows, J.F.; Chen, T. Identification of novel Amurin-2 variants from the skin secretion of Rana amurensis, and the design of cationicity-enhanced analogues. *Biochem. Biophys. Res. Commun.* **2018**, *497*, 943–949. [CrossRef] [PubMed]
12. Galanth, C.; Abbassi, F.; Lequin, O.; Ayala-Sanmartin, J.; Ladram, A.; Nicolas, P.; Amiche, M. Mechanism of Antibacterial Action of Dermaseptin B2: Interplay between Helix–Hinge–Helix Structure and Membrane Curvature Strain. *Biochemistry* **2008**, *48*, 313–327. [CrossRef] [PubMed]
13. Silva, L.P.; Leite, J.R.S.; Brand, G.D.; Regis, W.B.; Tedesco, A.C.; Azevedo, R.B.; Freitas, S.M.; Bloch, C., Jr. Dermaseptins from *Phyllomedusa oreades* and *Phyllomedusa distincta*: Liposomes fusion and/or lysis investigated by fluorescence and atomic force microscopy. *Comp. Biochem. Physiol. Part A Mol. Integr. Physiol.* **2008**, *151*, 329–335. [CrossRef] [PubMed]
14. Hoskin, D.W.; Ramamoorthy, A. Studies on anticancer activities of antimicrobial peptides. *Biochim. Biophys. Acta (BBA)-Biomembr.* **2008**, *1778*, 357–375. [CrossRef] [PubMed]
15. Giangaspero, A.; Sandri, L.; Tossi, A. Amphipathic α helical antimicrobial peptides. A systematic study of the effects of structural and physical properties on biological activity. *Eur. J. Biochem.* **2001**, *268*, 5589–5600. [CrossRef] [PubMed]
16. Timofeeva, L.; Kleshcheva, N. Antimicrobial polymers: Mechanism of action, factors of activity, and applications. *Appl. Microbiol. Biotechnol.* **2011**, *89*, 475–492. [CrossRef] [PubMed]
17. Fjell, C.D.; Hiss, J.A.; Hancock, R.E.; Schneider, G. Designing antimicrobial peptides: Form follows function. *Nat. Rev. Drug Discov.* **2012**, *11*, 37. [CrossRef] [PubMed]
18. Schweizer, F. Cationic amphiphilic peptides with cancer-selective toxicity. *Eur. J. Pharmacol.* **2009**, *625*, 190–194. [CrossRef] [PubMed]
19. Mor, A.; Nicolas, P. Isolation and structure of novel defensive peptides from frog skin. *Eur. J. Biochem.* **1994**, *219*, 145–154. [CrossRef] [PubMed]
20. Charpentier, S.; Amiche, M.; Mester, J.; Vouille, V.; Le Caer, J.P.; Nicolas, P.; Delfour, A. Structure, synthesis, and molecular cloning of dermaseptins B, a family of skin peptide antibiotics. *J. Biol. Chem.* **1998**, *273*, 14690–14697. [CrossRef] [PubMed]
21. Riedl, S.; Zweytick, D.; Lohner, K. Membrane-active host defense peptides–challenges and perspectives for the development of novel anticancer drugs. *Chem. Phys. Lipids* **2011**, *164*, 766–781. [CrossRef] [PubMed]
22. Van Zoggel, H.; Carpentier, G.; Dos Santos, C.; Hamma-Kourbali, Y.; Courty, J.; Amiche, M.; Delbé, J. Antitumor and angiostatic activities of the antimicrobial peptide dermaseptin B2. *PLoS ONE* **2012**, *7*, e44351. [CrossRef] [PubMed]

23. Belmadani, A.; Semlali, A.; Rouabhia, M. Dermaseptin-S1 decreases Candida albicans growth, biofilm formation and the expression of hyphal wall protein 1 and aspartic protease genes. *J. Appl. Microbiol.* **2018**, *125*, 72–83. [CrossRef] [PubMed]
24. Mor, A.; Nicolas, P. The NH2-terminal alpha-helical domain 1–18 of dermaseptin is responsible for antimicrobial activity. *J. Biol. Chem.* **1994**, *269*, 1934–1939. [PubMed]
25. Hancock, R.E.; Chapple, D.S. Peptide antibiotics. *Antimicrob. Agents Chemother.* **1999**, *43*, 1317–1323. [CrossRef]
26. Lequin, O.; Ladram, A.; Chabbert, L.; Bruston, F.; Convert, O.; Vanhoye, D.; Chassaing, G.; Nicolas, P.; Amiche, M. Dermaseptin S9, an α-helical antimicrobial peptide with a hydrophobic core and cationic termini. *Biochemistry* **2006**, *45*, 468–480. [CrossRef] [PubMed]
27. Andreev, K.; Bianchi, C.; Laursen, J.S.; Citterio, L.; Hein-Kristensen, L.; Gram, L.; Kuzmenko, I.; Olsen, C.A.; Gidalevitz, D. Guanidino groups greatly enhance the action of antimicrobial peptidomimetics against bacterial cytoplasmic membranes. *Biochim. Biophys. Acta (BBA)-Biomembr.* **2014**, *1838*, 2492–2502. [CrossRef] [PubMed]
28. Andreev, K.; Martynowycz, M.W.; Huang, M.L.; Kuzmenko, I.; Bu, W.; Kirshenbaum, K.; Gidalevitz, D. Hydrophobic interactions modulate antimicrobial peptoid selectivity towards anionic lipid membranes. *Biochim. Biophys. Acta (BBA)-Biomembr.* **2018**, *1860*, 1414–1423. [CrossRef] [PubMed]
29. Miltz, J.; Rydlo, T.; Mor, A.; Polyakov, V. Potency evaluation of a dermaseptin S4 derivative for antimicrobial food packaging applications. *Packag. Technol. Sci. Int. J.* **2006**, *19*, 345–354. [CrossRef]
30. Tyler, M.J.; Stone, D.J.; Bowie, J.H. A novel method for the release and collection of dermal, glandular secretions from the skin of frogs. *J. Pharmacol. Toxicol. Methods* **1992**, *28*, 199–200. [CrossRef]
31. Gao, Y.; Wu, D.; Wang, L.; Lin, C.; Ma, C.; Xi, X.; Zhou, M.; Duan, J.; Bininda-Emonds, O.R.; Chen, T.; et al. Targeted modification of a novel amphibian antimicrobial peptide from Phyllomedusa tarsius to enhance its activity against MRSA and microbial biofilm. *Front. Microbiol.* **2017**, *8*, 628. [CrossRef] [PubMed]
32. Chen, X.; Zhang, L.; Wu, Y.; Wang, L.; Ma, C.; Xi, X.; Bininda-Emonds, O.R.; Shaw, C.; Chen, T.; Zhou, M. Evaluation of the bioactivity of a mastoparan peptide from wasp venom and of its analogues designed through targeted engineering. *Int. J. Biol. Sci.* **2018**, *14*, 599. [CrossRef] [PubMed]
33. Micsonai, A.; Wien, F.; Bulyáki, É.; Kun, J.; Moussong, É.; Lee, Y.H.; Goto, Y.; Réfrégiers, M.; Kardos, J. BeStSel: A web server for accurate protein secondary structure prediction and fold recognition from the circular dichroism spectra. *Nucleic Acids Res.* **2018**, *46*, W315–W322. [CrossRef] [PubMed]

toxins

MDPI

Article

Burkholderia Lethal Factor 1, a Novel Anti-Cancer Toxin, Demonstrates Selective Cytotoxicity in MYCN-Amplified Neuroblastoma Cells

Aleksander Rust [1,2,*], Sajid Shah [2,3], Guillaume M. Hautbergue [3] and Bazbek Davletov [2,*]

1 Structural and Molecular Biology, Division of Biosciences, Faculty of Life Sciences, University College London, London WC1E 6BT, UK

2 Department of Biomedical Science, University of Sheffield, Firth Court, Western Bank, Sheffield S10 2TN, UK; sajidsw1@hotmail.com

3 Sheffield Institute for Translational Neuroscience, Department of Neuroscience, University of Sheffield, 385a Glossop Road, Sheffield S10 2HQ, UK; g.hautbergue@sheffield.ac.uk

* Correspondence: a.rust@ucl.ac.uk (A.R.); b.davletov@sheffield.ac.uk (B.D.); Tel.: +44-207-679-2021 (A.R.); +44-114-222-5111 (B.D.)

Received: 18 May 2018; Accepted: 20 June 2018; Published: 27 June 2018

Abstract: Immunotoxins are being investigated as anti-cancer therapies and consist of a cytotoxic enzyme fused to a cancer targeting antibody. All currently used toxins function via the inhibition of protein synthesis, making them highly potent in both healthy and transformed cells. This non-specific cell killing mechanism causes dose-limiting side effects that can severely limit the potential of immunotoxin therapy. In this study, the recently characterised bacterial toxin Burkholderia lethal factor 1 (BLF1) is investigated as a possible alternative payload for targeted toxin therapy in the treatment of neuroblastoma. BLF1 inhibits translation initiation by inactivation of eukaryotic initiation translation factor 4A (eIF4A), a putative anti-cancer target that has been shown to regulate a number of oncogenic proteins at the translational level. We show that cellular delivery of BLF1 selectively induces apoptosis in neuroblastoma cells that display MYCN amplification but has little effect on non-transformed cells. Future immunotoxins based on this enzyme may therefore have higher specificity towards MYCN-amplified cancer cells than more conventional ribosome-inactivating proteins, leading to an increased therapeutic window and decreased side effects.

Keywords: ribosome-inactivating protein; BLF1; eIF4A; MYCN; cancer; neuroblastoma; apoptosis

Key Contribution: This study highlights that neuroblastoma cells that exhibits MYCN amplification are particularly sensitive to BLF1-mediated inhibition of eIF4A potentially providing a novel therapeutic strategy against MYCN-amplified cancer cells.

1. Introduction

A number of protein synthesis-inhibiting toxins isolated from plants (e.g., saporin) or bacteria (e.g., diphtheria toxin) are being investigated for use in cancer therapy as they are highly toxic to mammalian cells [1,2]. One of the major characteristics of these enzymatic toxins is that their catalytic nature conveys an extraordinarily high potency not possible with small molecules [3]. However, the non-selective toxic mechanism of protein synthesis inhibition means that these toxins are highly potent to all cell types. Re-targeting to specific cancer cells is, therefore, necessary and requires fusion of the toxin to either an antibody (or antibody fragment), or ligand specific for receptors highly expressed on the cancer cell surface [4]. Despite such targeting, low level presence of the receptors on healthy cells and non-specific cellular uptake are known to cause side effects such as vascular leak syndrome and hepatotoxicity for every targeted toxin developed to date. Toxins that maintain high potency but have a toxic mechanism more selective towards transformed cells would, therefore, be of high interest in targeted toxin design.

Burkholderia Lethal Factor 1 (BLF1) is a monomeric toxin from the bacterium *Burkholderia pseudomallei* that was characterised in 2011 and has been proposed as a possible anti-cancer agent [5,6]. As with saporin and diphtheria toxin, BLF1 has been shown to cause cytotoxicity with high potency by irreversible inhibition of translation initiation and subsequent protein synthesis. BLF1 targets the translation initiation phase of translation via inactivation of the eukaryotic initiation translation factor 4A (eIF4A) through deamidation of the glutamine 339 [4]. Translation initiation is the rate limiting step of protein synthesis and is up-regulated in most cancers, contributing to increased levels proteins involved in a number of oncogenic processes [7]. During translation initiation, the eIF4F complex is assembled from the mRNA cap-binding protein eIF4E, the RNA-helicase eIF4A and the scaffold protein eIF4G. These compose part of the 43S pre-initiation complex involved in scanning the 5′ UTR for the translation start [8]. The eIF4F complex acts as a central node upon which a number of oncogenic signalling pathways (e.g., Ras, PI3K/AKT/mTOR and Myc) converge [9]. eIF4A is an RNA-helicase that resolves the secondary structures found in the 5′ UTR of mRNAs. This is necessary for scanning of the 5′ UTR by the pre-initiation complex to reach the translation start site [8]. It has recently been shown that a subset of mRNAs with long and complex 5′UTRs that contain G-quadruplex secondary structures have high dependence on eIF4A activity [10]. A number of important proto-oncogenes such as c-Myc, cell cycle regulators and survival proteins have been shown to be regulated by this mechanism and inhibition of eIF4A leads to the preferential down-regulation of these proteins, triggering growth arrest and cell death. Indeed, pre-clinical testing of small molecule inhibitors of eIF4A such as rocaglates and hippuristanol have shown efficacy in a number of cancer models [11–14]. The unique enzymatic inhibitory mechanism of BLF1 may therefore offer advantages over conventional toxins for targeted toxin therapy.

We have previously shown that delivery of recombinantly expressed BLF1 into mouse neuroblastoma cells using lipofectamine 3000 (LF3000) leads to cell growth arrest with high potency [15]. LF3000 was used as it increases the potency of toxins in cell lines by around 1000-fold, allowing assessment of activity at low nanomolar concentrations similar to that seen with targeted immunotoxins. In this study we look in greater detail at the anti-proliferative effect of BLF1 in neuroblastoma cells with an emphasis on MYCN amplification status. Amplification of MYCN, a gene belonging to the Myc family of transcription factors, is found in around 50% of advanced stage neuroblastoma patients and is a significant marker of poor prognosis [16,17]. Overexpression of this gene has been shown to primarily increase expression of a number of genes involved in protein synthesis and ribosome biogenesis, making translation initiation a promising target for intervention [18]. We demonstrate that LF3000-mediated delivery of BLF1into cells selectively induces apoptosis in MYCN-amplified neuroblastoma cell lines and preferentially down-regulates the translation of eIF4A dependent proteins (as has been seen with small molecule inhibitors of eIF4A). This highlights the potential for incorporation of BLF1 into targeted toxin design. Additionally, we show that the small molecule inhibitor of eIF4A rocaglamide A (RocA) demonstrates selectivity towards MYCN over-expressing cells, making eIF4A a novel target for neuroblastoma treatment.

2. Results

2.1. BLF1 Induces Apoptosis in MYCN-Amplified Neuroblastoma

To assess the importance of eIF4A in MYCN-driven neuroblastomas, we investigated the effects of eIF4A inactivation by BLF1 on cell growth in cell lines with or without MYCN amplification. BLF1 activity was compared to the ribosome-inactivating protein saporin. Saporin is an enzyme produced in *Saponaria officinalis* seeds that depurinates 28S ribosomal RNA leading to inactivation of the ribosome and a complete block of protein synthesis [19]. This makes saporin highly toxic to all cell types and a good positive control for intracellular protein delivery. The effect of these enzymes on cell growth was tested in four different neuroblastoma cell lines of which two were MYCN-amplified (IMR-32 and SK-N-BE(2)), and two were non-MYCN-amplified (SH-SY5Y and LA-N-6). MYCN expression in the different cell lines was confirmed by immunoblotting which showed substantially higher MYCN levels in the MYCN-amplified cell lines (Figure 1A). Blotting for eIF4A showed that eIF4A levels do not appear to be affected by MYCN expression (Figure 1A). Analysis of growth inhibition following delivery with LF3000 and a 72-h incubation showed that both BLF1 and saporin exhibit similar levels of potency with GI50s in the low nanomolar range (Figure 1B,C).

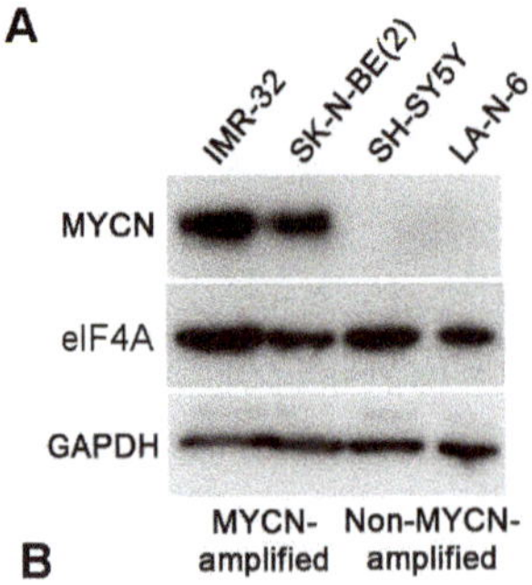

C

Cell line	BLF1 GI50 (95%CI)	Saporin GI50 (95%CI)
IMR-32	0.9 nM (0.4-2.2)	3.4 nM (2.2-5.1)
SK-N-BE(2)	0.8 nM (0.5-1.5)	10.7 nM (4.3-26.5)
SH-SY5Y	6.4 nM (1.7-24.4)	1.9 nM (1.0-3.5)
LAN-6	60.3 nM (20.3-178.6)	124 nM (33.1-460.3)

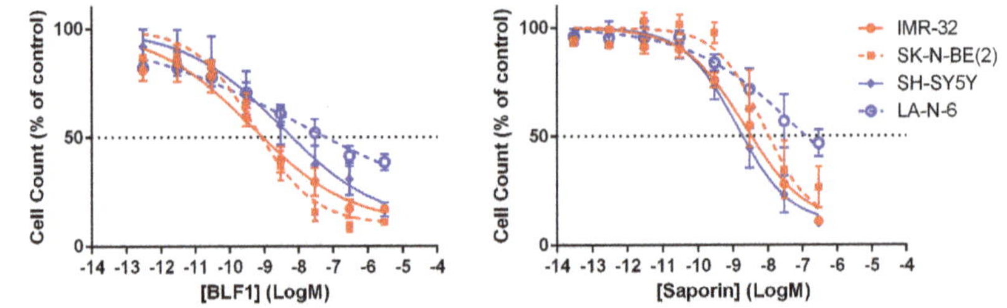

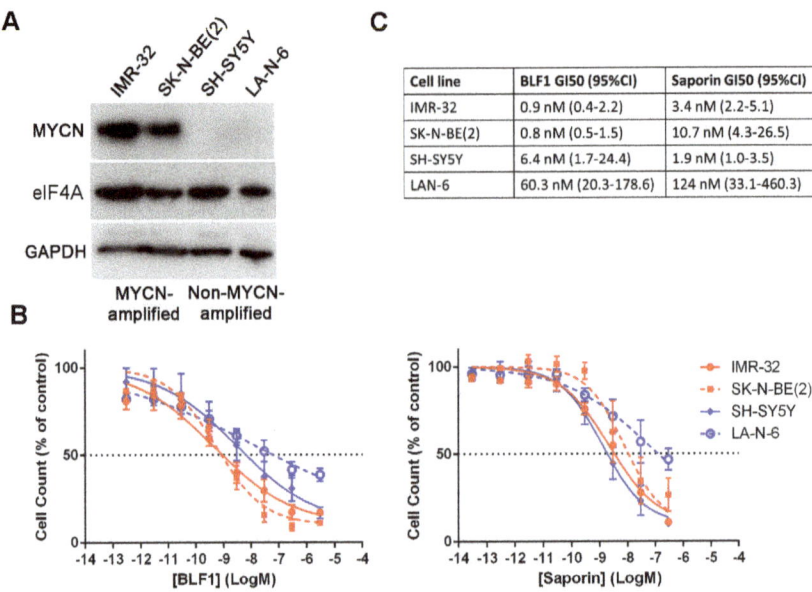

Figure 1. BLF1 causes growth inhibition preferentially in MYCN-amplified cell lines. (**A**) Immunoblot showing protein levels of MYCN and GAPDH in the MYCN-amplified cell lines IMR-32 and SK-N-BE(2), and the non-MYCN-amplified cell lines SH-SY5Y and LA-N-6; (**B**) Alamarblue assay following 72 h titration with BLF1 or saporin in the presence of LF3000. MYCN-amplified cell lines are shown in red and non-MYCN-amplified cell lines are shown in blue ($n \geq 3$, $\pm$ SEM); (**C**) Table showing calculated GI50s and 95% confidence intervals.

The effect of BLF1 on neuroblastoma cells was investigated in more detail by analysing cell viability using the nuclear dyes propidium iodide (which stains dead cells) and Hoechst 33342 (which stains all cells). The toxins were added to cells at a high concentration of 300 nM as this causes maximal growth inhibition of all the cell lines (Figure 1B). Staining following incubation for 72 h with 300 nM BLF1 or 300 nM saporin in the presence of LF3000 showed that BLF1 caused cell death at a similar level to saporin in MYCN-amplified cells (Figure 2A,B). However, in MYCN-non-amplified cells, BLF1 had no significant effect on cell viability, as evidenced by the lack of propidium iodide staining and normal nuclear morphology, whereas saporin still caused cell death (Figure 2A,B). This suggests that BLF1 is preferentially cytotoxic to MYCN amplified cell lines, but is only cytostatic in non-MYCN-amplified cells. Analysis of the mechanism of cell killing in the IMR-32 and SK-N-BE(2) MYCN-amplified cells revealed that both BLF1 and saporin cause caspase 3/7 activation and induce apoptosis (Figure 2C). Of note, SK-N-BE(2) cells have a non-functional p53 mutation which suggests that the induction of apoptosis occurs independently of p53 [20]. This is of potential therapeutic benefit because inactivating p53 mutations are often found in refractory tumours following relapse [21].

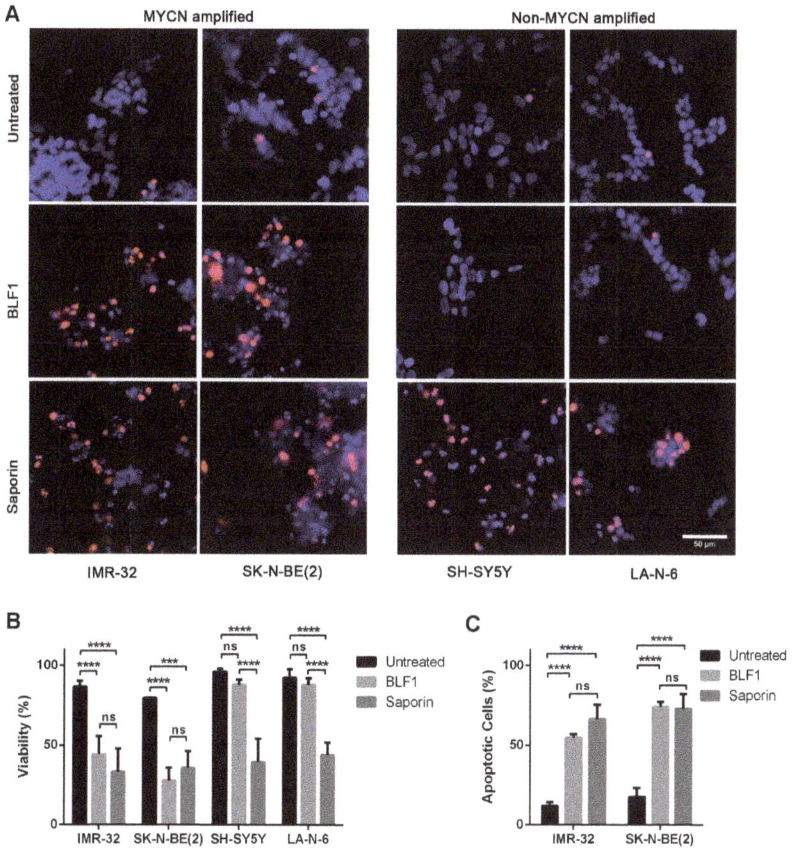

Figure 2. BLF1 induces apoptosis in MYCN-amplified neuroblastoma cell lines. (**A**) Fluorescence microscopy using Hoechst (blue) and propidium iodide (red) staining shows viability of neuroblastoma cell lines following 72 h incubation with 300 nM BLF1 or 300 nM saporin in the presence of LF3000; (**B**) Quantification of A shows that BLF1 exhibits similar cytotoxicity to saporin in MYCN-amplified cell lines but is not significantly cytotoxic to non-MYCN-amplified cells when compared to the untreated control (LF3000 alone), whereas saporin still exhibits high levels of cytotoxicity (n = 3, ± SEM). Data was analysed by two-way ANOVA and Tukey's multiple comparisons test with *** $p < 0.001$ and **** $p < 0.0001$; (**C**) CellEvent Caspase 3/7 activation assay shows that 300 nM BLF1 induces similar levels of apoptosis to saporin (300 nM) following a 72-h incubation in the presence of LF3000 (n = 3, ± SEM). Data was analysed by two-way ANOVA and Tukey's multiple comparisons test with **** $p < 0.0001$; ns: non significant.

2.2. The Cytotoxic Effect of eIF4A Inhibition in Neuroblastoma Is Directly Dependent on MYCN Expression

To confirm that cytotoxicity caused by BLF1 is MYCN dependent, the SHEP-21N cell line was used. This neuroblastoma cell line constitutively expresses high levels of MYCN which can be repressed by the addition of tetracycline [22]. Down-regulation of MYCN after a 72-h incubation with tetracycline was confirmed via immunoblotting (Figure 3A). Interestingly, down-regulation of MYCN also caused a decrease in levels of eIF4A, suggesting a regulatory role of MYCN in eIF4A expression (Figure 3A). This is in agreement with a previous study which utilised the SHEP-21N cell line and showed that

MYCN up-regulates eIF4A as well as a number of other proteins involved in ribosome biogenesis and protein synthesis [18].

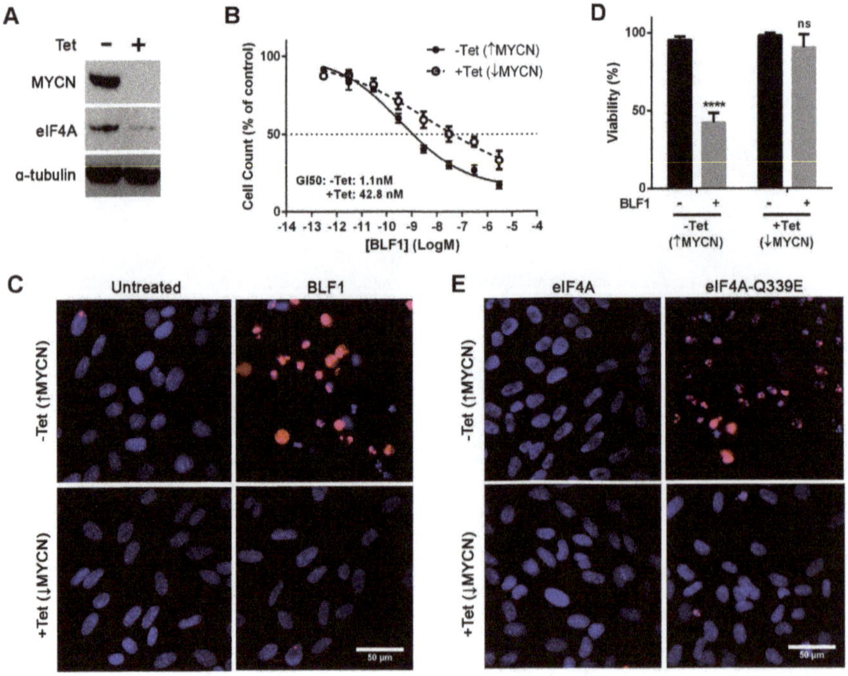

Figure 3. BLF1 exhibits increased potency in SHEP-21N cells expressing MYCN. (**A**) Immunoblot showing protein levels of MYCN, eIF4A and a loading control α-tubulin in SHEP-21N cells following 72 h of incubation in the presence or absence of tetracycline; (**B**) AlamarBlue assay following 72-h titrations with BLF1 in the presence of LF3000 demonstrates increased growth inhibition in SHEP-21N cells expressing MYCN (n = 3, ± SEM); (**C**) Fluorescence microscopy using Hoechst (blue) and propidium iodide (red) staining shows viability of SHEP-21N cells following 72-h incubation with 300 nM BLF1 in the presence of LF3000; (**D**) Quantification of the panel C results shows that BLF1 is significantly cytotoxic to SHEP-21N cells expressing MYCN but has no significant effect on cells not expressing MYCN when compared to the untreated control (LF3000 alone) (n = 3, ± SEM). Data was analysed by multiple unpaired t-tests with **** $p < 0.0001$; ns: non significant. (**E**) Fluorescence microscopy using Hoechst and propidium iodide staining shows viability of SHEP-21N cells following 72 h incubation with 1 µM eIF4A or eIF4A-Q339E dominant-negative mutant in the presence of LF3000.

Analysis of growth inhibition showed that BLF1 efficacy is decreased in the SHEP-21N cell line following tetracycline-induced down-regulation of MYCN (GI50: 1 nM without tetracycline and 42 nM with tetracycline) (Figure 3B). Propidium iodide staining showed that 300 nM BLF1 was only cytotoxic to the cells expressing high levels of MYCN (Figure 3C,D). BLF1 causes cytotoxicity by deamidation of a specific glutamine residue (Gln-339) in the eIF4A translation factor to glutamic acid [5]. This creates a dominant negative eIF4A mutant (eIF4A-Q339E) that can bind to mRNA and stall ribosome initiation complexes. To confirm that the cytotoxicity observed is due to eIF4A inhibition, the Gln-339 eIF4A mutant was generated and delivered into SHEP-21N cells using LF3000. Again, cytotoxicity was only observed in cells expressing high levels of MYCN (Figure 3E). No effect on cell viability was seen after the addition of the native eIF4A protein (Figure 3E).

The effect of small molecule inhibition of eIF4A in the SHEP-21N cell line was then tested to independently confirm the suitability of eIF4A as a therapeutic target. Rocaglamide A (RocA) is a small molecule isolated from Aglaia plants that has recently been shown to convert eIF4A into a sequence selective translational repressor [23]. Assessment of growth inhibition following titrations of RocA revealed that MYCN expressing SHEP-21N cells were over six times more sensitive compared to the tetracycline-treated SHEP-21N with MYCN being suppressed (GI50: 2.4 nM without tetracycline and 16 nM with tetracycline) (Figure 4A). Co-staining with propidium iodide and Hoechst revealed that 100 nM RocA was cytotoxic to cells expressing MYCN but had no significant effect on viability of cells with suppressed MYCN (Figure 4B,C). Taken together, these data indicate that the cytotoxicity observed in cells expressing high levels of MYCN is due to inhibition of eIF4A.

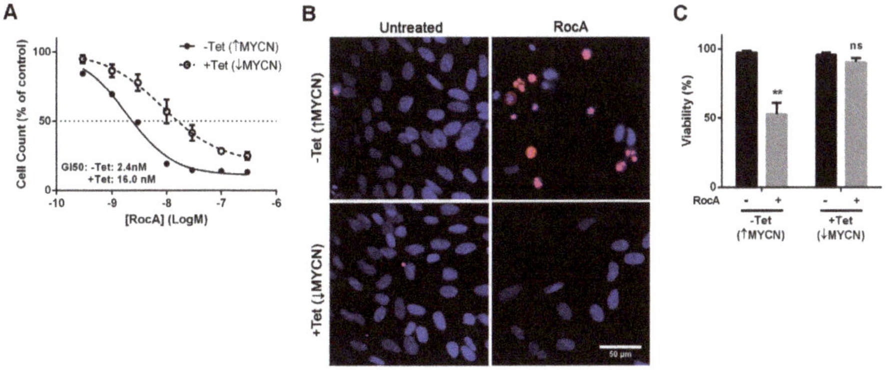

Figure 4. RocA exhibits increased potency in SHEP-21N cells expressing MYCN. (**A**) AlamarBlue cell assay shows more potent growth inhibition in SHEP-21N cells expressing MYCN, following 72 h titrations with RocA (n = 3, ± SEM); (**B**) Fluorescence micrographs showing cell viability following 72 h of incubation with 100 nM RocA. All nuclei are stained with Hoechst (blue) whereas the nuclei of dead cells are stained with propidium iodide (red); (**C**) Quantification of the panel C results shows that RocA is significantly cytotoxic to SHEP-21N cells expressing MYCN but has no effect on viability of cells not expressing MYCN when compared to the untreated control (DMSO alone) (n = 3, ± SEM). Data was analysed by multiple unpaired t-tests with ** $p < 0.01$; ns: non significant.

2.3. BLF1 Down-Regulates MYCN and Other Oncogenic Proteins

Pharmacological inactivation of eIF4A has been shown to down-regulate a number of oncogenic proteins which are encoded by mRNAs containing long and highly structured 5′ UTRs (Figure 5A) [10]. We therefore examined the effect of eIF4A inhibition on protein levels of MYCN and the cell-cycle regulator CDK4, which has previously been shown to be sensitive to the small molecule eIF4A inhibitors silvestrol and hippuristanol [13,24]. Two proteins that, according to ribosome-footprinting, should be insensitive to eIF4A inhibition, c-Jun and RhoA, were also used [10]. GAPDH and α-tubulin were used as controls, as their long half-lives mean levels should not be significantly affected by protein synthesis inhibition over 24 h. IMR-32 cells were incubated with 30 nM of protein as this concentration showed similarly high toxic activity for both BLF1 and saporin in growth assays (Figure 5B). Immunoblotting following a 24 h incubation with BLF1 or saporin in the presence of LF3000 revealed a dramatic decrease in MYCN and CDK4 protein levels after treatment with either enzyme (Figure 5B,C). However, only saporin caused down-regulation of the eIF4A-insensitive proteins c-Jun and RhoA. No effect was seen on levels of the house-keeping proteins GAPDH and α-tubulin. This shows that, as with small molecule inhibitors of eIF4A, BLF1 is able to preferentially down-regulate a subset of proteins. The specific down-regulation of MYCN provides a novel mechanistic insight into the cytotoxic activity of BLF1.

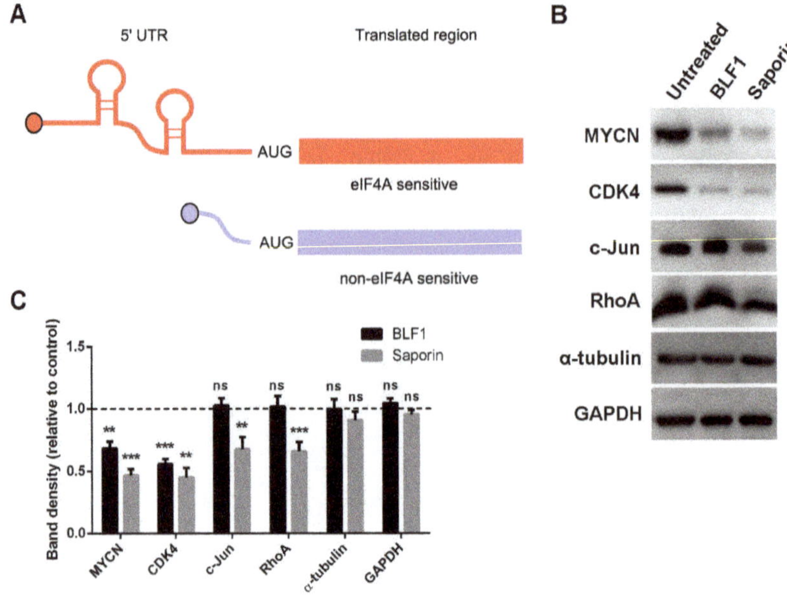

Figure 5. BLF1 down-regulates protein levels of MYCN and other eIF4A sensitive transcripts. (**A**) Schematic showing eIF4A sensitive (red) and non-eIF4A sensitive (blue) mRNAs. eIF4A sensitive mRNAs contain long and highly structured 5′ UTRs that require the RNA helicase of eIF4A to unwind secondary structures such as G-quadruplexes for scanning by the 43S pre-initiation translation complex. These coding transcripts usually encode for oncogenes and pro-survival proteins [10]; (**B**) Immunoblot showing protein levels of MYCN, CDK4, c-Jun, RhoA, α-tubulin and GAPDH in IMR-32 cells following a 24-h incubation with 30 nM BLF1 or saporin in the presence of LF3000; (**C**) Quantification of the panel B results shows the high level of down-regulation of MYCN and CDK4 following treatment with either toxin, whereas only saporin down-regulates the eIF4A insensitive transcripts c-Jun and RhoA. No effect on housekeeping genes GAPDH or α-tubulin was seen. Protein levels are shown as a ratio of the untreated control (LF3000 alone) ($n \geq 3$, ± SEM). Data was analysed by multiple one-sample t-tests to a normalised control of 1 with ** $p < 0.01$ and *** $p < 0.001$; ns: non significant.

2.4. BLF1 Has No Effect on Primary Mouse Fibroblasts

BLF1 shows cytostatic activity in non-MYCN-amplified cells which suggests that it may also have reduced efficacy in non-dividing healthy cells. This is an important characteristic when considering viable therapies for cancer treatment. To test this, primary fibroblasts were cultured from the mouse ear, grown until fully confluent and cultured in low serum media. This has previously been shown to inhibit cell division and cause cells to enter a quiescent state [25]. Analysis of cell numbers showed that BLF1 had no effect on viability, whereas saporin was still able to potently induce cell death (Figure 6A). This was confirmed by Hoechst 33,342 and propidium iodide staining which showed that 300 nM saporin caused nuclear condensation characteristic of apoptosis, whereas 300 nM of BLF1 had no effect on nuclear morphology (Figure 6B,C). The lack of efficacy of BLF1 in non-dividing cells suggests a wide therapeutic window that will be of benefit for future therapy development.

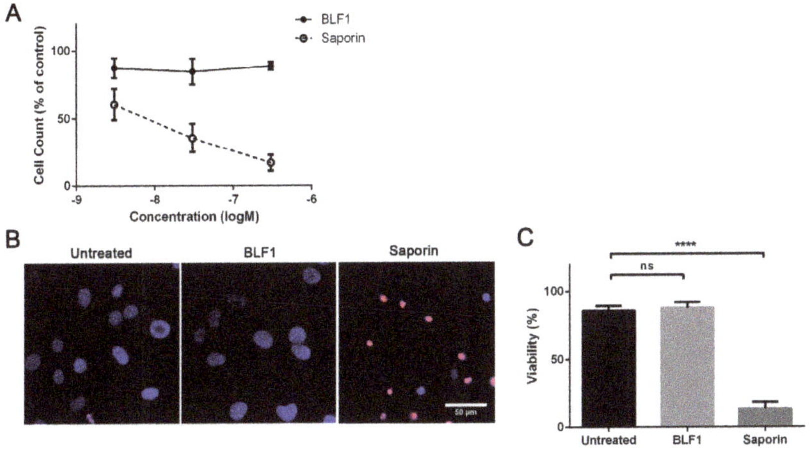

Figure 6. BLF1 has no effect on viability of primary mouse fibroblasts. (**A**) AlamarBlue assay following titrations of BLF1 and saporin in the presence of LF3000 in quiescent mouse fibroblasts shows that BLF1 has no effect on cell numbers whereas saporin causes a large reduction ($n = 3$, ± SEM); (**B**) Fluorescence microscopy images showing Hoechst 33342 (blue) and propidium iodide (red) staining following treatment of mouse fibroblasts with 300 nM BLF1 or saporin in the presence of LF3000; (**C**) Quantification of B shows that BLF1 has no effect on cells whereas saporin causes a significant amount of cell death ($n = 3$, ± SEM). Data was analysed by one-way ANOVA and Dunnet's multiple comparisons test with **** $p < 0.0001$; ns: non significant.

3. Discussion

In this paper, we have shown that inactivation of eIF4A by BLF1, following intracellular delivery with LF3000, inhibits the growth of neuroblastoma cells with high potency. Furthermore, this toxin exhibits increased efficacy and a specific cytotoxic action in MYCN amplified cells. Downregulation of MYCN via siRNA has previously been shown to induce caspase 3 activation and apoptosis in MYCN amplified neuroblastoma [26]. Caspase 3 activation and induction of apoptosis in a MYCN dependent manner was also observed following BLF1 treatment, demonstrating selective cytotoxicity of this toxin towards MYCN-amplified neuroblastoma cell lines. The exact mechanism by which BLF1 exerts cytotoxicity towards MYCN-amplified cells but is only cytostatic towards non-MYCN-amplified cells remains unclear. Levels of eIF4A do not appear to differ between MYCN-amplified and non-MYCN-amplified cells, which suggests that cytotoxicity is due to inhibition of cellular processes necessary for cell survival in a MYCN-driven setting and not an increased dependence on eIF4A itself. Ribosome profiling has previously highlighted hundreds of transcripts that are sensitive to eIF4A inhibition [10]. Importantly, a number of eIF4A sensitive transcripts have been shown to be critical for MYCN-dependent cell growth, including the p53 inhibitor MDM2 [27], the pro-survival protein BCL2 [28] and the transcription factor B-MYB [29]. BLF1 caused a down-regulation of the eIF4A-sensitive oncogenic protein CDK4 which has previously been shown to be important for maintaining an undifferentiated phenotype in neuroblastoma cell lines [30]. Inactivation of eIF4A appears to concomitantly target several key pathways involved in proliferation and cell survival of MYCN amplified neuroblastoma, making eIF4A a highly attractive target for the treatment of MYCN-amplified tumours.

The highly potent selective toxicity exhibited by BLF1 upon entry into the cytosol highlights this toxin as a potential future therapeutic. Commonly used toxins in immunotoxin therapies, such as saporin, work by the global inhibition of protein synthesis which leads to cell death in both healthy and transformed cells [31]. Off-target toxicity due to non-specific uptake of toxin by healthy cells has led to dose-limiting side effects in a number of trials and limited the success of this form of treatment [1]. BLF1 induced cell death in MYCN-amplified neuroblastoma with high potency similar to saporin, but had only a cytostatic effect on non-MYCN amplified cells at the concentrations tested. Additionally, no effect was seen following BLF1 treatment of quiescent fibroblasts. This makes BLF1 a promising candidate for immunotoxin development as it displays a degree of selectivity towards rapidly dividing cells. This may help to reduce common side effects observed during immunotoxin treatment and increase the therapeutic window. Although validated in MYCN-driven neuroblastoma cell lines, BLF1 may also be of use in more common c-Myc-driven tumours, as has been seen with small molecule inhibitors of eIF4A. This would greatly increase the applicability of this enzyme for cancer treatment as c-Myc is one of the key factors driving cancerous growth [32]. Further investigation of BLF1 as an immunotoxin and targeting of this enzyme to MYCN-amplified neuroblastoma or other Myc-driven cancers will therefore be of high interest.

MYCN-amplified neuroblastoma is a high-risk disease with very few treatment options and poor outcome. A common feature of these cancers is the acquisition of chemo-resistance and relapse, making identification of novel targets for treatment of this disease of high importance [17]. The selective cytotoxic effects of BLF1 and RocA observed in MYCN-expressing SHEP-21N cells suggests that future immunotoxins and eIF4A small molecule inhibitors have potential as future therapeutics in the treatment of neuroblastomas exhibiting MYCN amplification.

4. Materials and Methods

4.1. Recombinant Proteins, Toxins and Small Molecules

Saporin from *Saponaria officinalis* seeds and Rocaglamide A were obtained commercially from Sigma. Recombinant hexahistidine-tagged fusions of BLF1 and eIF4A encompassing amino-acids 20-406 were purified using Ion Metal Affinity Chromatography on TALON-Cobalt (Clontech, Fremont, CA, USA) and S200 gel filtration (GE Healthcare, Little Chalfont, UK) as described previously [5].

4.2. Cell Culture

SHEP-21N cells were a gift from Prof. Manfred Schwab, and IMR-32 cells were a gift from Prof. Peter Andrews. SK-N-BE(2) and LA-N-6 cells were obtained from the Childrens Oncology Group Cell Culture Repository, and SH-SY5Y cells were obtained from Sigma. SHEP-21N, IMR-32, SK-N-BE(2) and LA-N-6 cells were grown in RPMI media (Life Technologies, Carlsbad, CA, USA) supplemented with 10 per cent Fetal Bovine Serum (FBS) (Life Technologies). SH-SY5Y cells were grown in a 1:1 mix of MEM (Life Technologies) and F12 nutrient mix (Life Tehcnologies) supplemented with 15 per cent FBS and 1 per cent non-essential amino acids (Life Technologies). Cells were maintained at 37 °C, 5 per cent CO_2.

For primary mouse fibroblast cultures, C57BL/6 mice were sacrificed using a humane method as listed in Schedule 1 of the Animal (Scientific procedure) Act 1986. Small segments of tissue (approximately 3 mm^2) were taken from the ear and placed in Hank's balanced salt solution (HBSS) (Life Technologies). The tissue was then diced into small pieces and incubated with 2000 U/mL collagenase XI (Sigma) for 25 min at 37 °C. Following washing in HBSS, 0.05 per cent trypsin was added and the tissue was incubated for 20 min at 37 °C. The supernatant was discarded, and the tissue was re-suspended in fibroblast media (DMEM + 1 per cent NEAA + 1 per cent penicillin/streptomycin (P/S) (Sigma) + 10 per cent FBS), followed by trituration to break up cell aggregates. The suspension was plated in 3 cm dishes and incubated at 37 °C for 3 to 4 days to allow cell growth. Cells were expanded to a 25 cm^2 flask followed by a 75 cm^2 flask. Sub-culturing was carried out every 3–4 days when spent media was replaced or cells were passaged. For viability studies, cells were seeded in a 96-well plate and grown for 3–4 days until fully confluent. The media was then changed to low serum (0.1 per cent FBS) and cells were cultured for a further 1–2 days before treatment with proteins.

4.3. Cell Treatment and Protein Transduction

For pharmacological treatment, cells were seeded in 96-well plates at a density of 1×10^4 cells or in 48-well plates at a density of 2.5×10^4 cells and left for at least 5 h to attach. For immunoblotting experiments, cells were left for 24 h before treatment. Protein was delivered into cells by lipofection as described previously [15]. Briefly, protein was diluted to 10 times the final concentration in Optimem (Life Technologies) before the addition of LF3000 (Life Technologies) in a 1/50 dilution. The mix was incubated at 20 °C for 20 min before adding to cells in a 1/10 dilution to give the final concentrations.

4.4. Growth Inhibition Assay

Growth inhibition was measured using the alamarBlue assay (Thermofisher, Waltham, MA, USA) according to manufacturer's instructions. Following incubation with drugs or protein, the alamarBlue reagent was added directly to cells in a 1/10 dilution. Cells were incubated at 37 °C for 4 h before fluorescence was measured at 560 nm excitation/590 nm emission using a Fluorskan Ascent plate reader (Thermo Fisher Scientific, Loughborough, UK). GI50s were calculated as the concentration necessary to cause 50 per cent growth inhibition compared to vehicle (DMSO for RocA or LF3000 for protein) treated cells after normalisation to a blank control (cell-free medium with alamarBlue reagent).

4.5. Fluorescence Microscopy

For microscopy assays, cells were seeded in μClear 96-well plates (Greiner, Kremsmünster, Austria) for greater resolution and incubated for 72 h following treatment before imaging. Viability of cells was assessed by the addition of 1 μg/mL Hoechst 33,342 (Life Technologies), which stains the DNA of all cells, and 0.5 μg/mL Propidium iodide (Life Technologies), which only stains the DNA of dead cells. Cells were then incubated at 37 °C for 15 min before imaging using an epifluorescence microscope (CMIRB, Leica, London, UK) at a 40× objective. Experiments were performed in duplicate with at least 4 images taken per well. Cells were counted using ImageJ and the percentage of Propidium iodide to Hoechst stained cells was calculated to give viability.

Apoptosis was assessed using the CellEvent Caspase 3/7 Green detection kit (Life Technologies) according to manufacturer's instructions. The CellEvent reagent was added to cells at a final concentration of 5 μM and cells were incubated at 37 °C for 15 min. Hoechst 33,342 (1 μg/mL) was then added and cells were incubated for a further 15 min before imaging. Images were taken at a 40× objective and the number of apoptotic cells was then calculated as a percentage of the total cell number. Experiments were performed in duplicate and at least 4 images were taken per well.

4.6. Immunoblotting

Following 24 h treatment, cells were washed once with cold PBS before lysis using RIPA buffer (50 mM Tris-HCl pH 7.5, 150 mM NaCl, 1 mM EDTA, 1 mM EGTA, 0.1 per cent SDS, 0.5 per cent sodium deoxycholate and 1 per cent NP-40). Protein concentration was measured using the DC assay (Biorad, Hercules, CA, USA) according to manufacturer's instructions. Lysates were run on 12 per cent Bis-Tris sodium dodecyl sulphate-polyacrylamide gel electrophoresis (SDS-PAGE) gels (Life Technologies) and protein was transferred to a polyvinylidene difluoride membrane (Biorad) before probing with antibodies. Primary antibodies used include MYCN (Santa-Cruz: sc-56729, 1:500 [33]), eIF4A1 (Cell Signalling: 2490, 1:1000 [23]), CDK4 (Abcam: ab108357, 1:1000 [34]), c-Jun (Santa-Cruz: sc-74543, 1:500 [35]), RhoA (Santa-Cruz: sc-418, 1:500 [36]), GAPDH (Life Technologies: AM4300, 1:5000 [37]) and α-tubulin (Sigma: T6074, 1:5000 [38]). Following incubation with peroxidase conjugated sheep anti-mouse or donkey anti-rabbit secondary antibodies (1:24,000, GE Healthcare), proteins were visualised using the SuperSignal West Dura ECL reagent (Thermo Scientific, Waltham, MA, USA) by X-ray film exposure with signals quantified using ImageJ after film scanning.

4.7. Statistical Analysis

Data were expressed as mean ± SEM and specific statistical tests are indicated in the figure legends. Statistical analysis was carried out using Graphpad Prism 6. A p-value <0.05 was considered statistically significant. For single comparisons, unpaired t-tests were used. For single comparisons to a normalised control the one-sample t-test was used. For multiple comparisons, data was analysed by either one-way ANOVA followed by the Dunnet's method for comparison to a single control sample, or by two-way ANOVA followed by the Tukey range test.

Author Contributions: A.R. and S.S. conducted experiments; A.R., G.M.H. and B.D. designed the project and prepared the manuscript for publication.

Funding: This study was supported by Sheffield University grant 311469 and Medical Research Council grant U10578791.

Conflicts of Interest: The authors declare no conflict of interest.

References

1. Alewine, C.; Hassan, R.; Pastan, I. Advances in anticancer immunotoxin therapy. *Oncologist* **2015**, *20*, 176–185. [CrossRef] [PubMed]
2. Rust, A.; Partridge, L.J.; Davletov, B.; Hautbergue, G.M. The use of plant-derived ribosome inactivating proteins in immunotoxin development: Past, present and future generations. *Toxins* **2017**, *9*, 344. [CrossRef] [PubMed]
3. Shan, L.; Liu, Y.; Wang, P. Recombinant immunotoxin therapy of solid tumors: Challenges and strategies. *J. Basic Clin. Med.* **2013**, *2*, 1–6. [PubMed]
4. Pastan, I.; Hassan, R.; FitzGerald, D.J.; Kreitman, R.J. Immunotoxin treatment of cancer. *Annu. Rev. Med.* **2007**, *58*, 221–237. [CrossRef] [PubMed]
5. Cruz-Migoni, A.; Hautbergue, G.M.; Artymiuk, P.J.; Baker, P.J.; Bokori-Brown, M.; Chang, C.T.; Dickman, M.J.; Essex-Lopresti, A.; Harding, S.V.; Mahadi, N.M.; et al. A burkholderia pseudomallei toxin inhibits helicase activity of translation factor eIF4A. *Science* **2011**, *334*, 821–824. [CrossRef] [PubMed]
6. Walsh, M.J.; Dodd, J.E.; Hautbergue, G.M. Ribosome-inactivating proteins: Potent poisons and molecular tools. *Virulence* **2013**, *4*, 774–784. [CrossRef] [PubMed]
7. Bhat, M.; Robichaud, N.; Hulea, L.; Sonenberg, N.; Pelletier, J.; Topisirovic, I. Targeting the translation machinery in cancer. *Nature Rev. Drug Discov.* **2015**, *14*, 261–278. [CrossRef] [PubMed]
8. Jackson, R.J.; Hellen, C.U.; Pestova, T.V. The mechanism of eukaryotic translation initiation and principles of its regulation. *Nat. Rev. Mol. Cell Biol.* **2010**, *11*, 113–127. [CrossRef] [PubMed]
9. Pelletier, J.; Graff, J.; Ruggero, D.; Sonenberg, N. Targeting the eif4f translation initiation complex: A critical nexus for cancer development. *Cancer Res.* **2015**, *75*, 250–263. [CrossRef] [PubMed]
10. Wolfe, A.L.; Singh, K.; Zhong, Y.; Drewe, P.; Rajasekhar, V.K.; Sanghvi, V.R.; Mavrakis, K.J.; Jiang, M.; Roderick, J.E.; Van der Meulen, J.; et al. RNA G-quadruplexes cause eIF4A-dependent oncogene translation in cancer. *Nature* **2014**, *513*, 65–70. [CrossRef] [PubMed]
11. Alachkar, H.; Santhanam, R.; Harb, J.G.; Lucas, D.M.; Oaks, J.J.; Hickey, C.J.; Pan, L.; Kinghorn, A.D.; Caligiuri, M.A.; Perrotti, D.; et al. Silvestrol exhibits significant in vivo and in vitro antileukemic activities and inhibits flt3 and mir-155 expressions in acute myeloid leukemia. *J. Hematol. Oncol.* **2013**, *6*, 21. [CrossRef] [PubMed]
12. Lucas, D.M.; Edwards, R.B.; Lozanski, G.; West, D.A.; Shin, J.D.; Vargo, M.A.; Davis, M.E.; Rozewski, D.M.; Johnson, A.J.; Su, B.N.; et al. The novel plant-derived agent silvestrol has b-cell selective activity in chronic lymphocytic leukemia and acute lymphoblastic leukemia in vitro and in vivo. *Blood* **2009**, *113*, 4656–4666. [CrossRef] [PubMed]
13. Tsumuraya, T.; Ishikawa, C.; Machijima, Y.; Nakachi, S.; Senba, M.; Tanaka, J.; Mori, N. Effects of hippuristanol, an inhibitor of eIF4A, on adult T-cell leukemia. *Biochem. Pharmacol.* **2011**, *81*, 713–722. [CrossRef] [PubMed]
14. Moerke, N.J.; Aktas, H.; Chen, H.; Cantel, S.; Reibarkh, M.Y.; Fahmy, A.; Gross, J.D.; Degterev, A.; Yuan, J.; Chorev, M.; et al. Small-molecule inhibition of the interaction between the translation initiation factors eIF4E and eIF4G. *Cell* **2007**, *128*, 257–267. [CrossRef] [PubMed]

15. Rust, A.; Hassan, H.H.; Sedelnikova, S.; Niranjan, D.; Hautbergue, G.; Abbas, S.A.; Partridge, L.; Rice, D.; Binz, T.; Davletov, B. Two complementary approaches for intracellular delivery of exogenous enzymes. *Sci. Rep.* **2015**, *5*, 12444. [CrossRef] [PubMed]
16. Brodeur, G.M.; Seeger, R.C.; Schwab, M.; Varmus, H.E.; Bishop, J.M. Amplification of N-MYC in untreated human neuroblastomas correlates with advanced disease stage. *Science* **1984**, *224*, 1121–1124. [CrossRef] [PubMed]
17. Buechner, J.; Einvik, C. N-myc and noncoding rnas in neuroblastoma. *Mol Cancer Res* **2012**, *10*, 1243–1253. [CrossRef] [PubMed]
18. Boon, K.; Caron, H.N.; van Asperen, R.; Valentijn, L.; Hermus, M.C.; van Sluis, P.; Roobeek, I.; Weis, I.; Voûte, P.; Schwab, M.; et al. N-MYC enhances the expression of a large set of genes functioning in ribosome biogenesis and protein synthesis. *Embo J.* **2001**, *20*, 1383–1393. [CrossRef] [PubMed]
19. Stirpe, F.; Battelli, M.G. Ribosome-inactivating proteins: Progress and problems. *Cell Mol. Life Sci.* **2006**, *63*, 1850–1866. [CrossRef] [PubMed]
20. Goldschneider, D.; Horvilleur, E.; Plassa, L.F.; Guillaud-Bataille, M.; Million, K.; Wittmer-Dupret, E.; Danglot, G.; de The, H.; Benard, J.; May, E.; et al. Expression of c-terminal deleted p53 isoforms in neuroblastoma. *Nucleic Acids Res.* **2006**, *34*, 5603–5612. [CrossRef] [PubMed]
21. Tweddle, D.A.; Pearson, A.D.; Haber, M.; Norris, M.D.; Xue, C.; Flemming, C.; Lunec, J. The p53 pathway and its inactivation in neuroblastoma. *Cancer Lett.* **2003**, *197*, 93–98. [CrossRef]
22. Lutz, W.; Stohr, M.; Schurmann, J.; Wenzel, A.; Lohr, A.; Schwab, M. Conditional expression of N-MYC in human neuroblastoma cells increases expression of alpha-prothymosin and ornithine decarboxylase and accelerates progression into s-phase early after mitogenic stimulation of quiescent cells. *Oncogene* **1996**, *13*, 803–812. [PubMed]
23. Iwasaki, S.; Floor, S.N.; Ingolia, N.T. Rocaglates convert dead-box protein eif4a into a sequence-selective translational repressor. *Nature* **2016**, *534*, 558–561. [CrossRef] [PubMed]
24. Patton, J.T.; Lustberg, M.E.; Lozanski, G.; Garman, S.L.; Towns, W.H.; Drohan, C.M.; Lehman, A.; Zhang, X.; Bolon, B.; Pan, L.; et al. The translation inhibitor silvestrol exhibits direct anti-tumor activity while preserving innate and adaptive immunity against EBV-driven lymphoproliferative disease. *Oncotarget* **2015**, *6*, 2693–2708. [CrossRef] [PubMed]
25. Kuznetsov, G.; Xu, Q.; Rudolph-Owen, L.; Tendyke, K.; Liu, J.; Towle, M.; Zhao, N.; Marsh, J.; Agoulnik, S.; Twine, N.; et al. Potent in vitro and in vivo anticancer activities of des-methyl, des-amino pateamine a, a synthetic analogue of marine natural product pateamine a. *Mol. Cancer Ther.* **2009**, *8*, 1250–1260. [CrossRef] [PubMed]
26. Kang, J.H.; Rychahou, P.G.; Ishola, T.A.; Qiao, J.; Evers, B.M.; Chung, D.H. MYCN silencing induces differentiation and apoptosis in human neuroblastoma cells. *Biochem. Biophys. Res. Commun.* **2006**, *351*, 192–197. [CrossRef] [PubMed]
27. Gamble, L.D.; Kees, U.R.; Tweddle, D.A.; Lunec, J. Mycn sensitizes neuroblastoma to the MDM2-p53 antagonists nutlin-3 and MI-63. *Oncogene* **2012**, *31*, 752–763. [CrossRef] [PubMed]
28. Ham, J.; Costa, C.; Sano, R.; Lochmann, T.L.; Sennott, E.M.; Patel, N.U.; Dastur, A.; Gomez-Caraballo, M.; Krytska, K.; Hata, A.N.; et al. Exploitation of the apoptosis-primed state of MYCN-amplified neuroblastoma to develop a potent and specific targeted therapy combination. *Cancer Cell* **2016**, *29*, 159–172. [CrossRef] [PubMed]
29. Gualdrini, F.; Corvetta, D.; Cantilena, S.; Chayka, O.; Tanno, B.; Raschella, G.; Sala, A. Addiction of mycn amplified tumours to B-MYB underscores a reciprocal regulatory loop. *Oncotarget* **2010**, *1*, 278–288. [PubMed]
30. Molenaar, J.J.; Ebus, M.E.; Koster, J.; van Sluis, P.; van Noesel, C.J.; Versteeg, R.; Caron, H.N. Cyclin D1 and CDK4 activity contribute to the undifferentiated phenotype in neuroblastoma. *Cancer Res.* **2008**, *68*, 2599–2609. [CrossRef] [PubMed]
31. Schrot, J.; Weng, A.; Melzig, M.F. Ribosome-inactivating and related proteins. *Toxins* **2015**, *7*, 1556–1615. [CrossRef] [PubMed]
32. Hartl, M. The quest for targets executing MYC-dependent cell transformation. *Front. Oncol.* **2016**, *6*, 132. [CrossRef] [PubMed]
33. Zeid, R.; Lawlor, M.A.; Poon, E.; Reyes, J.M.; Fulciniti, M.; Lopez, M.A.; Scott, T.G.; Nabet, B.; Erb, M.A.; Winter, G.E.; et al. Enhancer invasion shapes MYCN-dependent transcriptional amplification in neuroblastoma. *Nat. Genet.* **2018**, *50*, 515–523. [CrossRef] [PubMed]

34. Shen, L.; Qu, X.; Li, H.; Xu, C.; Wei, M.; Wang, Q.; Ru, Y.; Liu, B.; Xu, Y.; Li, K.; et al. NDRG2 facilitates colorectal cancer differentiation through the regulation of SKP2-p21/p27 axis. *Oncogene* **2018**, *37*, 1759–1774. [CrossRef] [PubMed]
35. He, H.; Sinha, I.; Fan, R.; Haldosen, L.A.; Yan, F.; Zhao, C.; Dahlman-Wright, K. C-JUN/AP-1 overexpression reprograms eralpha signaling related to tamoxifen response in eralpha-positive breast cancer. *Oncogene* **2018**, *37*, 2586–2600. [CrossRef] [PubMed]
36. Shi, G.X.; Yang, W.S.; Jin, L.; Matter, M.L.; Ramos, J.W. RSK2 drives cell motility by serine phosphorylation of larg and activation of rho gtpases. *Proc. Natl. Acad. Sci. USA* **2018**, *115*, e190–e199. [CrossRef] [PubMed]
37. Shen, Y.; Liu, P.; Jiang, T.; Hu, Y.; Au, F.K.C.; Qi, R.Z. The catalytic subunit of DNA polymerase delta inhibits gammaturc activity and regulates golgi-derived microtubules. *Nat. Commun.* **2017**, *8*, 554. [CrossRef] [PubMed]
38. Antic, I.; Biancucci, M.; Zhu, Y.; Gius, D.R.; Satchell, K.J. Site-specific processing of RAS and RAP1 switch I by a martx toxin effector domain. *Nat. Commun.* **2015**, *6*, 7396. [CrossRef] [PubMed]

toxins MDPI

Article

A *Hylarana latouchii* Skin Secretion-Derived Novel Bombesin-Related Pentadecapeptide (Ranatensin-HLa) Evoke Myotropic Effects on the in vitro Rat Smooth Muscles

Yan Lin [1,2,*,†], Nan Hu [2,†], Haoyang He [2], Chengbang Ma [2], Mei Zhou [2], Lei Wang [2] and Tianbao Chen [2]

1 College of Bee Science, Fujian Agriculture and Forestry University, Fuzhou 350002, China
2 Natural Drug Discovery Group, School of Pharmacy, Queen's University, Belfast BT9 7BL, Northern Ireland, UK; nhu01@qub.ac.uk (N.H.); hhe06@qub.ac.uk (H.H.); c.ma@qub.ac.uk (C.M.); m.zhou@qub.ac.uk (M.Z.); l.wang@qub.ac.uk (L.W.); t.chen@qub.ac.uk (T.C.)
* Correspondence: ylin19@qub.ac.uk; Tel.: +86-188-5013-8341
† These authors contributed equally to this work.

Received: 21 February 2019; Accepted: 28 March 2019; Published: 5 April 2019

Abstract: Amphibians have developed successful defensive strategies for combating predators and invasive microorganisms encountered in their broad range of environments, which involve secretion of complex cocktails of noxious, toxic and diverse bioactive molecules from the skins. In recent years, amphibian skin secretions have been considered as an extraordinary warehouse for the discovery of therapeutic medicines. In this study, through bioactivity screening of the *Hylarana latouchii* skin secretion-derived fractions, a novel peptide belonging to ranatensin subfamily (ranatensin-HLa) was discovered, and structurally and pharmacologically-characterised. It consists of 15 amino acid residues, pGlu-NGDRAPQWAVGHFM-NH_2, and its synthetic replicate was found to exhibit pharmacological activities on increasing the contraction of the in vitro rat bladder and uterus smooth muscles. Corresponding characteristic sigmoidal dose-response curves with EC_{50} values of 7.1 nM and 5.5 nM were produced, respectively, in bladder and uterus. Moreover, the precursor of ranatensin-HLa showed a high degree of similarity to those of bombesin-like peptides from *Odorrana grahami* and *Odorrana schmackeri*. *Hylarana latouchii* skin continues to serve as a storehouse with diverse lead compounds for the development of therapeutically effective medicines.

Keywords: frog; mass spectrometry; molecular cloning; bombesin-related peptide; smooth muscle

Key Contribution: A novel pentadecapeptide falling within ranatensin subfamily, which possessed myotropic effects, was identified. It demonstrated moderate myoactivity on rat bladder and uterus, and its biosynthetic precursor exhibited high homology with that of some other bombesin-like peptides from Ranidae family.

1. Introduction

The word "amphibian", *amphi* meaning both and *bios* meaning life, tells us much about the amphibian lifestyle—a double life. More specifically, amphibians are a class of animals that many of them have a biphasic lifestyle, undergoing shift from strictly aquatic larvae to more terrestrial adults. They are widespread on the majority of continents apart from Antarctica [1,2]. In the present day, there are 7975 species of amphibians in total within the Class Amphibia, with 88% of anurans (frogs and toads) [3]. From aquatically to terrestrially, amphibians have gone through drastic environmental changes, in which the highly-specialised skin plays a multitude of vital roles in their adaptability, including respiration, thermoregulation, water absorption, camouflage, defensive barrier, and osmoregulation [4]. Amphibian skin is characterised by numerous dispersed mucous glands which excrete mucus continually to maintain the moisture of skin, and granular glands that discharge extremely toxic components to ward off predators and pathologic microorganisms [5,6].

Amphibian skin secretions contain a very high level of diversity of biochemical compounds such as bioactive peptides/proteins, biogenic amines, steroids and alkaloids, possessing various pharmacological properties like hallucinogenic, cardiotonic, analgesic, antibacterial, antifungal, antidiabetic and antineoplastic activities [7–9]. Peptides constitute the most significant molecular group in the secretions of frog skin, and can be broadly classified into two main groups—antimicrobial peptides and pharmacologically-active peptides [8,10]. The former group of peptides are the fundamental elements of frogs' innate immune system against pathogen attack, serving as the front line of an anti-infective defence barrier [11]; the second group of peptides execute irreplaceable responsibilities of regulating the physiological functions of the skin itself [12] and disrupting the homeostatic balance of predators [13]. Since analogues of a number of amphibian skin-derived peptides can be found in mammalian gastrointestinal tract, nervous tissues and endocrine organs but exist in extremely high quantities in amphibian skin, they have excited the interest of scientists in developing novel therapeutic drugs and elucidating relevant physiological processes in mammals [10,14].

Bombesin, identified in the skin of *Bombina bombina* [15], is one of the most important neuropeptides, whose discovery promoted the isolation of two counterparts in mammals—gastrin-releasing peptide (GRP) and neuromedin B (NMB) [16,17]. Bombesin-like peptides exhibit numerous pharmacological functions, such as stimulation of vascular and extra vascular smooth muscle contraction, renal circulation and gastric secretion [18]. They can be divided into three subfamilies: bombesin, ranatensin/litorin and phyllolitorin [19,20]; each member of them has a common structure in the C-terminal tetrapeptide amide. For bombesins, the final four amino acids are -Gly-His-Leu-Met-NH_2; -Gly-His-Phe-Met-NH_2 is at the C-termini of ranatensin/litorin subfamily; and phyllolitorin subfamily ends with -Gly-Ser-Phe/Leu-Met-NH_2 [21,22]. They function by binding to five closely-related bombesin-like G-protein coupled receptors, including the NMB receptor (BB1-R), the GRP receptor (BB2-R), the bombesin receptor subtype-3 (BB3-R), the bombesin receptor subtype-4 (BB4-R) and the bombesin receptor subtype-3.5 (BB3.5-R) [21,23–25].

Here, from the skin secretion of the broad-folded frog, *Hylarana latouchii*, a structurally novel ranatensin-related peptide was isolated and investigated for its pharmacological activity. It evoked contractile effects on rat bladder and uterus by binding to NMB receptor and GRP receptor in the in vitro smooth muscle assays. Furthermore, prepropeptides of ranatensin-HLa and other closely related peptides including previously discovered ranatensin-HL were analysed to illustrate the evolutionary strategies of some species in the Ranidae family.

2. Results

2.1. Bioactivity Screening Resulted in Discovery of Ranatensin-HLa

Following chromatographic fractionation of *H. latouchii* skin secretion, the bioactivity screening of sequential fractions resulted in the authentication of a component in fraction # 93 possessing considerable myoactivities towards rat urinary bladder and uterus. The active fraction was in accord with an absorbance peak in chromatogram with retention time of about 93 min as shown in Figure 1. Matrix-assisted laser desorption/ionisation time-of-flight (MALDI-TOF) mass spectrometric analysis indicated that this fraction comprised a single predominant peptide with molecular weight of 1695.32 Da. Its primary structure was further analysed by MS/MS fragmentation sequencing (Figure 2), which was followed by structural bioinformatic analysis inquiring the National Centre for Biotechnology Information-Basic Local Alignment Search Tool Protein Database (NCBI-BLASTp), indicating the highest sequence similarity with a proline rich bombesin-related peptide (PR-bombesin) from *Bombina maxima* and ranatensin from *Rana pipiens*. The sequence alignment of mature PR-bombesin, ranatensin and ranatensin-HLa was shown in Figure 3. Due to the identity of the C-terminal 4 residues (-Gly-His-Phe-Met-NH_2) with the typical ranatensin subfamily peptide, this peptide was regarded as a member of the ranatensin group and was assigned a name of ranatensin-HLa ("HL" represented *Hylarana latouchii* and "a" represented distinction from a previously described peptide named ranatensin-HL).

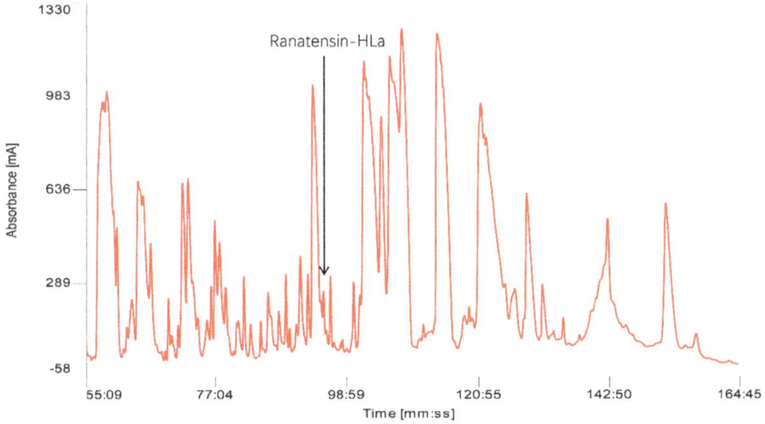

Figure 1. Chromatogram of *Hylarana latouchii* skin secretion indicating myoactive peptide (ranatensin-HLa). Ranatensin-HLa was eluted at around 93 min as pointed with an arrow.

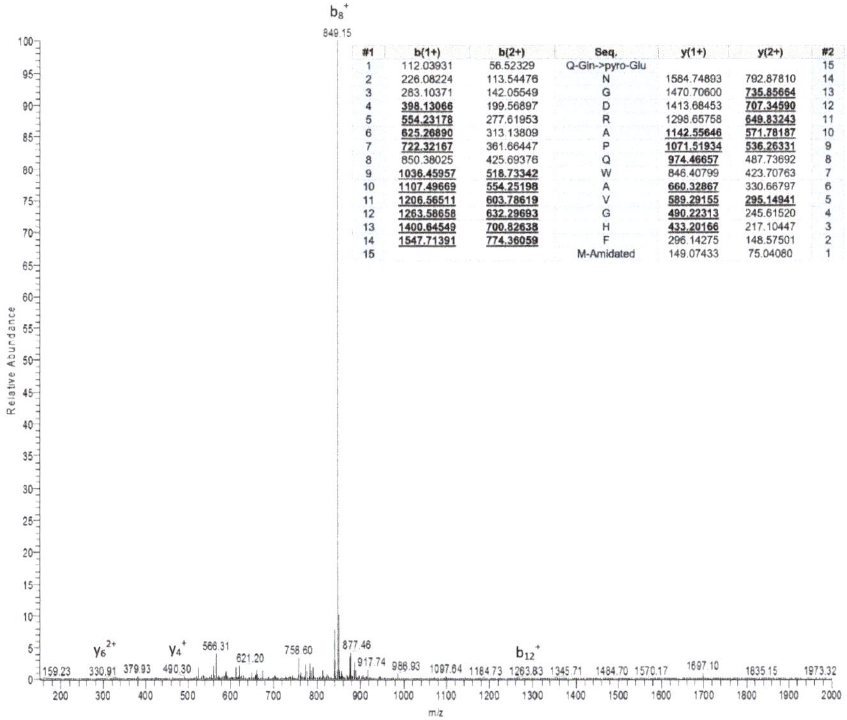

#1	b(1+)	b(2+)	Seq.	y(1+)	y(2+)	#2
1	112.03931	56.52329	Q-Gln->pyro-Glu			15
2	226.08224	113.54476	N	1584.74893	792.87810	14
3	283.10371	142.05549	G	1470.70600	**735.85664**	13
4	**398.13066**	199.56897	D	1413.68453	**707.34590**	12
5	**554.23178**	277.61953	R	1298.65758	**649.83243**	11
6	**625.26890**	313.13809	A	**1142.55646**	**571.78187**	10
7	**722.32167**	361.66447	P	**1071.51934**	**536.26331**	9
8	850.38025	425.69376	Q	**974.46657**	487.73692	8
9	**1036.45957**	**518.73342**	W	846.40799	423.70763	7
10	**1107.49669**	**554.25198**	A	**660.32867**	330.66797	6
11	**1206.56511**	**603.78619**	V	**589.29155**	**295.14941**	5
12	**1263.58658**	**632.29693**	G	**490.22313**	245.61520	4
13	**1400.64549**	**700.82638**	H	**433.20166**	217.10447	3
14	**1547.71391**	**774.36059**	F	296.14275	148.57501	2
15			M-Amidated	149.07433	75.04080	1

Figure 2. Mass spectrum of the primary structural analysis of the HPLC fraction possessing myotropic activity using MS/MS fragmentation sequencing. Predicted *b*- and *y*- MS/MS fragment ion series (both singly- and doubly-charged) are shown in panel. Observed ions are indicated in bold underlined typeface.

```
PR-bombesin        QKKPPRPPQWAVGHFM     16
Ranatensin         .....EVPQWAVGHFM     11
Ranatensin-HLa     .QNGDRAPQWAVGHFM     15
Consensus                 pqwavghfm
```

Figure 3. Sequences alignment of the mature peptides with the highest similarity in the bioinformatic analysis of ranatensin-HLa through inquiring NCBI-BLASTp database. Ranatensin-HLa and a bombesin-related peptide (PR-bombesin) (*Bombina maxima*) (Accession number: AAM10624.1) showed 62% identity. Meanwhile, Ranatensin-HLa and ranatensin (*Rana pipiens*) (Accession number: 701177B) showed 60% identity.

2.2. Molecular Cloning of the cDNA Encoding the Biosynthetic Precursor of Ranatensin-HLa

According to the primary structure of ranatensin-HLa, a degenerate primer was designed and used to clone the complete cDNA encoding the prepropeptide of ranatensin-HLa. Consequently, the full-length cDNA was iteratively cloned from the cDNA library established from *H. latouchii* skin secretion through RACE PCR. The translated amino acid sequence of the open-reading frame (ORF) of the precursor was composed of 72 residues which was constituted by a canonical architecture involving a putative signal peptide domain, an N-terminal spacer peptide domain flanked by a typical constructive dibasic amino acid cleavage site (-RR-), a single copy of mature peptide (15 amino acids) and a short C-terminal extended peptide containing another convertase cleavage site (-KK-) and a amide donor (glycine) (Figure 4). The sequence of ranatensin-HLa open-reading frame was analysed by searching for the nucleotide and protein database within NCBI-BLAST, indicating significant sequence homology with bombesin-like peptide precursors identified in *Odorrana grahami* (odorranain-BLP-4) and *Odorrana schmackeri* (bombesin-OS), respectively. The corresponding open-reading frame nucleotide and amino acid sequences alignments were shown in Figure 5. The cDNA sequence of ranatensin-HLa precursor with accession code LR032001 has been deposited in the EMBL Nucleotide Sequence Database.

```
     M  T  A   V  P  A  I   R  I  L   P  I  G   F  L  G  I
  1  ATGACTGCGG TTCCTGCCAT CAGAATCCTG CCCATTGGGT TCCTGGGTAT
     TACTGACGCC AAGGACGGTA GTCTTAGGAC GGGTAACCCA AGGACCCATA
     ·L  L  L   F  S  I   I  S  R  S   V  C  V   D  F  T  E
 51  TCTGCTGCTC TTCTCCATCA TCTCCCGCTC TGTTTGCGTA GATTTCACGG
     AGACGACGAG AAGAGGTAGT AGAGGGCGAG ACAAACGCAT CTAAAGTGCC
     · D  A  G   K  L  D   K  I  D   A  F [R  R]  E  A  Q
101  AAGATGCTGG AAAACTAGAC AAAATCGATG CGTTTCGGAG AGAAGCACAA
     TTCTACGACC TTTTGATCTG TTTTAGCTAC GCAAAGCCTC TCTTCGTGTT
     N  G  D   R  A  P  Q   W  A  V   G  H  F   M  G [K  K]
151  AATGGAGATC GAGCACCTCA GTGGGCAGTT GGACACTTCA TGGGTAAGAA
     TTACCTCTAG CTCGTGGAGT CACCCGTCAA CCTGTGAAGT ACCCATTCTT
     ·S  L  Q   E  D  *
201  GAGCCTACAA GAAGATTAGC GTATGCTCTG ACACCCAGCC AGATGCAAGA
     CTCGGATGTT CTTCTAATCG CATACGAGAC TGTGGGTCGG TCTACGTTCT
251  AGCACAGCGG ACACTTTTTG GAGAAGTATT TTAACATGTC CTGGAAGTAA
     TCGTGTCGCC TGTGAAAAAC CTCTTCATAA AATTGTACAG GACCTTCATT
301  AGAAGATATT TATTTTCCTG TGCTTATAAA ATAAAAAAAA AAAAAAAAAA
     TCTTCTATAA ATAAAAGGAC ACGAATATTT TATTTTTTTT TTTTTTTTTT
351  AAAAA
     TTTTT
```

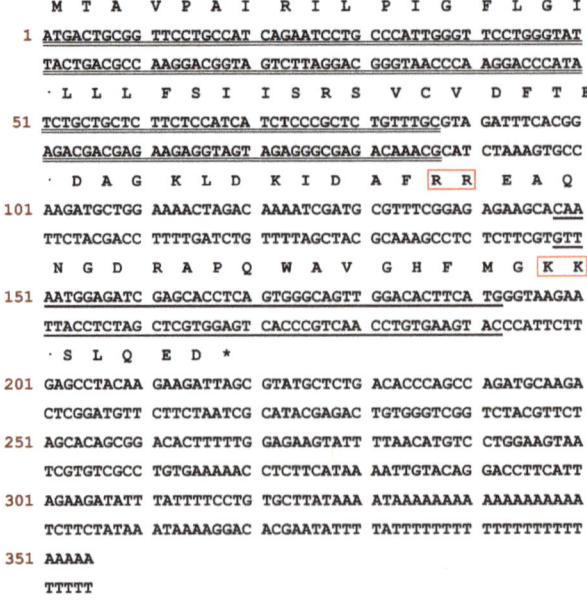

Figure 4. Nucleotide and amino acid sequences of the cDNA encoding the prepropeptide of ranatensin-HLa. Putative signal peptide domain is double-underlined, mature peptide sequence is single-underlined and stop codon is marked by an asterisk. The constructive convertase processing sites for the release of mature peptide are squared in red.

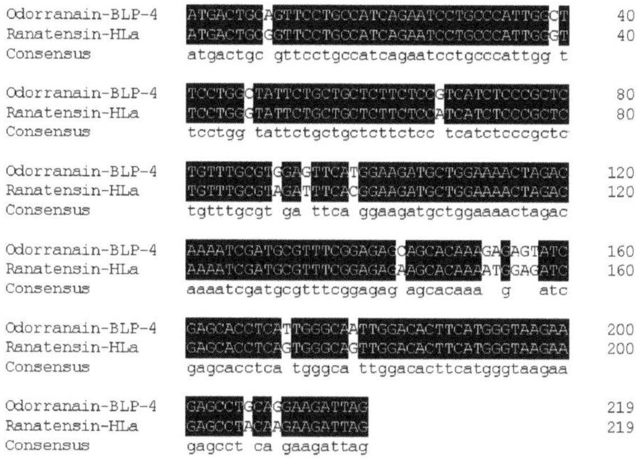

(a)

(b)

Figure 5. Sequence alignments of the full-length open-reading frames of prepro-ranatensin-HLa and the precursors with the highest sequence homology in the bioinformatic analysis using NCBI-BLAST nucleotide and protein database. (**a**) Ranatensin-HLa ORF and a bombesin-like peptide (odorranain-BLP-4) precursor (*Odorrana grahami*) (Accession number: DQ836112.1) showed 92% of sequence identity; (**b**) Ranatensin-HLa ORF and a bombesin (bombesin-OS) precursor (*Odorrana schmackeri*) (Accession number: ATP61827.1) showed 90% of sequence identity.

2.3. Pharmacological Effects of Ranatensin-HLa on Rat Smooth Muscles

The purified synthetic replicates of ranatensin-HLa were used to assess the biological activities on isolated bladder and uterus muscle strips, in which it exhibited pacific pharmacological activities on stimulating contraction of bladder and enhancing the periodicity of spontaneously contractive activity of uterus with EC_{50} values of 7.1 nM and 5.5 nM in a dose-dependent manner (Figure 6).

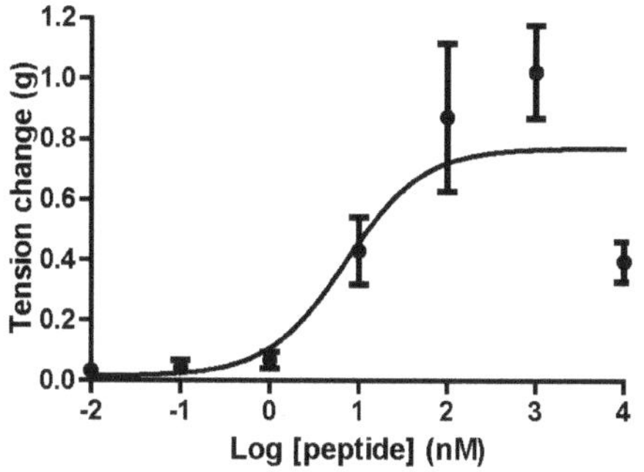

(a)

Figure 6. *Cont.*

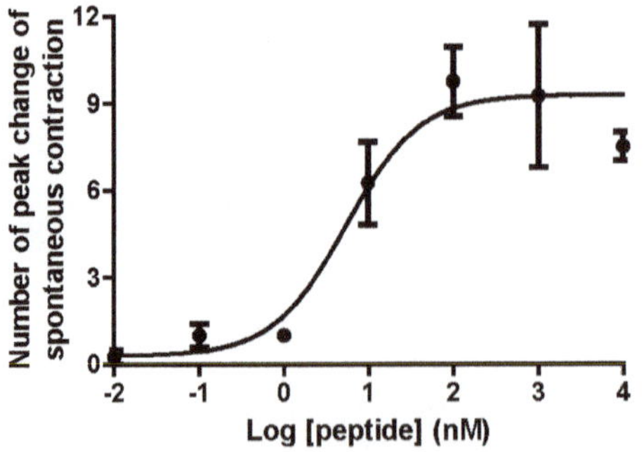

(b)

Figure 6. Myotropic effects of synthetic ranatensin-HLa on isolated smooth muscle preparations from rat bladder and uterus. (**a**) Dose-response curve of the effectiveness of ranatensin-HLa on bladder smooth muscles (EC_{50} = 7.1 nM); (**b**) Dose-response curve of the effectiveness of ranatensin-HLa on uterus smooth muscles (EC_{50} = 5.5 nM). Five determinations were conducted for the generation of each data point.

3. Discussion

In recent years, natural drug discovery has been intensified and extended to structurally- and functionally-diverse compounds of amphibian origin. In reality, since ancient times, amphibian-derived substances have been applied for many conditions; for instance, the skin secretion of poison-dart frogs have been used for the purpose of effective hunting by Native Americans up to the present [26,27], and Chan Su, processed from the dried skin secretion of *Bufo bufo gargarizans* Cantor or *Bufo melanostictus* Schneider with the efficacy of detoxification, has been an important traditional remedy for around 3000 years [28–31]. With the advancement in the technologies of isolation, identification and analysis, it has been recognised that the skin of amphibians is an extraordinary arsenal of a plethora of biologically-active components possessing diverse pharmacological properties [7–9]. These bioactive compounds are endogenously produced and deposited in granular glands, being discharged to the surface of integument upon stimulation to implement diverse physiological and defensive functions [14,32].

In this study, another bombesin-like peptide termed ranatensin-HLa with new structure and belonged to ranatensin subfamily was isolated from the skin secretion of *H. latouchii* following the discovery of ranatensin-HL in 2017. The overall structure of ranatensin-HLa precursor was consistent with other prepropeptides from frog skin, containing a characteristic construction of "signal peptide–N-terminal spacer peptide–mature peptide". Besides, as other neuropeptides from amphibian skin, including ranatensin-HL, the N-terminal glutamine of ranatensin-HLa was also converted into pyroglutamine, and it was also C-terminally amidated. Nevertheless, compared with ranatensin-HL which was released peculiarly from a 125-amino acid residue large precursor, the size of ranatensin-HLa precursor (72 amino acid residues) was similar to that of other ranid frogs-derived bombesin-related peptide precursors [33]. Moreover, like most biosynthetic precursors of bombesin-related peptide formerly cloned from Ranidae family, yet unlike ranatensin-HL precursor, the N-terminal of ranatensin-HLa was flanked by a common convertase proteolytic site made up of dibasic amino acids (Arg-Arg) (Figure 4). According to the sequence comparison of the nucleotides and amino acids of the open-reading frames of ranatensin-HL and ranatensin-HLa (Figure 7), the signal peptides and N-terminal spacer peptides were highly conserved, while the mature peptides were significantly distinct, implying that they might originate from a mutual ancestral gene and became C-terminally different due to the stress response to the threats encountered in which the functional region should evolve rapidly in order to overcome different predators and microorganisms existing in a certain ecological niche. It has been speculated that such diverse molecules in a single species may result from the associated effects of gene duplications, mutation and positive selection [34–36]. Consequently, it makes predators or pathogens more difficult to fight against the components in such complicated cocktails of skin secretions and protects frogs from predators' attack or invasive microorganisms to some extent.

```
Ranatensin-HL    MTAVPAIRILPIGFLANLLLFSFISLSVCVEFTEDTGKLG     40
Ranatensin-HLa   MTAVPAIRILPIGFLGILLLFSIISRSVCVDFTEDAGKLD     40
Consensus        mtavpairilpigfl  lllfs is svcv fted gkl

Ranatensin-HL    KINVLQRAGN......QWAIGHFMGKKSLQDTYRLREQDM     74
Ranatensin-HLa   KIDAFRREAQNGDRAPQWAVGHFMGKKSLQED........     72
Consensus        ki    r         qwa ghfmgkkslq

Ranatensin-HL    EEAAIFPPRSMENMRDTLLQEQRRALSPSQIQGAQRILKK    114
Ranatensin-HLa   ........................................     72
Consensus

Ranatensin-HL    ILEQYFNMSRK                                 125
Ranatensin-HLa   ...........                                  72
Consensus
```

(a)

```
Ranatensin-HL    ATGACTGCGGTCCCTGCCATCAGAATCCTGCCCATTGGCT     40
Ranatensin-HLa   ATGACTGCGGTTCCTGCCATCAGAATCCTGCCCATTGGGT     40
Consensus        atgactgcggt cctgccatcagaatcctgcccattgg t

Ranatensin-HL    TCCTGGCTAACCTGCTGCTCTTCTCCTTCATCTCGCTCTC     80
Ranatensin-HLa   TCCTGGGTATTCTGCTGCTCTTCTCCATCATCTCCCGCTC     80
Consensus        tcctgg ta  ctgctgctcttctcc tcatctc c ctc

Ranatensin-HL    TGTTTGCGTGGAGTTCACGGAGGATACTGGCAAACTGGGC    120
Ranatensin-HLa   TGTTTGCGTAGATTTCACGGAAGATGCTGGAAAACTAGAC    120
Consensus        tgtttgcgt ga ttcacgga gat ctgg aaact g c

Ranatensin-HL    AAAATCAATGTGCTTCAGAGAGCAGGGAATCAGTGGGCAA    160
Ranatensin-HLa   AAAATCGATGCGTTTCGGAGAGAAGCACA..AAATGGAGA    158
Consensus        aaaatc atg g ttc gagag ag   a  a   gg  a

Ranatensin-HL    TTGGACACTTTA.TGGGTAAGAAGAGCCTGCA.AGACACG    198
Ranatensin-HLa   TCGAGCACCTCAGTGGGCAGTTGGACACTTCATGGGTAAG    198
Consensus        t g  cac t a tggg a    ga  ct ca  g  a g

Ranatensin-HL    TACAGATTAAGAGAACAAGATATGGAAGAGGCAGCCATCT    238
Ranatensin-HLa   AAGAGCCTACAAGAAGATTAG...................    219
Consensus         a ag  ta  agaa a  a

Ranatensin-HL    TTCCACCACGCTCTATGGAGAACATGAGAGACACTCTACT    278
Ranatensin-HLa   ........................................    219
Consensus

Ranatensin-HL    ACAAGAGCAGCGAAGAGCACTGTCACCCAGCCAGATCCAA    318
Ranatensin-HLa   ........................................    219
Consensus

Ranatensin-HL    GGAGCACAGCGGATACTAAAGAAGATTCTGGAGCAGTATT    358
Ranatensin-HLa   ........................................    219
Consensus

Ranatensin-HL    TTAACATGTCCCGGAAGTAA                        378
Ranatensin-HLa   ....................                        219
Consensus
```

(b)

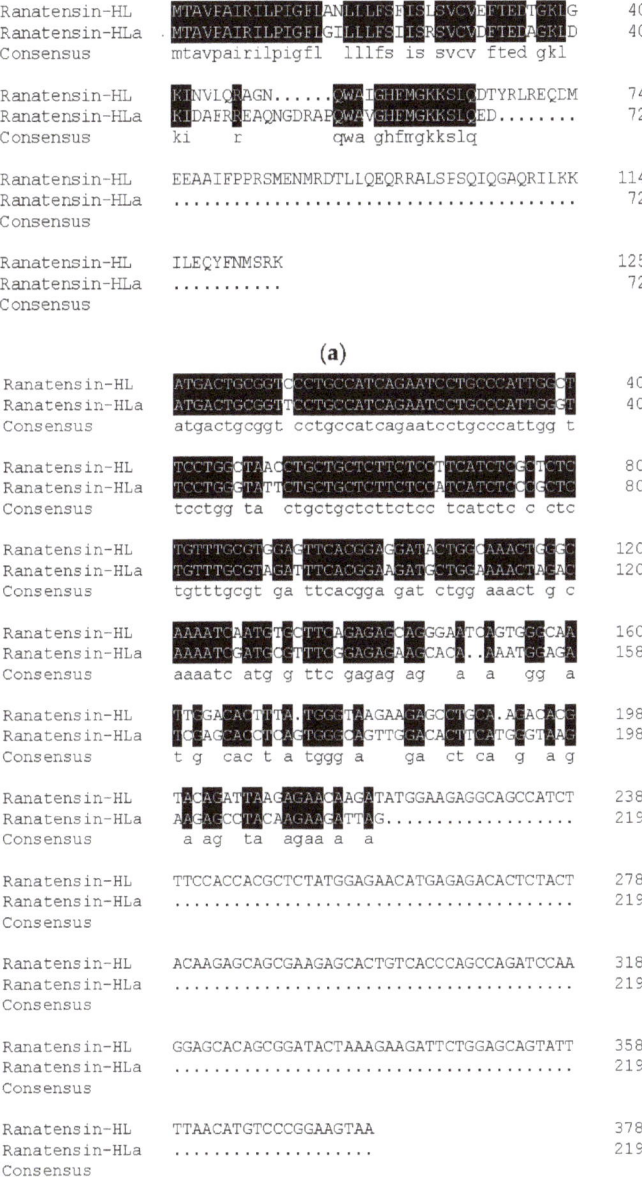

Figure 7. Comparison of the nucleotide and amino acid sequences of the biosynthetic precursors of ranatensin-HL [33] and ranatensin-HLa. (**a**) Comparison of the translated amino acid sequences of ranatensin-HL and ranatensin-HLa precursors; (**b**) Nucleotide sequences comparison of ranatensin-HL and ranatensin-HLa precursors.

The nucleotide and amino acid sequences of the biosynthetic precursor of ranatensin-HLa were highly homological to those of bombesin-like peptides from *Odorrana grahami* (odorranain-BLP-4) and *Odorrana schmackeri* (bombesin-OS) (Figure 5). Additionally, taking the high homology of ranatensin-HL from *H. latouchii* and another bombesin-related peptide from *Lithobates catesbeianus* [33] into consideration, we guessed that *H. latouchii*, *Odorrana graham*, *Odorrana schmackeri* and *Lithobates catesbeianus* were likely to have retained bombesin/ranatensin peptides in their skin secretion, or these peptides might be the outcome of convergent evolution [37].

In terms of the pharmacological activity of ranatensin-HLa, it stimulated similar contractile frequency in rat uterus as ranatensin-HL and was more potent in evoking rat bladder contraction (Table 1). Since NMB- and GRP-preferring subtype receptors are expressed in bladder and uterus respectively [38,39], it was speculated that both ranatensin-HL and ranatensin-HLa might cause myotropic contraction by binding to the NMB receptor in bladder and GRP receptor in uterus. The greater potency of ranatensin-HLa towards bladder may be attributed to either the length or the composition of the N-terminal amino acids (Figure 8). Furthermore, as both ranatensin-HL and ranatensin-HLa were more effective than ranatensin-HL-10 in tested smooth muscles, it further suggested that the longer stretch of N-terminal might be beneficial to the efficacy of peptides [40] in exposing functional cores or extending half-life [41] (Figure 8 and Table 1).

Table 1. Comparison of ranatensin-HLs on the myoactivities [33].

Peptides	Bladder (EC_{50}, nM)	Uterus (EC_{50}, nM)
Ranatensin-HLa	7.1	5.5
Ranatensin-HL	19.2	5.4
Ranatensin-HL-10	63.8	70.9

```
Ranatensin-HLa      .QNGDRAPQWAVGHFM   15
Ranatensin-HL       QRAGN...QWAIGHFM   13
Ranatensin-HL-10    ...GN...QWAIGHFM   10
Consensus              g    qwa ghfm
```

Figure 8. Comparison of ranatensin-HLs on the primary structures [33].

Undoubtedly, frog skin will continue to serve as a treasury with extraordinarily-diverse lead compounds for the development of therapeutically effective medicines. Moreover, the dissimilar bioactive peptides present in different species and the corresponding precursor-encoding cDNA sequences will offer an additional insight into the evolutionary relationships between different specie [32,42].

4. Conclusions

Through the combination of mass spectrometry and molecular cloning, another naturally-occurring bombesin-related peptide belonging to ranatensin subfamily was characterised in *H. latouchii* skin secretion. In the in vitro pharmacological investigation, it displayed moderate myoactivity on rat bladder and uterus. The precursor of ranatensin-HLa was compared with that of ranatensin-HL previously discovered from *H. latouchii*; as a result, the signal peptides and N-terminal spacer peptides were highly conserved, but remarkable mature peptide modification was observed, which provided the evidence that the diversity of bioactive substances within a single species might be ascribed to selective pressure and might eventually lead to the frogs' optimisation of adaptation.

5. Materials and Methods

5.1. Specimen Biodata and Skin Secretion Acquisition

Adults of the broad-folded Frog, *Hylarana latouchii* were captured in Fujian Province of the People's Republic of China. Frogs were stimulated by mild electricity on the dorsal skin surface to discharge the defensive skin secretion which was subsequently rinsed off with distilled deionised water to be collected. The detailed methods were previously described [43]. The collected skin secretion was then lyophilised and kept at −20 °C prior to usage. The procedure of secretion acquisition had been overseen by the Institutional Animal Care and Use Committee (IACUC) of Queen's University Belfast, and approved on 1 March 2011. It was carried out under the UK animal (Scientific Procedures) Act 1986, Project licence PPL 2694, which was issued by the Department of Health, Social Services and Public Safety, Northern Ireland.

5.2. Chromatographic Fractionation of H. latouchii Skin Secretion with Reverse Phase HPLC

A sample made from dissolving 5 mg of lyophilised *H. latouchii* skin secretion in 1 mL of trifluoroacetic acid (TFA)/water (0.05/99.95, *v*/*v*) was subjected to reverse phase HPLC chromatography. It was clarified and loaded onto a Jupiter C-5 semi preparative column (300 Å, 5 µm, 25 cm × 1 cm, Phenomenex, Macclesfield, Cheshire, UK) which was installed in a Cecil Adept CE4200 HPLC system (Amersham Biosciences, Buckinghamshire, UK) equipped with a Powerstream HPLC software for chromatographic fractionation. Chromatography was performed by eluting from 0.05/99.95 (*v*/*v*) TFA/water to 0.05/19.95/80.00 (*v*/*v*/*v*) TFA/water/acetonitrile over 240 min at a flow rate of 1 mL/min. The effluent fractions were detected at λ214 nm, and were collected automatically and continuously every minute. One hundred microliter of each collected fraction was transferred to be freeze-dried and stored at −20 °C for the subsequent screening of myoactivity employing smooth muscle bioassays.

5.3. Myoactivity Screening of Chromatographic Fractions

According to the institutional animal experimentation ethics and UK animal research guidelines, the method of carbon dioxide asphyxiation in combination of cervical dislocation was employed to realize euthanasia of female Wister rats with weights of 250-300 g for the in vitro smooth muscle pharmacological assays. Upon removal of the intact urinary bladder, uterine horns, ileum and proximal rat tail artery from each rat by dissection, they were kept in ice-cold Kreb's solution (118 mM NaCl, 4.7 mM KCl, 25 mM $NaHCO_3$, 1.15 mM NaH_2PO_4, 2.5 mM $CaCl_2$, 1.1 mM $MgCl_2$ and 5.6 mM glucose) spiritedly pneumatic with 95% O_2 and 5% CO_2 immediately. Then, the dissected tissue strips were carefully mounted on a transducer and immersed in organ baths filled with flowing (2 mL/min) Kreb's solution constantly bubbled with mixed 95% O_2 and 5% CO_2 at 37 °C. Prior to the application of sample solutions, the tissue preparations were allowed to be equilibrated under the normal physiological tension of 1.0 g (for bladder) and 0.5 g (for bladder, uterus, artery and ileum) for 1 h. Then, HPLC fractions dissolved in 22 µl of Kreb's solution were added sequentially to the organ baths to examine the myotropic activity of each fraction.

5.4. Structural Characterization of Peptide Possessing Bioactivity

The molecular weight of the HPLC fraction which displayed myoactivity was primarily analysed using MALDI-TOF mass spectrometry (Voyager DE, PerSeptive Biosystems, Foster City, CA, USA). Then, the predominant peptide in the bioactive fraction was further fragmented in an LCQ-Fleet electrospray ion-trap mass spectrometer (Thermo Fisher Scientific, San Francisco, CA, USA) for MS/MS sequencing to determine the primary structure.

5.5. Molecular Cloning of the cDNA Encoding the Peptide Biosynthetic Precursor of Ranatensin-HLa

Based on primary structure analysis result, a degenerate sense primer (S: 5′-CARAAYGGNGAYM GNGCNCC-3′) complementary to pGlu-N-G-D-R-A-P- was designed. One milliliter of cell lysis/mRNA protection buffer provided by manufacturer (Dynal Biotec, Wirral, UK) was used to dissolve a total of 5 mg of lyophilised *H. latouchii* skin secretion, which was followed by polyadenylated mRNA isolation with magnetic oligo-dT beads (Dynal Biotech, Merseyside, UK). The isolated mRNA was then reverse transcribed to synthesise cDNA that was applied to run 3′-rapid amplification of cDNA ends (RACE) procedure with the designed degenerate primer in conjunction with a Nested Universal Primer (NUP) provided by kit (SMART-RACE kit, Clontech, Palo Alto, CA, USA) to get the 3′-end of biosynthetic precursor of myoactive peptide. The resultant PCR products were ligated into a pGEM-T easy vector (Promega Corporation, Southampton, UK) and sequenced using an ABI 3100 automated sequencer (Applied Biosystems, Foster City, CA, USA). The sequence information was used to further design a specific antisense primer (AS: 5′-GCATCTGGCTGGGTGTCAGAGCATA-3′) complementary to the 3′-end of the DNA sequence. Subsequently, this primer along with the NUP primer were used to conduct 5′-RACE PCR whose products were purified, cloned and sequenced as described above.

5.6. Fmoc Chemistry Solid-Phase Peptide Synthesis of Ranatensin-HLa

On the strength of results obtained from primary structure analysis and molecular cloning, the explicit sequence of the bioactive peptide could be established. It was followed by automatic synthesis using PS3 solid-phase peptide synthesiser (Protein Technologies, Tucson, AZ, USA) with the methodology of Fmoc chemistry. Reverse phase HPLC was employed for the purification of the synthetic products, and MALDI-TOF MS and MS/MS fragmentation sequencing were used for the confirmation of the purity and structural authenticity of the peptide.

5.7. The Effects of Synthetic Peptide on Rat Smooth Muscles Tension

Purified peptide was prepared in Kreb's solution and diffused to the 2-mL organ baths with bladder and uterus tissue strips equilibrated under normal physiological tension to get final concentrations ranging from 10^{-2} to 10^{4} nM to produce dose-response curves. Between each dose, the smooth muscle preparations were subjected to periods of 10-min wash and 10-min equilibration. Tension changes of bladder tissue strips were magnified and recorded by pressure transducers hooked up to a PowerLab System (AD Instruments Pty Ltd., Oxford, UK), and the frequency changes of the spontaneous contraction of uterus were recorded by counting the contractile peaks. Peptide was investigated for the myoactivity using at least five muscle strips in each concentration.

5.8. Statistical Analysis

Data were analysed by GraphPad Prism 5 software (San Diego, CA, USA) to establish the EC_{50} values for the peptides and to generate dose-response curves featured by changes in tension (for bladder) or changes in contractive frequency (for uterus) against peptide concentrations using a "best-fit" algorithm. Data points represented mean values ± SEM from Student's *t*-test.

Author Contributions: Conceptualization, Y.L. and T.C.; Methodology, C.M., M.Z. and L.W.; Validation, N.H., H.H. and C.M.; Formal Analysis, Y.L., N.H., H.H., T.C., L.W. and M.Z.; Investigation, Y.L. and N.H.; Resources, T.C. and Y.L.; Data Curation, Y.L. and N.H.; Writing—Original Draft Preparation, Y.L. and N.H.; Writing—Review & Editing, Y.L., C.M. and T.C.; Visualization, Y.L.; Supervision, Y.L. and T.C.; Project Administration, Y.L. and T.C.; Funding Acquisition, Y.L.

Funding: This research was funded by the grants from the National Natural Science Foundation of China, grant number 31500753.

Acknowledgments: Yan Lin was in receipt of an Overseas Studentship at Queen's University, Belfast. This work was also supported by the Natural Drug Discovery Group, School of Pharmacy, Queen's University.

Conflicts of Interest: The authors declare no conflict of interest.

References

1. Pough, F.H. Amphibian biology and husbandry. *Inst. Lab. Anim. Resour.* **2007**, *48*, 203–213. [CrossRef]
2. Crump, M.L. Amphibian diversity and life history. In *Amphibian Ecology and Conservation: A Handbook of Techniques*; Dodd, C.K., Ed.; Oxford University Press: Oxford, UK, 2009; pp. 3–20.
3. Frost, D.R. Amphibian Species of the World: An Online Reference. Version 6.0. Available online: http://research.amnh.org/herpetology/amphibia/index.html (accessed on 20 February 2019).
4. König, E.; Bininda-Emonds, O.R.P.; Shaw, C. The diversity and evolution of anuran skin peptides. *Peptides* **2015**, *63*, 96–117. [CrossRef]

5. Bueno, C.; Navas, P.; Aijon, J.; Lopez-Campos, J.L. Glycoconjugates in the epidermis of *Pleurodeles waltlii*. *J. Ultrasructure Res.* **1981**, *77*, 354–359.
6. Toledo, R.C.; Jared, C. Cutaneous granular glands and amphibian venoms. *Comp. Biochem. Physiol. Part Physiol.* **1995**, *111*, 1–29. [CrossRef]
7. Lazarus, L.H.; Attila, M. The toad, ugly and venomous, wears yet a precious jewel in his skin. *Prog. Neurobiol.* **1993**, *41*, 473–507. [PubMed]
8. Clarke, B.T. The natural history of amphibian skin secretions, their normal functioning and potential medical applications. *Biol. Rev.* **1997**, *72*, 365–379.
9. Gomes, A.; Giri, B.; Saha, A.; Mishra, R.; Dasgupta, S.C.; Debnath, A.; Gomes, A. Bioactive molecules from amphibian skin: Their biological activities with reference to therapeutic potentials for possible drug development. *Indian J. Exp. Biol.* **2007**, *45*, 579–593. [PubMed]
10. Bevins, C.; Zasloff, M. Peptides from frog skin. *Annu. Rev. Biochem.* **1990**, *59*, 395–414.
11. Giuliani, A.; Pirri, G.; Nicoletto, S.F. Antimicrobial peptides: An overview of a promising class of therapeutics. *Cent. Eur. J. Biol.* **2007**, *2*, 1–33.
12. Samgina, T.Y.; Artemenko, K.A.; Gorshkov, V.A.; Lebedev, A.T. Bioactive peptides from the skin of ranid frogs: Modern approaches to the mass spectrometric de novo sequencing. *Russ. Chem. Bull.* **2008**, *57*, 1080–1091.
13. Calderon, L.; Stabeli, R. Anuran Amphibians: A huge and threatened factory of a variety of active peptides with potential nanobiotechnological applications in the face of amphibian decline. *Chang. Divers. Chang. Environ. Oscar Grillo Gianfranco Venora* **2011**, *7*, 211–242.
14. Simmaco, M.; Mignogna, G.; Barra, D. Antimicrobial peptides from amphibian skin: What do they tell us? *Biopolym. (Peptide Sci.)* **1998**, *47*, 435–450. [CrossRef]
15. Anastasi, A.; Erspamer, V.; Bucci, M. Isolation and structure of bombesin and alytesin, two analogous active peptides from the skin of the european amphibians *Bombina* and *Alytes*. *Experientia* **1971**, *27*, 166–167. [CrossRef] [PubMed]
16. McDonald, T.J.; Jörnvall, H.; Nilsson, G.; Vagne, M.; Ghatei, M.; Bloom, S.R.; Mutt, V. Characterization of a gastrin releasing peptide from porcine non-antral gastric tissue. *Biochem. Biophys. Res. Commun.* **1979**, *90*, 227–233. [CrossRef]
17. Minamino, N.; Kangawa, K.; Matsuo, H. Neuromedin B: A novel bombesin-like peptide identified in porcine spinal cord. *Biochem. Biophys. Res. Commun.* **1983**, *114*, 541–548. [CrossRef]
18. Erspamer, V.; Erspamer, G.F.; Inselvini, M. Some pharmacological actions of alytesin and bombesin. *J. Pharm. Pharmacol.* **1970**, *22*, 875–876. [CrossRef] [PubMed]
19. Nagalla, S.R.; Barry, B.J.; Falick, A.M.; Gibson, B.W.; Taylor, J.E.; Dong, J.Z.; Spindel, E.R. There are three distinct forms of bombesin: identification of [Leu^{13}] bombesin, [Phe^{13}] bombesin, and [Ser^3, Arg^{10}, Phe^{13}] bombesin in the frog *Bombina orientalis*. *J. Biol. Chem.* **1996**, *271*, 7731–7737. [CrossRef] [PubMed]
20. Lai, R.; Liu, H.; Lee, W.H.; Zhang, Y. A novel proline rich bombesin-related peptide (PR-bombesin) from toad *Bombina maxima*. *Peptides* **2002**, *23*, 437–442. [CrossRef]
21. Jensen, R.T.; Battey, J.F.; Spindel, E.R.; Benya, R.V. International Union of Pharmacology. LXVIII. Mammalian Bombesin Receptors: Nomenclature, distribution, pharmacology, signaling, and functions in normal and disease states. *Pharmacol. Rev.* **2008**, *60*, 1–42. [CrossRef]
22. Krane, I.M.; Naylor, S.L.; Helin-Davis, D.; Chin, W.W.; Spindel, E.R. Molecular cloning of cDNAs encoding the human bombesin-like peptide neuromedin B. Chromosomal localization and comparison to cDNAs encoding its amphibian homolog ranatensin. *J. Biol. Chem.* **1988**, *263*, 13317–13323.
23. Iwabuchi, M.; Ui-Tei, K.; Yamada, K.; Matsuda, Y.; Sakai, Y.; Tanaka, K.; Ohki-Hamazaki, H. Molecular cloning and characterization of avian bombesin-like peptide receptors: New tools for investigating molecular basis for ligand selectivity. *Br. J. Pharmacol.* **2003**, *139*, 555–566. [CrossRef] [PubMed]
24. Nagalla, S.R.; Barry, B.J.; Creswick, K.C.; Eden, P.; Taylor, J.T.; Spindel, E.R. Cloning of a receptor for amphibian [Phe^{13}] bombesin distinct from the receptor for gastrin-releasing peptide: Identification of a fourth bombesin receptor subtype (BB4). *Proc. Natl. Acad. Sci.* **1995**, *92*, 6205–6209. [CrossRef] [PubMed]
25. Katsuno, T.; Pradhan, T.K.; Ryan, R.R.; Mantey, S.A.; Jensen, R.T. Pharmacology and cell biology of the bombesin receptor subtype 4 (BB4-R). *Biochemistry* **1999**, *38*, 7307–7320. [CrossRef] [PubMed]
26. Philippe, G.; Angenot, L. Recent developments in the field of arrow and dart poisons. *J. Ethnopharmacol.* **2005**, *100*, 85–91. [CrossRef] [PubMed]

27. Dumbacher, J.P.; Wako, A.; Derrickson, S.R.; Samuelson, A.; Spande, T.F.; Daly, J.W. Melyrid beetles (*Choresine*): A putative source for the batrachotoxin alkaloids found in poison-dart frogs and toxic passerine birds. *Proc. Natl. Acad. Sci. USA* **2004**, *101*, 15857–15860. [CrossRef] [PubMed]
28. Qi, F.; Li, A.; Inagaki, Y.; Gao, J.; Li, J.; Kokudo, N.; Li, X.K.; Tang, W. Chinese herbal medicines as adjuvant treatment during chemo- or radio-therapy for cancer. *Biosci. Trends* **2010**, *4*, 297–307. [PubMed]
29. Wang, L.; Raju, U.; Milas, L.; Molkentine, D.; Zhang, Z.; Yang, P.; Cohen, L.; Meng, Z.; Liao, Z. Huachansu, containing cardiac glycosides, enhances radiosensitivity of human lung cancer cells. *Anticancer Res.* **2011**, *31*, 2141–2148. [PubMed]
30. Takai, N.; Kira, N.; Ishii, T.; Yoshida, T.; Nishida, M.; Nishida, Y.; Nasu, K.; Narahara, H. Bufalin, a traditional oriental medicine, induces apoptosis in human cancer cells. *Asian Pac. J. Cancer Prev.* **2012**, *13*, 399–402. [CrossRef] [PubMed]
31. Li, J.; Zhang, Y.; Lin, Y.; Wang, X.; Fang, L.; Geng, Y.; Zhang, Q. Preparative separation and purification of bufadienolides from chansu by high-speed counter-current chromatography combined with preparative HPLC. *Quimica Nova* **2013**, *36*, 1–5. [CrossRef]
32. Barra, D.; Simmaco, M. Amphibian skin: A promising resource for antimicrobial peptides. *Trends Biotechnol.* **1995**, *13*, 205–209. [CrossRef]
33. Lin, Y.; Chen, T.; Zhou, M.; Wang, L.; Su, S.; Shaw, C. Ranatensin-HL: A bombesin-related tridecapeptide from the skin secretion of the broad-folded frog, *Hylarana latouchii*. *Molecules* **2017**, *22*, 1110. [CrossRef] [PubMed]
34. Conlon, J.M.; Kolodziejek, J.; Nowotny, N. Antimicrobial peptides from ranid frogs: Taxonomic and phylogenetic markers and a potential source of new therapeutic agents. *Biochim. Biophys. Acta Proteins Proteomics* **2004**, *1696*, 1–14. [CrossRef]
35. Duda, T.F.; Vanhoye, D.; Nicolas, P. Roles of diversifying selection and coordinated evolution in the evolution of amphibian antimicrobial peptides. *Mol. Biol. Evol.* **2002**, *19*, 858–864. [CrossRef] [PubMed]
36. Vanhoye, D.; Bruston, F.; Nicolas, P.; Amiche, M. Antimicrobial peptides from hylid and ranin frogs originated from a 150-million-year-old ancestral precursor with a conserved signal peptide but a hypermutable antimicrobial domain. *Eur. J. Biochem.* **2003**, *270*, 2068–2081. [CrossRef] [PubMed]
37. Roelants, K.; Fry, B.G.; Norman, J.A.; Clynen, E.; Schoofs, L.; Bossuyt, F. Identical skin toxins by convergent molecular adaptation in frogs. *Curr. Biol.* **2010**, *20*, 125–130. [CrossRef] [PubMed]
38. Kortezova, N.; Mizhorkova, Z.; Milusheva, E.; Coy, D.H.; Vizi, E.S.; Varga, G. GRP-preferring bombesin receptor subtype mediates contractile activity in cat terminal ileum. *Peptides* **1994**, *15*, 1331–1333. [CrossRef]
39. Kilgore, W.R.; Mantyh, P.W.; Mantyh, C.R.; McVey, D.C.; Vigna, S.R. Bombesin/GRP-preferring and neuromedin B-preferring receptors in the rat urogenital system. *Neuropeptides* **1993**, *24*, 43–52. [CrossRef]
40. Erspamer, G.F.; Severini, C.; Erspamer, V.; Melchiorri, P.; Delle Fave, G.; Nakajima, T. Parallel bioassay of 27 bombesin-like peptides on 9 smooth muscle preparations. Structure-activity relationships and bombesin receptor subtypes. *Regul. Pept.* **1988**, *21*, 1–11. [CrossRef]
41. Zhou, X.; Ma, C.; Zhou, M.; Zhang, Y.; Xi, X.; Zhong, R.; Chen, T.; Shaw, C.; Wang, L. Pharmacological effects of two novel bombesin-like peptides from the skin secretions of Chinese piebald odorous frog (*odorrana schmackeri*) and European edible frog (*Pelophylax kl. esculentus*) on smooth muscle. *Molecules* **2017**, *22*, 1798. [CrossRef]
42. Pukala, T.L.; Bowie, J.H.; Maselli, V.M.; Musgrave, I.F.; Tyler, M.J. Host-defence peptides from the glandular secretions of amphibians: Structure and activity. *Nat. Prod. Rep.* **2006**, *23*, 368–393. [CrossRef]
43. Tyler, M.J.; Stone, D.J.M.; Bowie, J.H. A novel method for the release and collection of dermal, glandular secretions from the skin of frogs. *J. Pharmacol. Toxicol. Methods* **1992**, *28*, 199–200. [CrossRef]

toxins

MDPI

Article

Domain II of *Pseudomonas* Exotoxin Is Critical for Efficacy of Bolus Doses in a Xenograft Model of Acute Lymphoblastic Leukemia

Fabian Müller [1,2], Tyler Cunningham [1,†], Richard Beers [1], Tapan K. Bera [1], Alan S. Wayne [3] and Ira Pastan [1,*]

1 Laboratory of Molecular Biology, Center for Cancer Research, National Cancer Institute, National Institutes of Health, Bethesda, MD 20892, USA; fabian.mueller@uk-erlangen.de (F.M.); cunningham.tyler@gmail.com (T.C.); beersr@dc37a.nci.nih.gov (R.B.); berat@mail.nih.gov (T.K.B.)

2 Department of Hematology/Oncology, University Hospital Erlangen, Erlangen 91054, Germany

3 Children's Center for Cancer and Blood Diseases, Division of Hematology, Oncology and Blood and Marrow Transplantation, Children's Hospital Los Angeles, Norris Comprehensive Cancer Center, Keck School of Medicine, University of Southern California, Los Angeles, CA 90027, USA; awayne@chla.usc.edu

* Correspondence: pastani@mail.nih.gov; Tel.: +1-(240)-760-6470

† Current Address: MD/PhD Program, University of Miami, Miller School of Medicine, Miami, FL, USA.

Received: 26 April 2018; Accepted: 17 May 2018; Published: 21 May 2018

Abstract: Moxetumomab pasudotox is a fusion protein of a CD22-targeting antibody and *Pseudomonas* exotoxin. Minutes of exposure to Moxetumomab achieves similar cell killing than hours of exposure to a novel deimmunized variant against some acute lymphoblastic leukemia (ALL). Because blood levels fall quickly, Moxetumomab is more than 1000-fold more active than the deimmunized variant in vivo. We aimed to identify which part of Moxetumomab increases in vivo efficacy and generated five immunotoxins, tested time-dependent activity, and determined the efficacy in a KOPN-8 xenograft model. Full domain II shortened the time cells had to be exposed to die to only a few minutes for some ALL; deimmunized domain III consistently extended the time. Against KOPN-8, full domain II accelerated time to arrest protein synthesis by three-fold and tripled PARP-cleavage. In vivo efficacy was increased by more than 10-fold by domain II and increasing size, and therefore half-life enhanced efficacy two- to four-fold. In summary, in vivo efficacy is determined by the time cells have to be exposed to immunotoxin to die and serum half-life. Thus, domain II is most critical for activity against some ALL treated with bolus doses; however, immunotoxins lacking all but the furin-cleavage site of domain II may be advantageous when treating continuously.

Keywords: immunotoxin; Moxetumomab pasudotox; targeted therapy; CD22; B cell non-Hodgkin lymphoma; acute lymphoblastic leukemia; mantle cell lymphoma

Key Contribution: CD22-targeted immunotoxin containing domain II of *Pseudomonas* exotoxin is substantially more active against some acute lymphoblastic leukemia than immunotoxin lacking all but the furin-cleavage site of domain II when being administered as intravenous bolus doses. Lack of domain II; however; increases the cytotoxicity by two-fold when cells are treated continuously for three days in culture.

1. Introduction

Therapeutic antibodies have become the standard of care in the treatment of a variety of diseases, including cancer [1]. Therapeutic antibodies enable a specific attack against target cells, while the unrelated tissue remains unaffected. To further increase the efficacy of therapeutic antibodies, the constant fragment (Fc) of antibodies has been engineered to improve recruitment of complement and immune cells [2] and antibodies were armed with radionucleotides [3], chemotherapeutic drug conjugates [4,5], or plant and bacterial toxins [6]. The latter, also called immunotoxins, show high clinical activity against hematologic malignancies [6,7]. The immunotoxin Moxetumomab pasudotox consists of a CD22-targeting disulfide-stabilized antibody fragment (dsFv) and a 38 kDa fragment of *Pseudomonas* exotoxin (PE). Moxetumomab pasudotox (hereafter referred to as dsFv-PE38) produces response rates of 86% in patients with relapsed/refractory (r/r) hairy cell leukemia [8], and of 32% in r/r pediatric acute lymphoblastic leukemia (ALL) [9].

Because patients that are treated with immunotoxins often develop anti-drug antibodies which likely reduce activity [8,10], less immunogenic immunotoxins have been developed. Major B- and T-cell epitopes that were located in domain II of *Pseudomonas* exotoxin were removed by deletion of all but the furin cleavage site [11] resulting in a 24 kDa PE (PE24) which on average is 2-fold more active than PE38 [12]. To further reduce immunogenicity, seven mutations that disrupt B-cell epitopes were introduced in domain III, hereafter referred to as PE24(B) [13,14]. By extending the 25 kDa dsFv of PE24(B) to a 50 kDa Fab, serum half-life and activity of bolus doses greatly increased [14]. Fab-PE24(B) achieved sustained complete remissions in a subcutaneous CA46 xenograft model; however, dsFv-PE38 could not.

Opposite the CA46 model, a systemic xenograft with the ALL cell line KOPN-8 responded substantially better to dsFv-PE38 than to Fab-PE24(B) [15]. The KOPN-8 bone marrow (BM) infiltration was more than 1000-fold more efficiently reduced after five doses of the maximally tolerated dose (MTD) of 0.4 mg/kg dsFv-PE38 QOD than after five doses of the MTD of 2.0 mg/kg Fab-PE24(B) [15]. We showed that the reason for the substantial difference in the efficacy was the time that KOPN-8 cells had to be exposed to immunotoxin to die. While the cells had to be exposed to dsFv-PE38 for only a few minutes to induced more than 50% cell death, they had to be exposed to Fab-PE24(B) for more than 6 h in order to reach similar cytotoxicity. Because the serum concentration of immunotoxin after a bolus dose fall within hours to inactive levels, bolus doses of Fab-PE24(B) did not reduce KOPN-8 tumor burden in vivo.

Here, we determine which part of dsFv-PE38 is the most critical to achieve the 1000-fold stronger efficacy in the KOPN-8 animal model. Five distinct immunotoxin variants were generated and were tested for their time-dependent activity against various B-cell malignancies and patient-derived ALL blasts in vitro, and findings that were validated in the systemic KOPN-8 xenograft model.

2. Results

2.1. Wild-Type PE24 Immunotoxin Shows Highest Overall Activity

To test which protein domain was responsible for the substantial difference of activity of dsFv-PE38 and Fab-PE24(B) against the KOPN-8 cell line, we constructed a total of five molecules controlling for the three major differences, namely size of the antibody fragment, full domain II versus furin cleavage site, and mutated domain III versus wild-type (WT) domain III (Figure 1A). By removing all but the furin cleavage site of domain II of the WT PE38, the molecule dsFv-PE24 was generated [12], and by exchanging the dsFv with a Fab, the novel 23 kDa larger Fab-PE24 was achieved. By replacing PE24 with PE24(B), the two immunotoxins dsFv-PE24(B) and Fab-PE24(B) were generated [14]. Various B-cell malignancy cell lines were continuously treated for three days with immunotoxin, and the concentration at which 50% of cell growth was inhibited, as determined by the WST-8 assay (Supplementary Table S1). With an IC_{50} of 0.3 pM against the Reh cell line, dsFv-PE38 was the most active of the five immunotoxins. The dsFv-PE24 and Fab-PE24, however, showed the highest average activity overall cell lines with average IC_{50}s of 6.9 pmol/L and 10.9 pmol/L, respectively. In addition to highest average activity, the PE24 variants showed the least variability with an approximately a three-fold lower variance (σ^2) than PE38 or PE24(B) variants. We mathematically determined the fold-difference of cytotoxicity of the immunotoxin variants relative to dsFv-PE38 (Figure 1B). The larger Fab-containing immunotoxins were generally less cytotoxic in vitro than their respective smaller, dsFv-containing molecule. This finding was consistent for both WT and the mutant PE24. Against the Reh and the KOPN-8 cell line, dsFv-PE38 was more active than any other molecule, while for the other cell lines, WT PE24 without domain II was the most active. Confirming previous results [14], the activity of mutated PE24(B) was, on average, similar to that of dsFv-PE38 in this assay.

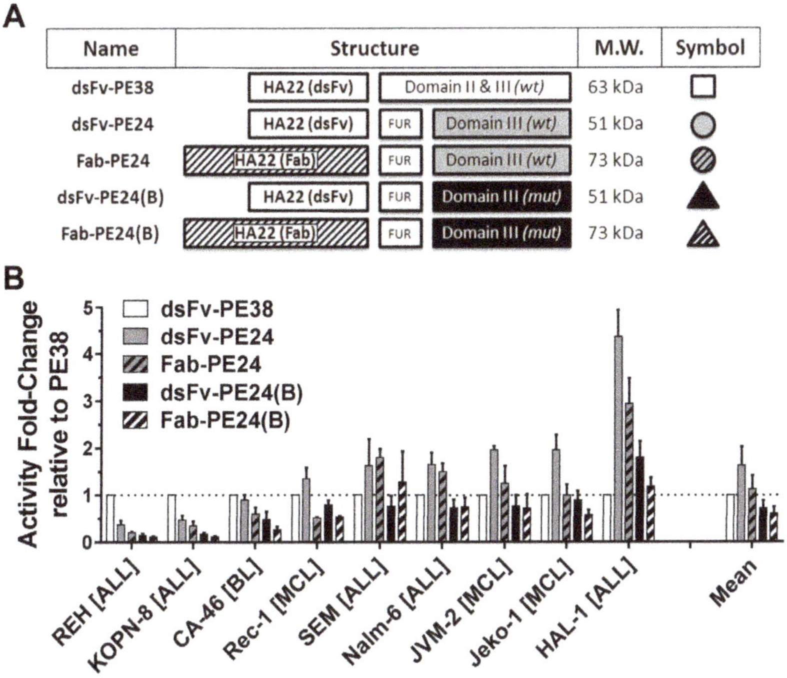

Figure 1. Variants of CD22-targeted immunotoxins are active in vitro. (**A**) Cartoon depicting the differences among five immunotoxin variants. Indicated are the names of the molecules, the antibody structure as disulfide-stabilized Fv (dsFv) or Fab, the respective PE-structure either as wild-type (WT) PE or a PE containing B-cell epitope-depleting mutations (mut) [14], the respective molecular weight (MW), and the symbol/color-code used throughout the manuscript; (**B**) The activity ($1/IC_{50}$) of the indicated immunotoxin variants against various malignant B-cell lines was determined by WST8. Shown are the average fold-changes relative to dsFv-PE38 of at least three independent experiments, errors as standard error of mean (SEM).

2.2. Immunotoxin Variants Show Highly Variable Time to Reach Maximal Cytotoxicity

Previously, we found that cells have to be exposed to CD22-targeted immunotoxins for a highly variable time for them to die [15]. Accordingly, we tested the time cells that had to be exposed to the five immunotoxins to induce cell death. We chose an equimolar concentration of 2.8 nmol/L immunotoxin, because it correlates with the amount of Fab-PE24(B) that reaches KOPN-8 cells growing in the murine BM after an intravenous single dose [15]. After treating with 2.8 nmol/L, the cells were washed at indicated times, replated in fresh medium, and three days after start of the assay, enough time for a cell that was exposed to a lethal dose of immunotoxin to die, cell viability was determined. The time that is needed to reach more than 50% cell death varied widely from less than 30 min to more than four days (Figure 2A, Supplementary Table S2). For the Reh and the KOPN-8 cells, the PE38-variant needed much less exposure time than the PE24 or the PE24(B) variants to induce cell death. All of the other cell lines were killed after shorter exposure time to the WT PE24 than to the mutated PE24(B) or the domain II-containing PE38. This was most pronounced for HAL-01, where the PE24-variant induced more than 50% cell death after 6 h, the PE24(B) after 9 h, and the PE38 after 24 h, suggesting that full domain II increased the exposure time needed to kill HAL-01 cells by 4-fold. Cells consistently had to be exposed longer to the PE24(B) variants than to WT PE24 to induce cell death. As suggested by the slightly higher activity in WST-8 assays, the dsFv-immunotoxins always needed shorter exposure time than their respective Fab variant to reach similar cell killing (Supplementary Figures S1 and S2). Whether an immunotoxin variant was more cytotoxic than any other variant after 1 h of exposure did not predict whether it remained as the most active variant after 72 h of exposure. To enable a comparison, we mathematically generated the relative cell escape as the ratio of living cells after treatment with two distinct immunotoxins. Comparing relative cell escape showed that, even though being slower initially, treatment with Fab-PE24 was more potent than dsFv-PE38 at later time points against most cell lines (Figure 2B). Cell lines, which showed less cell escape after 1 h of treatment with Fab-PE24 than after dsFv-PE38, consistently showed less escaping cells when being treated with Fab-PE24 for 72 h. On the other hand, two of three cell lines showing less relative cell escape after 1 h of treatment with dsFv-PE38 than with Fab-PE24 showed less escaping cells after 72 h of exposure to Fab-PE24 than to dsFv-PE38.

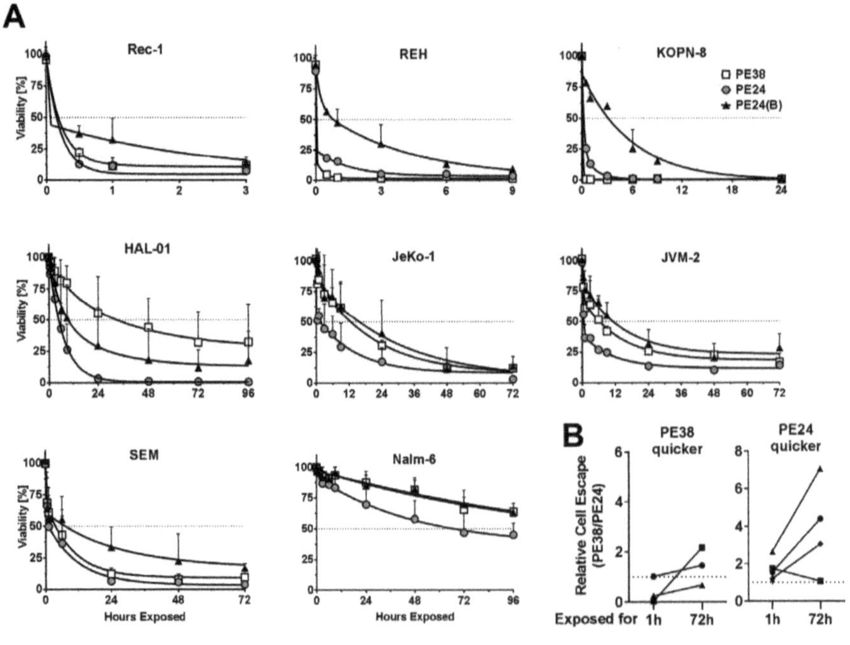

Figure 2. Immunotoxin variants induce exposure time-dependent cell death. (**A**) Indicated B-cell lymphoma cell lines were treated with 2.8 nmol/L immunotoxin for the indicated times, washed, and replated. Three days after assay initiation, cell viability was determined by flow cytometry. The symbols indicate the mean % of living cells at each data point of at least three independent experiments, errors are shown as SEM, curve fitting was done using two-phase decay regression analysis using GraphPad; (**B**) The rate of relative cell escape was determined as % surviving cells after PE38/% surviving cells after PE24 after 1 h and 72 h of drug exposure, respectively. Cell lines from A were divided into two groups, one group showing less escape after 1 h of PE38 (PE38 quicker) and one group showing less escaping cells after 1 h of PE24 (PE24 quicker). Results of the same cell line after 1 h and after 72 h of exposure are connected by lines.

2.3. *In Vivo Efficacy Emphasizes Importance of Domain II against KOPN-8*

We tested which of the molecular differences of dsFv-PE38 and Fab-PE24(B) most critically reduced the efficacy of intravenous bolus doses of immunotoxins. KOPN-8-bearing mice were treated with five doses immunotoxin at their MTD QOD and KOPN-8 BM infiltration rate was determined three days after the last dose (Figure 3A). Mice before treatment at day 8 showed, on average, a KOPN-8 BM infiltration of 6.1%, which rose to 59% at day 15 in untreated mice. Five doses of 0.4 mg/kg dsFv-PE38 resulted in 0.006%, of 1.0 mg/kg Fab-PE24 in 0.02%, of 2.0 mg/kg dsFv-PE24 in 0.08%, of 2.5 mg/kg Fab-PE24(B) in 24%, and of 4.0 mg/kg dsFv-PE24(B) in 41% KOPN-8 BM infiltration at day 15. Similar to previous results, five bolus doses of dsFv-PE38 reduced KOPN-8 BM infiltration 4000-fold more efficiently than Fab-PE24(B). Because the Fab-variants produced better responses than the dsFv variants in vivo, we focused on Fab-PE24, Fab-PE24(B), and dsFv-PE38, and tested next their effects on the survival of KOPN-8 bearing mice (Figure 3B). In line with the rate of reduction of KOPN-8 BM infiltration early after treatment, the median animal survival of KOPN-8 bearing mice treated with vehicle was 21.5 days, with Fab-PE24(B) 28 days, with Fab-PE24 37 days, and with dsFv-PE38 40 days.

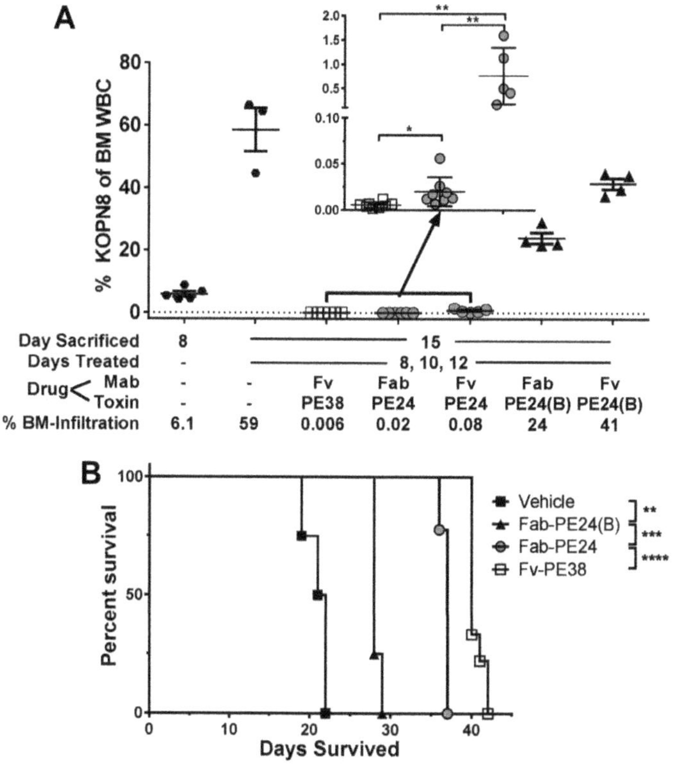

Figure 3. CD22-targeted immunotoxin variants show highly variable efficacy in vivo. (**A**) Mice that were intravenously injected with KOPN-8 on day 1 were either sacrificed on day 8 or treated with the indicated immunotoxin on days 8, 10, and 12 and sacrificed on day 15. BM was extracted and analyzed for KOPN-8 BM infiltration. Each symbol represents an individual mouse, bars indicate the mean of the respective group, errors are shown as SEM, *p*-values were determined by unpaired *t*-tests as * < 0.05 and ** < 0.01; (**B**) KOPN-8-bearing mice were treated with three doses of the indicated immunotoxin or vehicle and followed until disease progression. Shown is a Kaplan-Mayer plot, *p*-values determined by log-rank tests as ** < 0.01, *** < 0.001, and **** < 0.0001.

2.4. Biochemical Events Reflect Differences in Immunotoxin Activity

To test whether biochemical events reflected their distinct activity in vitro, we analyzed the critical steps of the immunotoxin intoxication processes. First, we compared the rate of internalization using Alexa647-conjugated immunotoxins (Figure 4A). For the three variants, the mean fluorescence intensity (MFI) of Alexa647 increased almost linearly over 6 h, and there was no significant difference between the MFIs of any of the immunotoxins. We next determined the rate of protein synthesis over time of immunotoxin treated cells (Figure 4B). Correlating with the shorter needed exposure time to kill efficiently, the dsFv-PE38 reduced protein synthesis in an exponential manner, and much faster than PE24 and PE24(B), which reduced protein synthesis in a linear manner. The rate of arresting protein synthesis was not different for PE24 and PE24(B). Thus, the advantage of the domain II containing immunotoxin against KOPN-8 occurred after internalization, but before the arrest of protein synthesis. Additionally, the time to reach the late-apoptotic PARP-cleavage was determined (Figure 4C). Six hours after the treatment started, the ratio of cleaved over uncleaved PARP was 1.12 for dsFv-PE38, 0.48 for Fab-PE24, and 0.13 for Fab-PE24(B). The time that is needed to induce the late-apoptotic event of PARP-cleavage in KOPN-8 cells was the shortest after treatment with dsFv-PE38, longer after Fab-PE24, and the longest after Fab-PE24(B), thereby also confirming differences in the efficacy of inducing cell death biochemically.

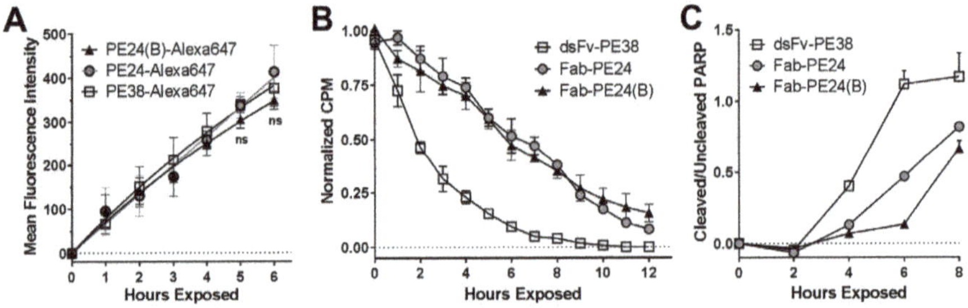

Figure 4. Differences of immunotoxin activity is reflected by distinct biochemical events. (**A**) The rate of internalization of immunotoxin molecules into KOPN-8 cells was determined by flow cytometry. The graph summarizes the mean fluorescence intensity of Alexa647 over time from three independent experiments, each done in triplicates, errors as SEM, significance determined by *t*-tests. The mean fluorescent intensity was corrected for the relative amount of Alexa647 conjugated with the respective immunotoxin variant; (**B**) Protein synthesis was quantified by [^{3}H]-leucine-incorporation, whereas the counts per minute (CPM) after treating KOPN-8 cells with immunotoxin for the indicated times were measured following an additional 1 h [^{3}H]-leucine pulse. Each data point was acquired in quadruplicates and the graph summarizes two independent experiments; errors shown as SEM; and, (**C**) The levels of loading control and whole and cleaved PARP were detected from KOPN-8 cells that were treated with immunotoxin for the indicated times by western blot and quantified. Shown are the background-subtracted and normalized ratios of cleaved/whole PARP over time. The graph summarizes the results from biological duplicates, each done in technical duplicates; symbols indicate mean values, errors as SEM.

2.5. Lack of Domain II Is Advantageous against Primary ALL Blasts

Also, primary patient-derived ALL blasts had to be exposed to 2.8 nmol/L immunotoxin for a highly variable time for them to die (Figure 5A). More than 50% of ALL blasts of patients 1, 2, and 3 died after only 1 h of exposure to dsFv-PE38 or Fab-PE24, of patients 4–7 after at least 24 h, and 50% cell killing was not yet reached at 72 h of immunotoxin treatment for patient 8. As for KOPN-8 and Reh cells, blasts of patients 1, 2, 3, and 5 were killed more efficiently after shorter exposure to dsFv-PE38 than after exposure to Fab-PE24 or Fab-PE24(B). On the other hand blasts of patients

4, 6, 7, and 8 died more efficiently after the short exposure to Fab-PE24 than after short exposure to dsFv-PE38 or Fab-PE24(B). Comparing the relative cell escape showed that even though being slower initially, treatment with Fab-PE24 was more potent than dsFv-PE38 at later time points against patient-derived ALL blasts (Figure 4B). ALL blasts that showed less cell escape after 1 h of treatment with Fab-PE24 than after dsFv-PE38 consistently showed less escaping cells when being treated with Fab-PE24 for 72 h. On the other hand, three of four primary ALL-blasts that showed less relative cell escape after 1 h of treatment with dsFv-PE38 than with Fab-PE24 showed less escaping cells after 72 h of exposure to Fab-PE24 than to dsFv-PE38 at 72 h. We then compiled the data on relative cell escape of cell lines and patient samples after 72 h of continuous treatment (Figure 5C). After 72 h of continuous treatment, on average, 2.2 (±0.44) times more cells escaped treatment with dsFv-PE38 than with Fab-PE24, 3.40 (±0.55) times more cells escaped after PE24(B) than after PE24, and the relative cell escape was similar after dsFv-PE38 and Fab-PE24(B) (0.95 ± 0.20) demonstrating the highest overall-activity after 72 h of continuous treatment with the WT PE24 immunotoxin.

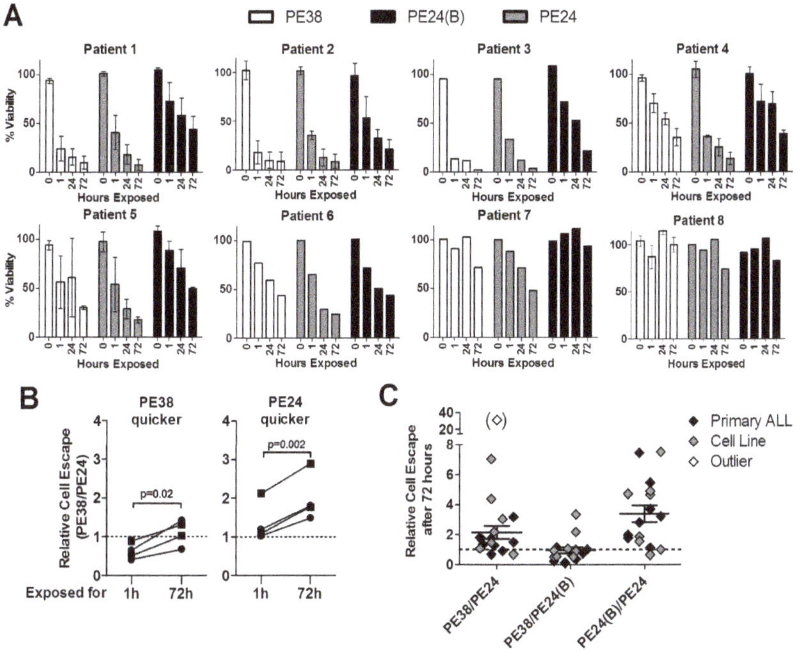

Figure 5. Patient-derived primary ALL cells are on average most sensitive to the PE24-variant. (**A**) Primary patient derived B-ALL cells were co-cultured on murine OP-9 stromal cells and treated with 2.8 nmol/L immunotoxin for the indicated times, washed, and replated on the same feeder layer. Three days after assay initiation, cell viability of the human ALL cells was determined by flow cytometry. Bars indicate the mean % of living cells at each time point, errors shown as SEM; and, (**B**) The relative % of cell killing was normalized as cell killing by PE24/cell killing by PE38 after 1 h and after 72 h of drug exposure. All of the cell lines were divided into two groups, one where PE24 was less active after only 1 h (PE38 quicker) and one where PE24 was more active after 1 h (PE24 quicker). Results of the same cell line after 1 h and after 72 h of exposure are connected by a line, p-values determined by paired *t*-tests; and, (**C**) Results from cell lines (grey rhombs) and patient cells (black rhombs) were pooled and the relative cell escape at 72 h of continuous exposure mathematically determined by dividing the fraction of living cells after the indicated treatment. The mean relative cell escape is indicated by a black line, error as SEM, clear rhomb in brackets was defined as outlier and was excluded from the mean relative escape.

3. Discussion

Testing five immunotoxin variants, we determined which difference between dsFv-PE38 and the deimmunized variant Fab-PE24(B) was responsible for a 4000-fold difference in efficacy in a KOPN-8 xenograft model and found WT domain II of PE38 to be most important for the highest activity after bolus dose administration. In vitro, domain II shortened the time KOPN-8 cells had to be exposed to immunotoxin for them to die to a few minutes, while the lack of domain II combined with deimmunized domain III extended the time that is needed to kill cells to several hours. Because immunotoxin serum concentration falls quickly after intravenous bolus doses, the drug exposure time that is needed to induce cell death was the most critical to predict in vivo efficacy of bolus doses of immunotoxin, whereas the current method of determining cytotoxicity using three-day continuous treatment was not. Even though PE38 was most active in the KOPN-8 xenograft model when being administered as bolus doses, continuous exposure to any immunotoxin produced substantially higher rates of cell death in vitro. When continuously exposed, WT PE24 was on average more cytotoxic than both PE38 and PE24(B), suggesting that the PE24-variant may be most favorable when treating CD22-expressing malignancies continuously in vivo.

3.1. *How Domain II Influences Immunotoxin Efficacy*

All five immunotoxin variants were internalized at a similar rate, indicating that a similar number of immunotoxin molecules entered the cells, and thus, presumably the endosomal compartment. A crucial step of immunotoxin intoxication is endosomal escape of domain III which then ADP-ribosylates elongation factor 2 in the cytosol [16]. No method can directly measure immunotoxin trafficking currently. Because a defined number of PE-molecules needs to reach the cytosol to kill a cell [17], we believe that the transport of PE38 is more efficient than the transport of PE24 and of PE24(B). In line with more efficient trafficking, arrest of protein synthesis by PE38 with full domain II follows a much faster, exponential time course, while PE24 arrests protein synthesis slower and proportionally over time. The distinct kinetics may suggest an additional function of domain II beyond a furin cleavage site. Whether domain II plays a role in directed vesicular cargo-transport towards the Trans-Golgi network is possible, but it has not been previously described [18–20]. Furin-cleavage of domain II was shown to be the most efficacious at the acidic pH of late endosomes [21], suggesting that domain II may allow for a timely liberation of domain III, whereas the furin-cleavage site of PE24 may be accessible in a pH independent manner [12]. It was shown that domain II of dsFv-PE38 is cleaved more efficiently by furin than PE24(B) when being internalized into KOPN-8 cells [15], and that the alterations of domain II and the furin-cleavage site greatly influence cytotoxicity [22,23]. Intriguingly, CD22-targeted immunotoxins that cannot be cleaved by furin loose approximately three-fold cytotoxicity, whereas mesothelin-targeted immunotoxins that cannot be cleaved by furin are inactive, indicating that the effects of furin-cleavage on cytotoxicity is target receptor and target cell dependent [24]. This is in line with highly variable endosomal (toxin) transport [25,26], and with a substantial variability of PE-trafficking depending on the target receptor and the target cell type [27–31].

A cell type specific transport of domain II-containing immunotoxin is also in line with the more efficient killing of some ALL, but not lymphoma cells after the short exposure shown in this study and is further supported by a higher clinical response rate to bolus dose Moxetumomab of 34% in pediatric ALL [9], while the response rate in adult ALL is 13% [32] and adult lymphoma patients fail to respond to domain II containing immunotoxin [33]. We cannot currently predict which ALL patients may respond better to bolus doses of domain II-containing immunotoxin.

3.2. *Efficacy In Vivo Is Influenced by Needed Exposure Time and Half-Life*

Generally, larger Fab-containing immunotoxins needed a longer exposure time than dsFv-containing immunotoxins in vitro. However, in vivo, the larger Fab-variants showed two-to four-fold higher efficacies, which is likely due to the two-fold longer half-life of larger Fab-molecules [14]. That Fab-variants were less active in vitro, but more active in vivo, suggests that in vivo efficacy is predicted by (i) the needed exposure time to kill target cells and (ii) serum half-life. These data add to our novel understanding that exposure time is a crucial component of in vivo efficacy of immunotoxins [15,34]. To prolong serum half-life, novel immunotoxins containing an albumin-binding domain were constructed, which, in line with our results, show improved efficacy in vivo [35]. Whether smaller dsFv containing immunotoxins given continuously to compensate for the short half-life may be more efficacious due to improved tissue penetration [36,37] is a question of ongoing research.

In summary, the present study suggests that approximately 1/3 of ALL are killed better after short exposure to domain II containing PE than after short exposure to the PE lacking domain II. Cytotoxicity of CD22-targeted immunotoxins substantially increase the longer cells that are exposed, thus emphasizing that patients with B-cell malignancies may respond better to continuous rather than to bolus dose administration. After 72 h, PE24 is on average two-fold more cytotoxic than PE38 suggesting that PE24 may be more efficacious than PE38 when being administered continuously.

4. Materials and Methods

4.1. Reagents

The immunotoxins dsFv-PE38, also Moxetumomab pasudotox [38], dsFv-PE24, also LR [12], the dsFv-PE24(B), and the Fab-PE24(B) [14] were produced, as previously described. The immunotoxin Fab-PE24 was cloned by exchanging the dsFv of LR with the Fab of Fab-PE24(B). The resulting protein was produced following standard procedures [39].

Secondary western blot antibodies were purchased from Santa Cruz Biotechnology (Santa Cruz, CA, USA), primary antibodies (Actin, GAPDH, and PARP) from Cell Signaling (Danvers, MA, USA), flow cytometry antibodies, and Annexin V-PE/7-AAD from Becton Dickinson (Franklin Lakes, NJ, USA).

4.2. Cell Lines

The cell lines were described previously [15,34]. All of the cells were grown in RPMI, supplemented with 10% fetal bovine serum, 100 U penicillin, and 100 mg streptomycin (Invitrogen, Carlsbad, CA, USA).

4.3. Cell Assays

Cytotoxicity was determined by WST-8, as described [15]. 5000 Cells/well were incubated with various recombinant immunotoxin (rIT) concentrations for 72 h. WST-8 reagent was added, absorbance measured 2 h later, and the values were normalized between Cycloheximide (10 μg/mL final, Sigma-Aldrich, Saint Louis, MO, USA) and untreated control. Non-linear regression to obtain IC_{50} concentrations was done using GraphPad Prism v6.01.

For in vitro apoptosis assays by flow cytometry, 1 million cells/mL were incubated with 2.8 nmol/L immunotoxin for various times, cells were washed twice, and were transferred to a new plate in complete RPMI. Seventy-two hours after assay initiation, cells were stained with 7-AAD/Annexin-PE, and measured with a FACS Calibur (Becton Dickinson, Franklin Lakes, NJ, USA). Results were analyzed with FlowJo software (Tree Star, v10.2, St. Ashland, OR, USA, 2016).

For primary patient cell assays, 20,000 OP-9 stromal cells were plated per well in a 24-well plate in α-MEM (1% P/S, 20% FBS) on Day 0. On Day 1, 300,000 patient cells were added and 2.8 nM rIT 4 h later. At the indicated times, cells and wells were washed twice with PBS and the cells were re-plated in the same co-culture wells. Seventy-two hours after initiation, the cells were stained with anti-hu-CD19 and Annexin V-PE/7-AAD and analyzed by flow.

For internalization assays immunotoxins were conjugated with Alexa647, following the manufacturer's instructions (Invitrogen). One million cells were incubated with Alexa647-conjugated immunotoxin at saturating concentrations (3 μg/mL) at 37 °C for the indicated times. Surface bound molecules were stripped for 10 min in 0.2 M Glycine pH 2.5, cells washed twice with phosphate buffered saline, and analyzed by flow cytometry (FACS Calibur, Becton Dickinson).

Inhibition of protein synthesis was determined by [^{3}H]-leucine incorporation, as described [12]. One million cells/mL were plated in complete RPMI and were treated for indicated times. Cells were pulsed with 2 μCi [^{3}H]-leucine for 1 h, frozen on dry ice for 30 min, thawed at 37%, and CPM was read using a liquid scintillation counter.

4.4. Patient Samples

Primary ALL samples from eight patients with CD22-positive pre-B-ALL treated on a phase I trial of HA22 (NCT01891981) were collected with informed consent before the first HA22 dose under protocol 04-C-0102, and was approved by the NCI Institutional Review Board

4.5. Animal Studies

Animals were handled according to NIH guidelines; studies under protocol #LMB-063 were approved 08/04/2010 by the NCI Animal Care and Use Committee.

Five million KOPN-8 cells were injected on Day 1 via tail vein into 6- to 8- week-old NSG mice (NOD.Cg-*Prkdc*^{scid} *Il2rg*^{tm1Wjl}/SzJ). Immunotoxin was given intravenously QOD at the maximal tolerated dose of 0.4 mg/kg (dsFv-PE38), 1 mg/kg (Fab-PE24), 2 mg/kg (dsFv-PE24), 2.5 mg/kg (Fab-PE24(B)), and 4 mg/kg (dsFv-PE24(B)). To assess early response, mice were euthanized three days after the last dose. For survival studies, mice were followed until disease progression (weight loss > 10%). BM was extracted by flushing femurs. Human ALL was stained with anti-human-CD19-FITC. Experiments on mouse-derived tissue were F_c-receptor blocked with anti-murine CD16/32.

4.6. Statistics

Statistical analyses were performed using Graph Pad Prism v6.01 (La Jolla, CA, USA, 2012) as *t*-tests (two group comparison) or log-rank tests for animal survival as indicated.

Supplementary Materials: The following are available online at http://www.mdpi.com/2072-6651/10/5/210/s1, Figure S1: Immunotoxin variants induce exposure time-dependent cell death, Figure S2: Immunotoxin variants indcue exposure time-dependent cell death, Table S1: IC_{50} pM for growth inhibition by WST8 assays, Table S2: 2-phase regression fitting of Figure 1.

Author Contributions: F.M., T.C., R.B., T.K.B. and I.P. designed and/or performed experiments and analyzed data. A.S.W. contributed reagents. F.M., A.S.W. and I.P. wrote the manuscript.

Acknowledgments: The content is solely the responsibility of the authors and does not necessarily represent the official views of the National Cancer Institute or the National Institutes of Health. The authors wish to thank the building 37 animal facility for their support and animal care taking and Anna Mazzuca. The work was supported in part by the Intramural Research Program of the NIH, National Cancer Institute, Center for Cancer Research. F.M. was supported in part by the German Research Foundation, award number MU 3619/1-1 and by the Interdisciplinary Center for Clinical Research, award number IZKF-J59.

Conflicts of Interest: T.K.B., A.S.W. and I.P. are co-inventors on patents assigned to the NIH for the investigational products. A.S.W. has received from Medimmune: research support, honorarium, travel support; Pfizer, Kite Pharma, and Spectrum Pharmaceuticals: honorarium and travel support. The other authors have declared that no conflict of interest exists.

References

1. Scott, A.M.; Wolchok, J.D.; Old, L.J. Antibody therapy of cancer. *Nat. Rev. Cancer* **2012**, *12*, 278–287. [CrossRef] [PubMed]
2. Beck, A.; Wurch, T.; Bailly, C.; Corvaia, N. Strategies and challenges for the next generation of therapeutic antibodies. *Nat. Rev. Immunol.* **2010**, *10*, 345–352. [CrossRef] [PubMed]

3. Sharkey, R.M.; Goldenberg, D.M. Cancer radioimmunotherapy. *Immunotherapy* **2011**, *3*, 349–370. [CrossRef] [PubMed]
4. Schrama, D.; Reisfeld, R.A.; Becker, J.C. Antibody targeted drugs as cancer therapeutics. *Nat. Rev. Drug Discov.* **2006**, *5*, 147–159. [CrossRef] [PubMed]
5. Senter, P.D.; Sievers, E.L. The discovery and development of brentuximab vedotin for use in relapsed hodgkin lymphoma and systemic anaplastic large cell lymphoma. *Nat. Biotechnol.* **2012**, *30*, 631–637. [CrossRef] [PubMed]
6. FitzGerald, D.J.; Wayne, A.S.; Kreitman, R.J.; Pastan, I. Treatment of hematologic malignancies with immunotoxins and antibody-drug conjugates. *Cancer Res.* **2011**, *71*, 6300–6309. [CrossRef] [PubMed]
7. Wayne, A.S.; Fitzgerald, D.J.; Kreitman, R.J.; Pastan, I. Immunotoxins for leukemia. *Blood* **2014**, *123*, 2470–2477. [CrossRef] [PubMed]
8. Kreitman, R.J.; Tallman, M.S.; Robak, T.; Coutre, S.; Wilson, W.H.; Stetler-Stevenson, M.; Fitzgerald, D.J.; Lechleider, R.; Pastan, I. Phase I trial of anti-CD22 recombinant immunotoxin moxetumomab pasudotox (CAT-8015 or HA22) in patients with hairy cell leukemia. *J. Clin. Oncol.* **2012**, *30*, 1822–1828. [CrossRef] [PubMed]
9. Wayne, A.S.; Shah, N.N.; Bhojwani, D.; Silverman, L.B.; Whitlock, J.A.; Stetler-Stevenson, M.; Sun, W.; Liang, M.; Yang, J.; Kreitman, R.J.; et al. Phase I study of the anti-CD22 immunotoxin moxetumomab pasudotox for childhood acute lymphoblastic leukemia. *Blood* **2017**, *130*, 1620–1627. [CrossRef] [PubMed]
10. Hassan, R.; Sharon, E.; Thomas, A.; Zhang, J.; Ling, A.; Miettinen, M.; Kreitman, R.J.; Steinberg, S.M.; Hollevoet, K.; Pastan, I. Phase 1 study of the antimesothelin immunotoxin SS1P in combination with pemetrexed and cisplatin for front-line therapy of pleural mesothelioma and correlation of tumor response with serum mesothelin, megakaryocyte potentiating factor, and cancer antigen 125. *Cancer* **2014**, *120*, 3311–3319. [PubMed]
11. Mazor, R.; Vassall, A.N.; Eberle, J.A.; Beers, R.; Weldon, J.E.; Venzon, D.J.; Tsang, K.Y.; Benhar, I.; Pastan, I. Identification and elimination of an immunodominant t-cell epitope in recombinant immunotoxins based on *Pseudomonas* exotoxin a. *Proc. Natl. Acad. Sci. USA* **2012**, *109*, E3597–E3603. [CrossRef] [PubMed]
12. Weldon, J.E.; Xiang, L.; Chertov, O.; Margulies, I.; Kreitman, R.J.; FitzGerald, D.J.; Pastan, I. A protease-resistant immunotoxin against CD22 with greatly increased activity against CLL and diminished animal toxicity. *Blood* **2009**, *113*, 3792–3800. [CrossRef] [PubMed]
13. Onda, M.; Beers, R.; Xiang, L.; Lee, B.; Weldon, J.E.; Kreitman, R.J.; Pastan, I. Recombinant immunotoxin against B-cell malignancies with no immunogenicity in mice by removal of B-cell epitopes. *Proc. Natl. Acad. Sci. USA* **2011**, *108*, 5742–5747. [CrossRef] [PubMed]
14. Bera, T.K.; Onda, M.; Kreitman, R.J.; Pastan, I. An improved recombinant Fab-immunotoxin targeting CD22 expressing malignancies. *Leuk. Res.* **2014**, *38*, 1224–1229. [CrossRef] [PubMed]
15. Muller, F.; Cunningham, T.; Liu, X.F.; Wayne, A.S.; Pastan, I. Wide variability in the time required for immunotoxins to kill B lineage acute lymphoblastic leukemia cells: Implications for trial design. *Clin. Cancer Res.* **2016**, *22*, 4913–4922. [CrossRef] [PubMed]
16. Pastan, I.; Hassan, R.; Fitzgerald, D.J.; Kreitman, R.J. Immunotoxin therapy of cancer. *Nat. Rev. Cancer* **2006**, *6*, 559–565. [CrossRef] [PubMed]
17. Kreitman, R.J.; Pastan, I. Accumulation of a recombinant immunotoxin in a tumor in vivo: Fewer than 1000 molecules per cell are sufficient for complete responses. *Cancer Res.* **1998**, *58*, 968–975. [PubMed]
18. Johannes, L.; Popoff, V. Tracing the retrograde route in protein trafficking. *Cell* **2008**, *135*, 1175–1187. [CrossRef] [PubMed]
19. Pfeffer, S.R. Multiple routes of protein transport from endosomes to the trans golgi network. *FEBS Lett.* **2009**, *583*, 3811–3816. [CrossRef] [PubMed]
20. Eaton, S.; Martin-Belmonte, F. Cargo sorting in the endocytic pathway: A key regulator of cell polarity and tissue dynamics. *Cold Spring Harb. Perspect. Biol.* **2014**, *6*, a016899. [CrossRef] [PubMed]
21. Chiron, M.F.; Fryling, C.M.; FitzGerald, D. Furin-mediated cleavage of *Pseudomonas* exotoxin-derived chimeric toxins. *J. Biol. Chem.* **1997**, *272*, 31707–31711. [CrossRef] [PubMed]
22. Kaplan, G.; Lee, F.; Onda, M.; Kolyvas, E.; Bhardwaj, G.; Baker, D.; Pastan, I. Protection of the furin cleavage site in low-toxicity immunotoxins based on *Pseudomonas* exotoxin A. *Toxins (Basel)* **2016**, *8*, 217. [CrossRef] [PubMed]

23. Mazor, R.; Kaplan, G.; Park, D.; Jang, Y.; Lee, F.; Kreitman, R.; Pastan, I. Rational design of low immunogenic anti CD25 recombinant immunotoxin for T cell malignancies by elimination of T cell epitopes in PE38. *Cell. Immunol.* **2017**, *313*, 59–66. [CrossRef] [PubMed]
24. Weldon, J.E.; Xiang, L.; Zhang, J.; Beers, R.; Walker, D.A.; Onda, M.; Hassan, R.; Pastan, I. A recombinant immunotoxin against the tumor-associated antigen mesothelin reengineered for high activity, low off-target toxicity, and reduced antigenicity. *Mol. Cancer Ther.* **2013**, *12*, 48–57. [CrossRef] [PubMed]
25. Stenmark, H. Rab gtpases as coordinators of vesicle traffic. *Nat. Rev. Mol. Cell Biol.* **2009**, *10*, 513–525. [CrossRef] [PubMed]
26. Moreau, D.; Kumar, P.; Wang, S.C.; Chaumet, A.; Chew, S.Y.; Chevalley, H.; Bard, F. Genome-wide RNAi screens identify genes required for ricin and PE intoxications. *Dev. Cell* **2011**, *21*, 231–244. [CrossRef] [PubMed]
27. Pasetto, M.; Antignani, A.; Ormanoglu, P.; Buehler, E.; Guha, R.; Pastan, I.; Martin, S.E.; FitzGerald, D.J. Whole-genome RNAi screen highlights components of the endoplasmic reticulum/Golgi as a source of resistance to immunotoxin-mediated cytotoxicity. *Proc. Natl. Acad. Sci. USA* **2015**, *112*, E1135–E1142. [CrossRef] [PubMed]
28. Verdurmen, W.P.R.; Mazlami, M.; Pluckthun, A. A quantitative comparison of cytosolic delivery via different protein uptake systems. *Sci. Rep.* **2017**, *7*, 13194. [CrossRef] [PubMed]
29. Du, X.; Beers, R.; Fitzgerald, D.J.; Pastan, I. Differential cellular internalization of anti-CD19 and -CD22 immunotoxins results in different cytotoxic activity. *Cancer Res.* **2008**, *68*, 6300–6305. [CrossRef] [PubMed]
30. O'Reilly, M.K.; Tian, H.; Paulson, J.C. CD22 is a recycling receptor that can shuttle cargo between the cell surface and endosomal compartments of B cells. *J. Immunol.* **2011**, *186*, 1554–1563. [CrossRef] [PubMed]
31. Michalska, M.; Wolf, P. *Pseudomonas* exotoxin A: Optimized by evolution for effective killing. *Front. Microbiol.* **2015**, *6*, 963. [CrossRef] [PubMed]
32. Short, N.J.; Kantarjian, H.; Jabbour, E.; Cortes, J.E.; Thomas, D.A.; Rytting, M.E.; Daver, N.; Alvarado, Y.; Konopleva, M.; Kebriaei, P.; et al. A phase I study of moxetumomab pasudotox in adults with relapsed or refractory B-cell acute lymphoblastic leukaemia. *Br. J. Haematol.* **2017**. [CrossRef] [PubMed]
33. Kreitman, R.J.; Squires, D.R.; Stetler-Stevenson, M.; Noel, P.; FitzGerald, D.J.; Wilson, W.H.; Pastan, I. Phase I trial of recombinant immunotoxin RFB4(dsfv)-PE38 (BL22) in patients with B-cell malignancies. *J. Clin. Oncol.* **2005**, *23*, 6719–6729. [CrossRef] [PubMed]
34. Müller, F.; Stookey, S.; Cunningham, T.; Pastan, I. Paclitaxel synergizes with exposure time adjusted CD22-targeting immunotoxins against B-cell malignancies. *Oncotarget* **2017**, *8*, 30644–30655. [CrossRef] [PubMed]
35. Wei, J.; Bera, T.K.; Liu, X.F.; Zhou, Q.; Onda, M.; Ho, M.; Tai, C.H.; Pastan, I. Recombinant immunotoxins with albumin-binding domains have long half-lives and high antitumor activity. *Proc. Natl. Acad. Sci. USA* **2018**, *115*, E3501–E3508. [CrossRef] [PubMed]
36. Minchinton, A.I.; Tannock, I.F. Drug penetration in solid tumours. *Nat. Rev. Cancer* **2006**, *6*, 583–592. [CrossRef] [PubMed]
37. Thurber, G.M.; Schmidt, M.M.; Wittrup, K.D. Factors determining antibody distribution in tumors. *Trends Pharmacol. Sci.* **2008**, *29*, 57–61. [CrossRef] [PubMed]
38. Salvatore, G.; Beers, R.; Margulies, I.; Kreitman, R.J.; Pastan, I. Improved cytotoxic activity toward cell lines and fresh leukemia cells of a mutant anti-CD22 immunotoxin obtained by antibody phage display. *Clin. Cancer Res.* **2002**, *8*, 995–1002. [PubMed]
39. Pastan, I.; Beers, R.; Bera, T.K. Recombinant immunotoxins in the treatment of cancer. *Methods Mol. Biol.* **2004**, *248*, 503–518. [PubMed]

toxins MDPI

Article

Alternagin-C (ALT-C), a Disintegrin-Like Cys-Rich Protein Isolated from the Venom of the Snake *Rhinocerophis alternatus*, Stimulates Angiogenesis and Antioxidant Defenses in the Liver of Freshwater Fish, *Hoplias malabaricus*

Diana Amaral Monteiro *, Heloisa Sobreiro Selistre-de-Araújo, Driele Tavares, Marisa Narciso Fernandes, Ana Lúcia Kalinin and Francisco Tadeu Rantin

Department of Physiological Sciences, Federal University of São Carlos (UFSCar), São Carlos, SP 13565-905, Brazil; hsaraujo@ufscar.br (H.S.S.-d.-A.); driele.tavares@gmail.com (D.T.); dmnf@ufscar.br (M.N.F.); akalinin@ufscar.br (A.L.K.); ftrantin@ufscar.br (F.T.R.)

* Correspondence: dianaamonteiro@yahoo.com.br; Tel.: +55-16-3351-9775

Academic Editor: Hang Fai (Henry) Kwok
Received: 29 August 2017; Accepted: 26 September 2017; Published: 28 September 2017

Abstract: Alternagin-C (ALT-C) is a disintegrin-like protein isolated from *Rhinocerophis alternatus* snake venom, which induces endothelial cell proliferation and angiogenesis. The aim of this study was to evaluate the systemic effects of a single dose of alternagin-C (0.5 $mg \cdot kg^{-1}$, via intra-arterial) on oxidative stress biomarkers, histological alterations, vascular endothelial growth factor (VEGF) production, and the degree of vascularization in the liver of the freshwater fish traíra, *Hoplias malabaricus*, seven days after the initiation of therapy. ALT-C treatment increased VEGF levels and hepatic angiogenesis. ALT-C also enhanced hepatic antioxidant enzymes activities such as superoxide dismutase, catalase, glutathione peroxidase, and glutathione reductase, decreasing the basal oxidative damage to lipids and proteins in the fish liver. These results indicate that ALT-C improved hepatic tissue and may play a crucial role in tissue regeneration mechanisms.

Keywords: disintegrin; blood vessel formation; VEGF; antioxidant enzymes; oxidative stress biomarkers

1. Introduction

Snake venoms contain a complex pool of proteins (more than 90% of the dry weight), organic compounds with a low molecular mass, and inorganic compounds [1]. Among these compounds are acetylcholinesterases, ADPases, phospholipases, hialuronidases, and hemostasis active compounds such as metalloproteases, named the snake venom metalloproteases (SVMPs), and serinoproteases [2].

Disintegrins of snake venoms are mostly derived from proteolytically processed precursor forms having a metalloprotease (SVMP) domain [3]. Alternagin-C (ALT-C) is an ECD (Glu-Cys-Asp sequence)-containing disintegrin-like/cysteine-rich domain released from metalloprotease alternagin isolated from the crude venom of the snake *Rhinocerophis alternatus*, popularly known in South America as urutu. Disintegrin-like proteins trigger integrin-mediated intracellular signal transduction events that modify gene expression and cell responses and interfere with cell-cell and cell-matrix interactions in a bi-directional manner across cell membranes [4,5].

Previous studies demonstrated that ALT-C induces endothelial cell proliferation and angiogenesis both in vitro and in vivo by up-regulating the expression of vascular endothelial growth factor (VEGF) and its receptors [4,6]. ALT-C binds to $\alpha_2\beta_1$ integrin, a major collagen receptor, competitively inhibiting cell adhesion to collagen, triggering downstream signaling molecules, and inducing a significant increase in several genes related to cell cycle control, including VEGF and other growth factors such as inducible early growth response, interleukin 11, early growth response 2 and 3, and the insulin-induced gene [7]. ALT-C also induced significant cytoskeleton dynamic changes with the polymerization of F-actin, focal adhesion kinase (FAK), and phosphoinositol 3-kinase (PI3K) activation, as well as erk-2 translocation [3].

An integrin-binding peptide, such as ALT-C, able to up-regulate the expression of growth factors, may be considered an interesting tool in experimental studies of tissue regeneration. Therefore, the aim of this study was to evaluate the effects of ALT-C on oxidative stress biomarkers, histopathological alterations, VEGF production, and the degree of vascularization in the liver of traíra, *Hoplias malabaricus* (Erythrinidae), a Neotropical freshwater fish species. The three major liver functions essential for life are: (a) uptake, metabolism, storage, and redistribution of nutrients; (b) metabolism of lipophilic compounds, including xenobiotics; (c) formation and excretion of bile. All of these functions have been shown to be involved not only in physiological states, but also in diseases leading to alterations in hepatic morphology and physiology [8]. The maintenance of structure and function is relevant for the liver itself and may also ensure the integrity and homeostasis of other organs crucial for fish survival. Furthermore, the use of the fish as a model for drug-induced liver injury is promising and may support better choices taken in the early stages of drug discovery, before a compound is tested in mammals [9]. The results indicated that ALT-C improved antioxidant defenses of fish liver by decreasing the level of oxidative stress biomarkers and by increasing the activity of antioxidant enzymes. As far as we know this is the first report of such effects for a disintegrin-like/cysteine-rich protein.

2. Results

ALT-C treatment increased the degree of liver vascularization. Figure 1 shows histological sections of the liver of fish of both experimental groups (Control and ALT-C), in which a larger number and/or size of the blood vessels present in the hepatic parenchyma of the ALT-C treated fish can be evidenced.

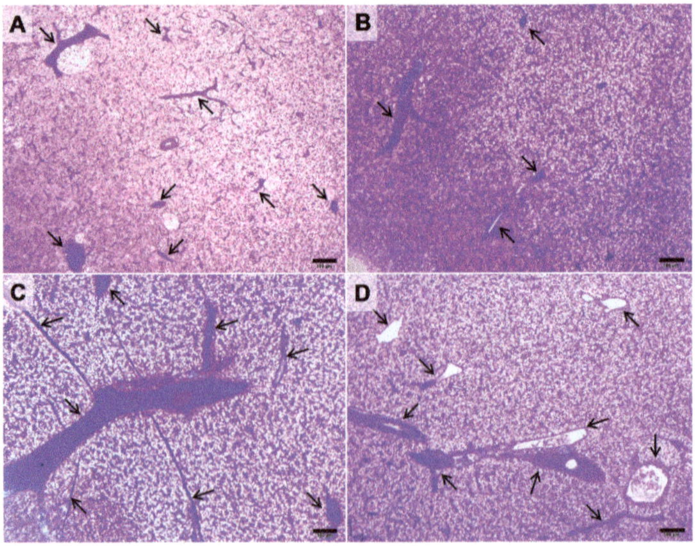

Figure 1. Light micrographs of sections through the liver of traíra (*H. malabaricus*) from the control group (n = 10, **A** and **B**) and after seven days of treatment with alternagin-C, in a single dose of 0.5 mg·kg^{-1}, intra-arterial (n = 10, **C** and **D**). Arrows indicate blood vessels. Samples were stained with toluidine-blue/basic fuchsin. *Bar* = 100 μm.

Histologically, polygonal hepatocytes with spherical and centralized nuclei clearly organized in cords surrounding sinusoid capillaries were observed in the liver of this species, characterizing the normal aspect of the tissue (Figure 2). Although the Control group exhibited normal aspect to the hepatic tissue, some structural changes were observed in some areas, such as: cytoplasmic degeneration and architectural/structural alterations, where it was not possible to see the format and the cellular delimitation, as well as the cord arrangement (Figure 2A), and cellular atrophy (Figure 2B). Other changes such as the accumulation of intracellular substances (eosinophilic-like granules, Figure 2B), the formation of cytoplasmic vacuoles (Figure 2B), and the presence of melano-macrophage centers (Figure 2C) were also observed. The ALT-C group also exhibited characteristics of normal hepatic tissue with some histopathological alterations but in lower frequencies. The liver parenchyma was homogeneous with polygonal shaped hepatocytes having a spherical nucleus and showed rare pathological features (Figure 2C,D). Few areas of morphological damages were observed like cytoplasmic degeneration in association with architectural/structural alterations. Additionally, in a smaller quantity, the melano-macrophage centers and accumulation intracellular substances were detected. Overall, the tissue of treated animals showed a smaller frequency of alterations when compared to the control group (Table 1).

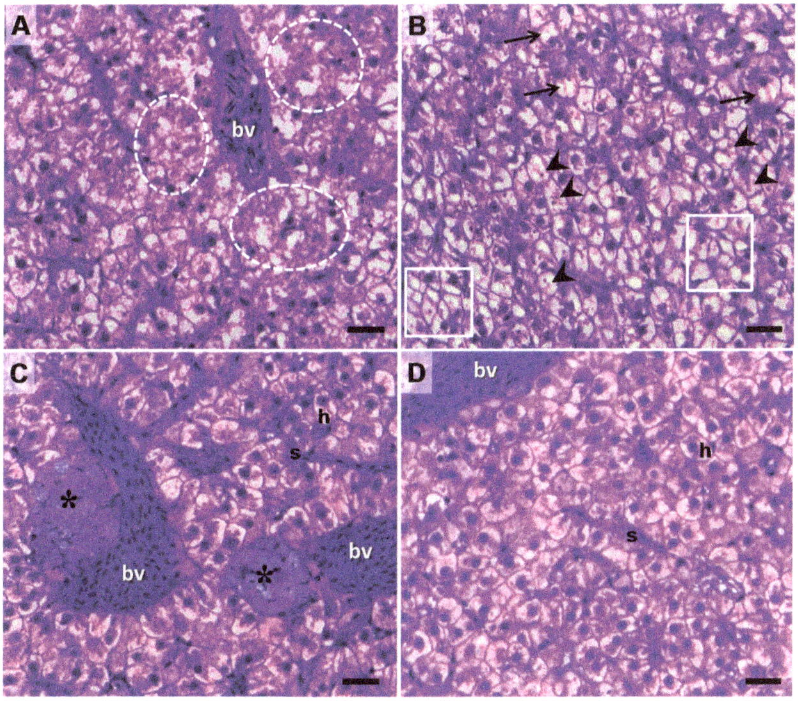

Figure 2. Light micrographs of sections through the liver of traíra (*H. malabaricus*) from the control group (**A**,**B**) and after seven days of treatment with alternagin-C, in a single dose of 0.5 mg·kg^{-1}, intra-arterial (**C**,**D**). h: hepatocytes; s: sinusoids; bv: blood vessel; black arrow: cytoplasmic vacuoles; asterisks: melano-macrophage centers; dotted circle: cytoplasmic degeneration and architectural/structural alterations; arrow head: eosinophilic granules in cytoplasm; squar: atrophy. Samples were stained with toluidine-blue/basic fuchsin and photomicrographs were taken using 400 × magnification. *Bar* = 20 μm.

Table 1. Liver histopathology of *H. malabaricus* after seven days of treatment with alternagin-C (single dose of 0.5 mg·kg^{-1}, intra-arterial).

Lesion	Control	ALT-C
Aneurism/Haemorrhage/Hyperaemia	A	A
Architectural and structural alterations	+	0+
Atrophy	+	A
Cytoplasmic vacuoles	0+	0+
Cytoplasmic degeneration	++	+
Eosinophilic granules in cytoplasm	0+	0+
Hypertrophy	A	A
Melano-macrophages centers	+	0+
Necrosis	A	A
Nuclear alterations	A	A

A: absent; 0+: very low frequency; +: low frequency; ++: frequent.

The treatment with 0.5 mg·kg^{-1} of ALT-C induced hepatic angiogenesis by up-regulating the expression of VEGF. The liver tissue of fish from the ALT-C group displayed elevated ($P < 0.05$) VEGF levels (31%, Figure 3A) and a higher ($P < 0.05$) percentage of area occupied by blood vessels (1.46 fold) than the hepatic tissue of animals from the Control group (Figure 3B).

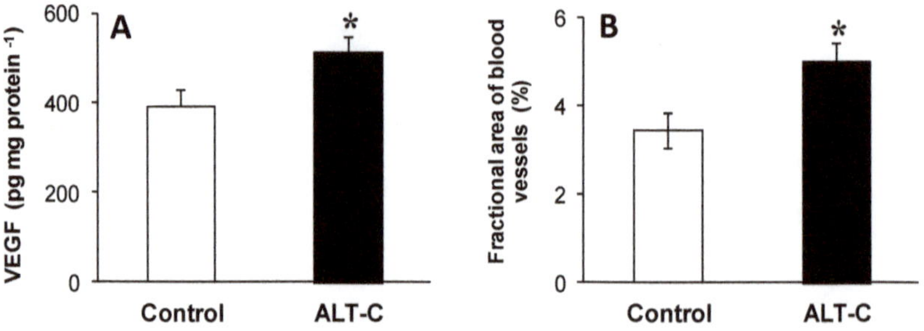

Figure 3. (**A**) Hepatic VEGF levels and (**B**) fractional area of the blood vessels in the liver histological sections of traíra (*H. malabaricus*) under control conditions ($n = 10$) and after seven days of treatment with alternagin-C ($n = 10$, single dose of 0.5 mg·kg^{-1}, intra-arterial). Data are presented as means ± S.E.M. Asterisks indicate significant difference ($P < 0.05$) between fish groups.

After seven days following a single-dose of ALT-C, no fish died and no changes in hepatic protein levels were observed (Control = 72. 4 ± 4.3 and ALT-C = 74.4 ± 4.1 mg·g tissue^{-1}). ALT-C treatment induced significant ($P < 0.05$) increases in the hepatic superoxide dismutase (SOD), catalase (CAT), glutathione peroxidase (GPx), and glutathione reductase (GR) activities (76%, 60%, 158%, and 31%, respectively). On the other hand, glutathione S-transferase (GST) activity and reduced glutathione (GSH) content remained unaffected (Figure 4).

Additionally, the liver of fish from the ALT-C group showed significantly ($P < 0.05$) reduced levels of lipid peroxidation and protein carbonyl content (17 and 32%, respectively), when compared to the control values (Figure 5).

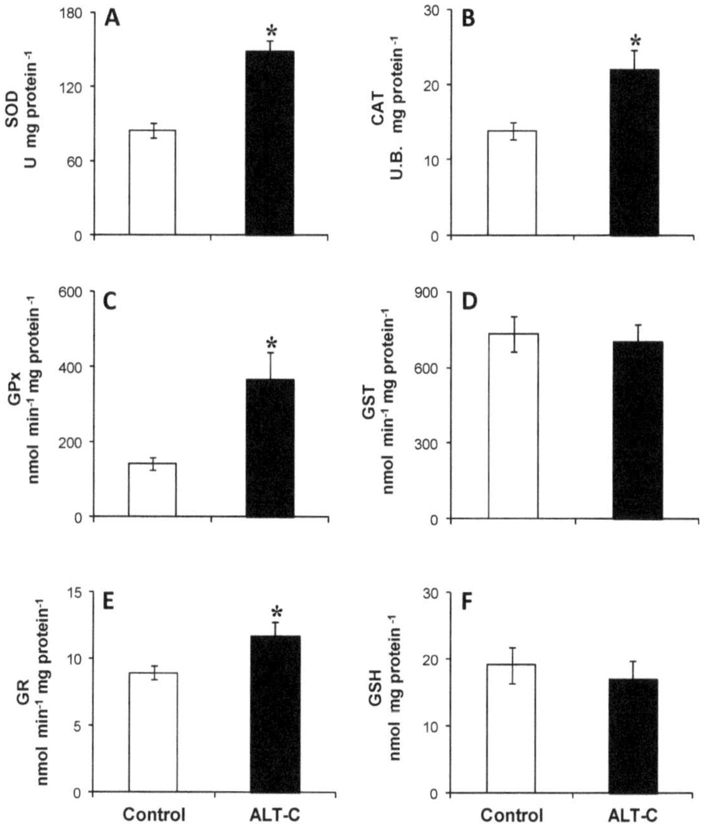

Figure 4. Activities of antioxidant enzymes (**A**) superoxide dismutase (SOD), (**B**) catalase (CAT), (**C**) glutathione peroxidase (GPx), (**D**) glutathione S-transferase (GST), (**E**) glutathione reductase (GR), and (**F**) reduced glutathione (GSH) levels in the liver of traíra, *H. malabaricus*, under control conditions (n = 10) and after seven days of treatment with alternagin-C (n = 10, single dose of 0.5 mg·kg^{-1}, intra-arterial). Data are presented as means ± S.E.M. Asterisks indicate a significant difference ($P < 0.05$) between fish groups.

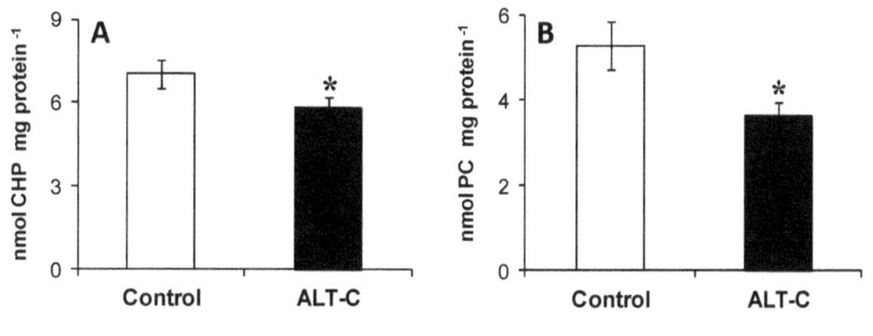

Figure 5. Oxidative stress indices: levels of (**A**) lipid hydroperoxide (cumene hydroperoxide—CHP—equivalents) and (**B**) protein carbonyl (PC) levels in the liver of traíra, *H. malabaricus*, under control conditions (n = 10) and after seven days of treatment with alternagin-C (n = 10, single dose of 0.5 mg·kg^{-1}, intra-arterial). Data are presented as means ± S.E.M. Asterisks indicate significant difference ($P < 0.05$) between fish groups.

3. Discussion

This study has revealed that the administration of a single dose of ALT-C (0.5 mg·kg^{-1}) improved fish hepatic tissue. When compared to controls, the liver of fish from the ALT-C group displayed higher VEGF levels and degree of vascularization, increases of the antioxidant defenses with a concomitant reduction of the oxidative damages, and a decrease in the incidence of some liver histopathological findings, usually present at a low frequency in controls, mainly cytoplasmic degeneration, hepatocyte atrophy, and architectural/structural alterations.

The fish liver is a highly vascularized organ composed of two afferent vessels, the hepatic artery and portal vein, and a single efferent vessel, the hepatic vein. The sinusoid capillaries are present among the hepatocytes and contain arterial and afferent venous blood. The treatment with ALT-C induced fish hepatic angiogenesis leading to enhanced blood flow. This process is defined as a dynamic and growth factor-dependent process leading to the formation of new blood vessels from preexisting ones and it is essential in many physiological and pathological conditions [10]. Previous studies demonstrated that ALT-C up-regulates VEGF expression and induces endothelial cell proliferation in vitro [4,6]. ALT-C competitively interacts with the $\alpha_2\beta_1$ integrin, the major collagen receptor, triggering intracellular signaling typical of integrin-activated pathways and inducing the expression of several growth factors, mainly VEGF, one of the most effective angiogenic peptides [11].

Angiogenesis, the formation and maintenance of blood vessel structures, is essential for the physiological functions of tissues and VEGF plays crucial roles in the formation of new blood vessels and microvascular permeability not only in physiological, but also in pathological angiogenesis [12]. ALT-C successfully induces protein kinase B (PKB) phosphorylation, an essential signaling pathway for endothelial cell proliferation which is activated by many angiogenic factors, including VEGF [7]. ALT-C also stimulated the formation of new vessels, increased the expression of growth factors, mainly VEGF and fibroblast growth factor 1 (FGF1), and augmented the fibroblast density and collagen deposition in vivo during the healing of wounded rat skin [13,14]. VEGF stimulates endothelial cell proliferation and differentiation, increases vascular permeability, supports endothelial cell survival, adhesion, and migration, and induces endothelial cell gene expression [15].

ALT-C is also able to induce in vivo angiogenesis, in a dose-dependent way, in the wounded rat skin model after topic treatment into the wound [13]. These authors evidenced that ATL-C induced the formation of new vessels and the expression of VEGF in the injured tissue. As previously demonstrated [16], the administration of a single dose of ALT-C (0.5 mg·kg^{-1}), via intra-arterial injection in *Hoplias malabaricus*, significantly increased myocardial VEGF levels after seven days. Following this line of evidence, the results of the present work pointed out the usefulness and effectiveness of ALT-C, as a pro-angiogenic desintegrin-like protein, by increasing the hepatic VEGF levels and the area occupied by blood vessels, after intra-arterial administration in fish, an alternative animal model. The use of fish in scientific research is growing due to both the expansion of the fish farming industry and an emergent awareness of questions concerning the humane use of mammalian models in basic research and chemical testing [17].

The changes that ALT-C induced in the hepatic microcirculation could result in a better oxygen and fuel supply to the fish liver. Oxygen supply is crucial to cell metabolism and reactive oxygen species (ROS) production mostly depends on appropriated oxygen levels [18,19]. ROS generation is a physiological process due to the oxidative metabolism of the cell or the activity of specific enzymatic complexes. Both the enzymatic and non-enzymatic antioxidant system are essential for the cellular response in order to deal with oxidative stress under a physiological condition [20]. Oxidative stress occurs when the critical balance between oxidants and antioxidants is disrupted due to the depletion of antioxidants or excessive accumulation of the ROS, or both [21,22], and can cause damage to lipids, protein, and DNA, leading to a disruption of cellular function and tissue injury [23].

In the present study, the enhanced hepatic fish tissue blood flow could have stimulated ROS production, such as the superoxide anion ($O_2^{\bullet-}$), hydrogen peroxide (H_2O_2), and hydroxyl radical ($HO^{\bullet}$). Increased ROS in cells may lead to an elevation of antioxidant enzymes as an adaptive response to neutralize the harmful effects of free radicals in the liver tissue. SOD and CAT enzymes have connected functions since SOD catalyses the conversion of the superoxide anion radical to H_2O and H_2O_2, which is detoxified by both CAT and GPx activity. The SOD-CAT system provides the first defense line against oxygen toxicity [24] and is usually used as a biomarker of ROS production [25,26]. The increased SOD and CAT activities induced by ALT-C in the liver of traíra indicate an elevated antioxidant status attempting to neutralize the impact of the ROS and, consequently, contributing to the reduction of basal protein and lipid oxidation levels. ROS are generated in a wide range of normal physiological conditions, resulting in basal levels of lipid and protein oxidation. When the cellular production of ROS overwhelms its antioxidant capacity, damage to cellular macromolecules such as lipids, proteins, and DNA above basal levels may occur, characterizing oxidative stress.

Traíras treated with ALT-C also displayed significant increases in the hepatic GPx and GR activities, while the GST activity remained unchanged. GPx protects tissues from oxidative damage by reducing H_2O_2 and a wide range of organic hydroperoxides that form an important group of toxic compounds produced by oxygen metabolism, therefore preventing lipid peroxidation in the membranes, and acts as a ROS scavenger [27,28]. The higher increase in the hepatic GPx activity in fish treated with ALT-C was probably enough to efficiently detoxify the H_2O_2 and the lipid hydroperoxides, leading to lower lipid peroxidation levels. The major detoxification function of GPx is the termination of radical chain propagation by quick reduction to yield ROS [29].

The detoxification of ROS and hydroperoxides by GPx involves a concomitant oxidation of reduced GSH to its oxidized form (GSSG). This GSSG is then reduced to GSH by GR at the expense of nicotinamide adenine dinucleotide phosphate (NADPH), which is recycled by the pentose phosphate pathway [30]. To maintain suitable GSH levels to the GPx and GST activities, a higher GR activity was detected in the fish of the ALT-C group when compared to the control group. GR plays a critical role in maintaining the larger glutathione pool in the reduced form, while GSH is the dominant non-protein thiol and is essential for protecting cells from oxidative damage and the toxicity of xenobiotic electrophiles, as well as maintaining redox homeostasis [31].

GSTs constitute a large multigene family of phase II detoxification enzymes involved in the conjugation of glutathione (GSH) to electrophilic compounds, such as xenobiotics, through thioether linkages, leading to the formation of conjugates that are more readily excreted and typically less toxic [32,33]. No differences were observed between experimental groups regarding hepatic GST activity indicating that ALT-C did not activate the liver detoxification pathways. These results reinforce the hypothesis that ALT-C, even though a disintegrin-like protein extracted from the snake venom, can be an important therapeutic tool. The liver is particularly susceptible to chemical injury because of its strategic location, prominent blood supply and prevalent role in the biotransformation of xenobiotics [34]. The formation of electrophilic reactive metabolites is considered to be an undesirable trait of drug candidates on the grounds of evidence linking this liability with drug-drug interactions, end-organ toxicity, and genotoxicity [35].

It is worth pointing out that the VEGF can induce Nrf2 (nuclear factor erythroid-2-related factor 2) expression by the activation of downstream effectors, including the protein kinase C (PKC), serine/threonine-protein kinase (Raf), and extracellular signal-regulated kinase (ERK1/2)/phosphatidylinositol 3-kinase (PI3K)-focal adhesion kinase pathways [36]. Nrf2 is a redox-sensitive transcription factor that binds to the antioxidant response element (ARE), leading to an upregulation of antioxidant gene expression that controls the elimination of ROS and electrophiles [37]. The results from this study indicate a possible activation of VEGF-Nrf2 signaling by ALT-C.

Two of the more obvious differences between fish and mammals are that fish have a lower perfusion rate and 50-fold slower bile flow rate [38,39], suggesting that the liver tissues of fish are more susceptible to damage by chemical agents. Due to these anatomic and physiologic considerations, histological changes often occur in the liver from control fish. The morphologic features of liver toxicity are often exacerbations of findings that may be observed in normal or control fish [38]. In the present investigation, the induction of hepatic angiogenesis and antioxidant defense systems by ALT-C with concomitant reductions in the basal oxidative damages was able to lead to an improvement of the major liver histological features. Future research directions may also be highlighted.

4. Conclusions

In summary, ALT-C treatment offered a considerable beneficial effect on fish liver vascularization supporting in vivo pro-angiogenic activity of this desintegrin-like peptide extracted from snake venom. The results point out that ALT-C induces angiogenesis through increased VEGF levels producing better hepatic tissue morphology and indicating the potential application of ALT-C in therapies for tissue repair. Furthermore, these ALT-C-induced changes in the hepatic tissue improved antioxidant defense systems. However, the action of ALT-C on different hepatic cell populations and the specific signaling pathways involved in its effects remains to be elucidated in future studies, as well as to validate the translation of the model to humans.

5. Materials and Methods

This study was performed under the approval of the Animal Ethics Committee of the Federal University of São Carlos (CEUA/UFSCar—Approval #049/14 September 2012) and in accordance with the Guide for Care and Use of Laboratory Animals published by National Institutes of Health and the ethical guidelines.

5.1. Animals

Adult specimens of *Hoplias malabaricus* (traíra, 131.3 $\pm$ 5.4 g, mean $\pm$ E.P.M.) were obtained from the Santa Candida Fish Farm (Santa Cruz da Conceição, São Paulo, Brazil). Fish were acclimated for 60 days prior to experimentation in 500 L holding tanks equipped with a continuous supply of well-aerated ($P_wO_2 > 130$ mmHg) and dechlorinated water at a constant temperature (24 °C) under a natural photoperiod (~12 h:12 h). During this period, fish were fed weekly with small live fishes.

5.2. Alternagin-C (ALT-C)

ALT-C was isolated from *Rhinocerophis alternatus* lyophilized venom (provided by the Institute Butantan, São Paulo, Brazil) by two steps of gel filtration followed by anion exchange chromatography. ALT-C is purified as a precursor peptide with a metalloprotease domain, from which it is released after proteolytic processing, resulting in a form with disintegrin and cysteine-rich domains according to the procedures previously described [11].

5.3. Experimental Design and Treatment

Fish were divided into two groups ($n = 10$ in each group): one group received a single 0.5 mg·kg^{-1} intra-arterial injection of alternagin-C (ALT-C group), and the other group was treated with sterile saline (Control group). Fish were anaesthetized in a 0.1% (w/v) solution of benzocaine and then placed onto an operating table. An indwelling cannula was implanted into the third afferent branchial artery on the left side using polyethylene tubing (PE 10) allowing drug injection according to the procedures previously described [40]. This cannula was filled with saline and heparin solution (NaCl 0.9%, 100 IU·mL^{-1} of heparin) and used to deliver injections of 0.2 mL of sterile saline or ALT-C (0.5 mg·kg^{-1}). To ensure complete drug delivery, after each injection, the cannula was cleaned with a new solution of saline (0.1 mL) and sutured in place. The chosen dose of ALT-C was based on our previous study [16]. After seven days, fish were euthanized through cerebral concussion followed by anterior spinal cord section. The fish liver was dissected and samples were either fixed in 4% paraformaldehyde (PFA) in 0.1 M phosphate buffer pH 7.2 for morphological analyses or frozen into liquid nitrogen and stored at −80 °C until an analysis of the VEGF levels or stress oxidative biomarkers.

5.4. Liver Morphological Analysis

For histomorphology, liver samples fixed in buffered PFA were dehydrated in ethanol crescent series, embedded in historesin (Leica, Wetzlar, Germany), and the sections (3 μm thickness) were stained with Toluidine blue and basic fuchsin. The slides were analyzed using an Olympus BX51 light microscope (Olympus, Ballerup, Denmark) equipped with a camera connected to a computer using Olympus DP2-BSW software (Version 2.2, Olympus, Ballerup, Denmark, 2008). For the quantification of liver vascularization, the total area occupied by the histological section and the total area occupied by the blood vessels (arteries and veins) were measured with the help of the image analysis software (Motic Image Plus 2.0, Motic China Group Co., Ltd., Hong Kong, China, 2006). The fractional area of the blood vessels was calculated through the ratio between the total area occupied by these vessels and the total area occupied by the histological section (six-eight non-contiguous fields/section/fish), multiplied by 100. The mean values for each fish were calculated and these values were used to determine the average of each experimental condition. The presence of histological alterations was also evaluated by a randomized blind method.

5.5. Quantification of VEGF

Frozen livers were homogenized in RIPA buffer (10 mM Tris pH 7.4, 100 mM NaCl, 1 mM EDTA, 1 mM EGTA, 1% Triton X-100, 10% glycerol, 0.1% SDS, 0.5% Na deoxycholate, and protease inhibitor cocktail from Sigma—P8340) and centrifuged at 10,000 rpm for 40 min at 4 °C. VEGF levels were quantified using the Murine VEGF Mini ELISA Development kit (cat. no. 900-M99, Peprotech, Rocky Hill, NJ, USA), according to the manufacturer's instructions. Briefly, 100 uL of standards (recombinant murine VEGF) or liver homogenates were added in triplicate on each well, previously coated overnight with polyclonal rabbit anti-murine VEGF capture antibody. After 2 h incubation, wells were washed and incubated with biotinylated rabbit anti-murine VEGF and avidin-HRP conjugate. Following another wash, ABTS (2,2′-Azino-bis 3-ethylbenzothiazoline-6-sulfonic acid) substrate solution was added to wells and color developed in proportion to the amount of VEGF bound in the initial step. The plate was read on a Spectra Max M5 plate reader (Molecular Devices, LLC, Sunnyvale, CA, USA) with an absorbance of 405 nm. Data are expressed as the VEGF content in pg of mg $protein^{-1}$.

5.6. Biomarkers for Antioxidant Defense and Oxidative Damage

5.6.1. Antioxidant Defenses

Samples of frozen tissue were homogenized in 0.1 M Na^+/K^+ phosphate buffer pH 7.0 at 18,000 rpm. Homogenates were centrifuged at 14,000 rpm for 30 min at 4 °C. The supernatants were used for the oxidative stress biomarker assays described below.

Superoxide dismutase (SOD) activity was evaluated based on the determination of the cytochrome *c* reduction rate by superoxide anions, monitored at 550 nm [41]. The reaction mixture (1 mL) contained 50 mM Na^+/K^+ phosphate buffer (pH 7.8), 0.1 mM EDTA, 1 mM xanthine, 20 mM cytochrome *c*, tissue homogenates, and a sufficient amount of xanthine oxidase to produce a rate of cytochrome *c* reduction of 0.025 absorbance units/min. One unit of SOD activity was calculated as the amount of enzyme causing 50% of the maximum inhibition of the cytochrome *c* reduction.

Catalase (CAT) activity was measured as the decrease of H_2O_2 concentration at 240 nm [42]. Decays in absorbance were recorded for 15 s in 50 mM sodium phosphate buffer (pH 7.0) containing 15 mM H_2O_2 and tissue homogenates. CAT, together with glutathione peroxidase, acts by removing the hydrogen peroxide. However, CAT is responsible for the detoxification of elevated levels of H_2O_2 (as the concentration contained in the reaction mixture). The non-enzymatic oxidation of H_2O_2, obtained using water instead of enzyme samples, was subtracted from tissue sample decay rates. CAT activity was expressed as Bergmeyer units (B.U.) per mg of protein [43].

Glutathione peroxidase (GPx) activity was assessed by a coupled assay with glutathione reductase (GR)-catalyzed oxidation of NADPH [44]. The consumption of NADPH was recorded at 340 nm in media containing 50 mM sodium phosphate buffer (pH 7.0), 1 mM EDTA, 0.2 mM NADPH, 1 mM sodium azide, 1 U·mL^{-1} GR, 1 mM GSH, 0.2 mM H_2O_2, and tissue homogenates. Sodium azide was used to block catalase activity. The non-enzymatic oxidation of GSH was determined by using water instead of the enzyme fraction, and its reaction rate was subtracted from the rates of liver homogenates in order to determine the true enzymatic activity. The activity of GPx was expressed as mU mg protein^{-1} and 1 mU was defined as 1 nmol of NADPH consumed min^{-1}·mL^{-1} of the sample, using NADPH molar extinction coefficient (ε_{340} = 6.2 mM^{-1}·cm^{-1}).

Glutathione-S-transferase (GST) activity was evaluated using 1-chloro-2, 4-dinitrobenzene (CDNB) as a substrate [45]. The assay mixture contained 1 mM CDNB in ethanol, 1mM GSH, 100 mM potassium phosphate buffer (pH 7.0), and tissue homogenates. The formation of adduct S-2, 4-dinitrophenyl glutathione was monitored by the increase in absorbance at 340 nm against a blank. The activity was measured as the amount of enzyme catalyzing the formation of 1 nmol of the product min^{-1}·mg^{-1} of protein, using the CDNB-GSH conjugate molar extinction coefficient (ε_{340} = 9.6 mM^{-1}·cm^{-1}).

Glutathione reductase (GR) activity was measured at 340 nm [46]. The reaction system of 1 mL contained: 1 mM oxidized glutathione (GSSG), 0.2 mM NADPH, 0.5 mM EDTA 50 mM K^+ phosphate buffer (pH 7.0), and a suitable amount of the glutathione reductase sample to change the absorbance from 0.05 to 0.30 per min at 340 nm. The GR activity in the samples was calculated by using the extinction coefficient of NADPH. The net rate for each sample was obtained by subtracting the rate obtained for the blank (the blank measures the spontaneous oxidation of NADPH). Under these conditions, the oxidation of 1 μmol of NADPH min^{-1}·mg^{-1} of protein was used as a unit of GR activity.

The estimate of reduced glutathione (GSH) content was analyzed according to using Elmann's reagent (DTNB) [47]. Supernatants of the acid extracts (1:1 *v*/*v* with 12% TCA) were added to 0.25 mM DTNB in a 0.1 sodium phosphate buffer pH 8.0, and a thiolate anion formation was determined at 412 nm against a GSH standard-curve. DTNB-reactive thiols levels (as GSH equivalents) were expressed as nmol·mg protein^{-1}.

5.6.2. Oxidative Damage

Lipid peroxidation (LPO) was determined by a FOX (ferrous oxidation-xylenol orange) assay for lipid hydroperoxide [48]. The principle of the FOX method is based on the oxidation of ferrous ions to ferric by the hydroperoxide activity under acidic conditions. Liver samples were homogenized in 0.1 M Na^+/K^+ phosphate buffer pH 7.0 at 18,000 rpm. Homogenates were centrifuged at 14,000 rpm for 30 min at 4 °C. The lipid hydroperoxide (LHP) was determined with 100 μL of supernatant samples (previously treated with 10% trichloroacetic acid—TCA) and 900 μL of reaction mixture containing 0.25 mM ammonium ferrous sulfate, 25 mM H_2SO_4, 0.1 mM xylenol orange, and 4 mM butylated hydroxytoluene in 90% (*v*/*v*) methanol. Blanks contained all components without supernatant. Mixtures were incubated for 30 min at room temperature prior to measurements at 560 nm. LHP levels were quantified using a calibration curve obtained with cumene hydroperoxide (CHP—2 to 200 nmol) in the corresponding reaction medium. Data are normalized by the amount of CHP equivalents per mg protein in the liver homogenate.

Protein carbonyl (PC) content was determined by colorimetric 2,4-dinitrophenylhydrazine (DNPH) [49]. Liver samples were homogenized in 50 mM K_2PO_4 buffer containing 1 mM EDTA and 40 μg·mL^{-1} phenylmethylsulfonyl fluoride (PMSF) and then centrifuged at 10,000 rpm for 10 min. Supernatant samples were incubated with 10 mM DNPH in 2.5 M hydrochloric acid at room temperature for 1 h, in a dark environment with a 15 min interval of vortexing. After that, these reactive mixtures were precipitated with 50% TCA and centrifuged for 10 min at 10,000 rpm. The pellets were washed three times with 1 mL of ethanol/ethyl acetate (1:1 *v*/*v*) mixture and dissolved in 6 M guanidine hydrochloride. The carbonyl content was measured spectrophotometrically at 370 nm.

Carbonyl content was calculated using the molar absorption coefficient of 22,000 $M^{-1} \cdot cm^{-1}$ relative to protein concentration.

5.7. Total Protein

The concentration of total protein in liver homogenates was carried out with Coomassie Brilliant Blue G-250 [50] adapted to a microplate reader [51] using bovine albumin as a standard.

5.8. Statistical Analysis

Data are shown as means ± SEM. Biochemical and morphometric data of the two experimental groups (Control and ALT-C) were compared and analyzed using a Student *t*-test or the non-parametric Mann-Whitney test, respectively. All tests were performed using GraphPad Prism software (Version 5.00, GraphPad Software, Inc., San Diego, CA, USA, 2007) and the data significance was designated at $P < 0.05$.

Acknowledgments: This study was supported by São Paulo State Research Foundation (FAPESP, #2012/10993-4; #2013/00798-2). The authors are thankful to Santa Cândida Fish Farm (Santa Cruz da Conceição-SP) which provided the fish and to Instituto Butantan for the venom samples.

Author Contributions: D.A.M. conceived and designed the experiments; D.A.M. performed the experiments and biochemical analysis; D.T. performed the histological analysis; D.A.M. analyzed the data, H.S.S.-d.-A. and F.T.R. supervised the study; A.L.K., H.S.S.-d.-A. and M.N.F. contributed reagents/materials/analysis tools; D.A.M. and D.T. wrote the paper; All authors read and approved the final manuscript.

Conflicts of Interest: The authors declare no conflict of interest.

References

1. Huancahuire-Vega, S.; Ponce-Soto, L.A.; Martins-de-Souza, D.; Marangoni, S. Biochemical and pharmacological characterization of PhTX-I a new myotoxic phospholipase A isolated from snake venom. *Comp. Biochem. Physiol. C* **2011**, *154*, 108–119. [CrossRef]
2. Ramos, O.H.P.; Selistre-de-Araujo, H.S. Snake venom metalloproteases—Structure and function of catalytic and disintegrin domains. *Comp. Biochem. Physiol. C* **2006**, *142*, 328–346. [CrossRef] [PubMed]
3. Selistre-de-Araujo, H.S.; Pontes, C.L.; Montenegro, C.F.; Martin, A.C. Snake venom disintegrins and cell migration. *Toxins* **2010**, *2*, 2606–2621. [CrossRef] [PubMed]
4. Ramos, O.H.; Terruggi, C.H.; Ribeiro, J.U.; Cominetti, M.R.; Figueiredo, C.C.; Berard, M.; Crepin, M.; Morandi, V.; Selistre-de-Araujo, H.S. Modulation of in vitro and in vivo angiogenesis by alternagin-C, a disintegrin-like protein from *Bothrops alternatus* snake venom and by a peptide derived from its sequence. *Arch. Biochem. Biophys.* **2007**, *461*, 1–6. [CrossRef] [PubMed]
5. Teklemariam, T.; Seoane, A.I.; Ramos, C.J.; Sanchez, E.E.; Lucena, S.E.; Perez, J.C.; Mandal, S.A.; Soto, J.G. Functional analysis of a recombinant PIII-SVMP, GST-acocostatin; an apoptotic inducer of HUVEC and HeLa, but not SK-Mel-28 cells. *Toxicon* **2011**, *57*, 646–656. [CrossRef] [PubMed]
6. Cominetti, M.R.; Terruggi, C.H.; Ramos, O.H.; Fox, J.W.; Mariano-Oliveira, A.; De Freitas, M.S.; Figueiredo, C.C.; Morandi, V.; Selistre-de-Araujo, H.S. Alternagin-C, a disintegrin-like protein, induces vascular endothelial cell growth factor (VEGF) expression and endothelial cell proliferation in vitro. *J. Biol. Chem.* **2004**, *279*, 18247–18255. [CrossRef] [PubMed]
7. Selistre-de-Araujo, H.S.; Cominetti, M.R.; Terruggi, C.H.B.; Mariano-Oliveira, A.; De Freitas, M.S.; Crepin, M.; Figueiredo, C.C.; Morandi, V. Alternagin-C, a disintegrin-like protein from the venom of *Bothrops alternatus*, modulates $\alpha_2\beta_1$ integrin-mediated cell adhesion, migration and proliferation. *Braz. J. Med. Biol. Res.* **2005**, *38*, 1505–1511. [CrossRef] [PubMed]
8. Hinton, D.E.; Segner, H.; Au, D.W.T.; Kullman, S.W.; Hardman, R.C. Liver toxicology. In *The Toxicology of Fishes*; Di Giulio, R.T., Hinton, D.E., Eds.; CRC Press: Boca Raton, FL, USA, 2008; pp. 327–400.
9. Vliegenthart, A.D.B.; Tucker, C.S.; Del Pozo, J.; Dear, J.W. Zebrafish as model organisms for studying drug-induced liver injury. *Br. J. Clin. Pharmacol.* **2014**, *78*, 1217–1227. [CrossRef] [PubMed]
10. Bocca, C.; Novo, E.; Miglietta, A.; Parola, M. Angiogenesis and fibrogenesis in chronic liver diseases. *Cell. Mol. Gastroenterol. Hepatol.* **2015**, *1*, 477–488. [CrossRef] [PubMed]

11. Souza, D.H.; Iemma, M.R.; Ferreira, L.L.; Faria, J.P.; Oliva, M.L.; Zingali, R.B.; Niewiarowski, S.; Selistre-de-Araujo, H.S. The disintegrin-like domain of the snake venom metalloprotease alternagin inhibits alpha2beta1 integrin-mediated cell adhesion. *Arch. Biochem. Biophys.* **2000**, *384*, 341–350. [CrossRef] [PubMed]
12. Shibuya, M. Vascular endothelial growth factor (VEGF) and its receptor (VEGFR) signaling in angiogenesis: A crucial target for anti- and pro-angiogenic therapies. *Genes Cancer* **2011**, *2*, 1097–1105. [CrossRef] [PubMed]
13. Sant'Ana, E.M.; Gouvea, C.M.; Nakaie, C.R.; Selistre-de-Araujo, H.S. Angiogenesis and growth factor modulation induced by alternagin C, a snake venom disintegrin-like, cysteine-rich protein on a rat skin wound model. *Arch. Biochem. Biophys.* **2008**, *479*, 20–27. [CrossRef] [PubMed]
14. Sant'Ana, E.M.; Gouvêa, C.M.; Durigan, J.L.; Cominetti, M.R.; Pimentel, E.R.; Selistre-de-Araújo, H.S. Rat skin wound healing induced by alternagin-C, a disintegrin-like, Cys-rich protein from *Bothrops alternatus* venom. *Int. Wound J.* **2011**, *8*, 245–252. [CrossRef] [PubMed]
15. Senger, D.R.; Perruzzi, C.A.; Streit, M.; Koteliansky, V.E.; de Fougerolles, A.R.; Detmar, M. The $\alpha 1\beta 1$ and $\alpha_2\beta_1$ integrins provide critical support for vascular endothelial growth factor signaling, endothelial cell migration, and tumor angiogenesis. *Am. J. Pathol.* **2002**, *160*, 195–204. [CrossRef]
16. Monteiro, D.A.; Kalinin, A.L.; Selistre-de-Araujo, H.S.; Vasconcelos, E.S.; Rantin, F.T. Alternagin-C (ALT-C), a disintegrin-like protein from snake venom promotes positive inotropism and chronotropism in fish heart. *Toxicon* **2016**, *110*, 1–11. [CrossRef] [PubMed]
17. Schaeck, M.; Van den Broeck, W.; Hermans, K.; Decostere, A. Fish as research tools: Alternatives to in vivo experiments. *Altern. Lab. Anim.* **2013**, *41*, 219–229. [PubMed]
18. Gao, L.; Laude, K.; Cai, H. Mitochondrial pathophysiology, reactive oxygen species, and cardiovascular diseases. *Vet. Clin. N. Am. Small Anim. Pract.* **2008**, *38*, 137–155. [CrossRef] [PubMed]
19. Hao, Y.; Cheng, D.; Ma, Y.; Zhou, W.; Wang, Y. The relationship between oxygen concentration, reactive oxygen species and the biological characteristics of human bone marrow hematopoietic stem cells. *Transplant. Proc.* **2011**, *43*, 2755–2761. [CrossRef] [PubMed]
20. Li, S.; Tan, H.Y.; Wang, N.; Zhang, Z.J.; Lao, L.; Wong, C.W.; Feng, Y. The role of oxidative stress and antioxidants in liver diseases. *Int. J. Mol. Sci.* **2015**, *16*, 26087–26124. [CrossRef] [PubMed]
21. Scandalios, J.G. Oxidative stress: Molecular perception and transduction of signals triggering antioxidant gene defenses. *Braz. J. Med. Biol. Res.* **2005**, *38*, 995–1014. [CrossRef] [PubMed]
22. Storey, K.B. Oxidative stress: Animal adaptation in nature. *Braz. J. Med. Res.* **1996**, *29*, 1715–1733.
23. Halliwell, B. Free radicals, antioxidants, and human disease: Curiosity, cause, or consequence? *Lancet* **1994**, *344*, 721–724. [CrossRef]
24. Pandey, S.; Parvez, S.; Sayeed, I.; Haque, R.; Bin-Hafeez, B.; Raisuddin, S. Biomarkers of oxidative stress: A comparative study of river Yamuna fish *Wallago attu* (Bl. & Schn.). *Sci. Total Environ.* **2003**, *309*, 105–115. [CrossRef] [PubMed]
25. Regoli, F.; Winston, G.W.; Gorbi, S.; Frenzilli, G.; Nigro, M.; Corsi, I.; Focardi, S. Integrating enzymatic responses to organic chemical exposure with total oxyradical absorbing capacity and DNA damage in the European eel *Anguilla anguilla*. *Environ. Toxicol. Chem.* **2003**, *22*, 2120–2129. [CrossRef] [PubMed]
26. Monteiro, D.A.; Rantin, F.T.; Kalinin, A.L. Inorganic mercury exposure: Toxicological effects, oxidative stress biomarkers and bioaccumulation in the tropical freshwater fish matrinxã, *Brycon amazonicus* (Spix and Agassiz, 1829). *Ecotoxicology* **2010**, *19*, 105–123. [CrossRef] [PubMed]
27. Arthur, J.R. The glutathione peroxidases. *Cell. Mol. Life Sci.* **1994**, *57*, 1825–1835. [CrossRef]
28. Miyamoto, Y.; Koh, Y.H.; Park, Y.S.; Fujiwara, N.; Sakiyama, H.; Misonou, Y.; Ookawara, T.; Suzuki, K.; Honke, K.; Taniguchi, N. Oxidative stress caused by inactivation of glutathione peroxidase and adaptive responses. *Biol. Chem.* **2003**, *384*, 567–574. [CrossRef] [PubMed]
29. Van der Oost, R.; Beyer, J.; Vermeulen, N.P.E. Fish bioaccumulation and biomarkers in environmental risk assessment: A review. *Environ. Toxicol. Pharmacol.* **2003**, *13*, 57–149. [CrossRef]
30. Zhang, J.; Shen, H.; Wang, X.; Wu, J.; Xue, Y. Effects of chronic exposure of 2,4-dichlorophenol on the antioxidant system in liver of freshwater fish *Carassius auratus*. *Chemosphere* **2004**, *55*, 167–174. [CrossRef] [PubMed]
31. Forman, H.J.; Zhang, H.; Rinna, A. Glutathione: Overview of its protective roles, measurement, and biosynthesis. *Mol. Aspects Med.* **2009**, *30*, 1–12. [CrossRef] [PubMed]

32. Pierrick, L.; Haoues, A.; Luc, D.; Nicole, P.; Mylène, W. Evolution of resistance to insecticide in disease vectors. In *Genetics and Evolution of Infectious Disease*; Tibayrenc, M., Ed.; Elsevier: London, UK, 2011; pp. 363–409. [CrossRef]
33. Townsend, D.M.; Manevich, Y.; He, L.; Hutchens, S.; Pazoles, C.J.; Tew, K.D. Novel role for glutathione S-transferase pi. Regulator of protein S-Glutathionylation following oxidative and nitrosative stress. *J. Biol. Chem.* **2009**, *284*, 436–445. [CrossRef] [PubMed]
34. Gu, X.; Manautou, J.E. Molecular mechanisms underlying chemical liver injury. *Expert Rev. Mol. Med.* **2012**, *14*, e4. [CrossRef] [PubMed]
35. Kalgutkar, A.S. Handling reactive metabolite positives in drug discovery: What has retrospective structure-toxicity analyses taught us? *Chem. Biol. Interact.* **2011**, *192*, 46–55. [CrossRef] [PubMed]
36. Izuta, H.; Matsunaga, N.; Shimazawa, M.; Sugiyama, T.; Ikeda, T.; Hara, H. Proliferative diabetic retinopathy and relations among antioxidant activity, oxidative stress, and VEGF in the vitreous body. *Mol. Vis.* **2010**, *16*, 130–136. [PubMed]
37. Kweider, N.; Fragoulis, A.; Rosen, C.; Pecks, U.; Rath, W.; Pufe, T.; Wruck, C.J. Interplay between vascular endothelial growth factor (VEGF) and nuclear factor erythroid 2-related factor-2 (Nrf2): Implications for preeclampsia. *J. Biol. Chem.* **2011**, *286*, 42863–42872. [CrossRef] [PubMed]
38. Kleinow, K.M.; Nichols, J.W.; Hayton, W.L.; McKim, J.M.; Barron, M.G. Toxicokinetics, in fishes. In *The Toxicology of Fishes*; Di Giulio, R.T., Hinton, D.E., Eds.; CRC Press: Boca Raton, FL, USA, 2008; pp. 55–152.
39. Wolf, J.C.; Wolfe, M.J. A brief overview of nonneoplastic hepatic toxicity in fish. *Toxicol. Pathol.* **2005**, *33*, 75–85. [CrossRef] [PubMed]
40. Axelsson, M.; Fritsche, R. Cannulation techniques. In *Analytical Techniques. Biochemistry and Molecular Biology of Fishes*; Hochachka, P.W., Mommsen, T.P., Eds.; Elsevier Science: Amsterdam, The Netherlands, 1994; pp. 17–36.
41. Flohé, L.; Ötting, F. Superoxide dismutase assays. *Methods Enzymol.* **1984**, *105*, 93–105. [PubMed]
42. Aebi, H. Catalase. In *Methods of Enzymatic Analysis*; Bergmayer, H.U., Ed.; Academic Press: London, UK, 1874; pp. 671–684.
43. Bergmeyer, H.U. Measurement of catalase activity. *Biochem. Z.* **1955**, *327*, 255–258. [PubMed]
44. Nakamura, W.; Hosoda, S.; Hayashi, K. Purification and properties of rat liver glutathione peroxidase. *Biochem. Biophys. Acta* **1974**, *358*, 251–261. [CrossRef]
45. Habig, W.H.; Pabst, M.J.; Jakoby, W.B. Glutathione S-transferase, the first enzymatic step in mercapturic acid formation. *J. Biol. Chem.* **1974**, *249*, 7130–7139. [CrossRef] [PubMed]
46. Carlberg, I.; Mannervik, B. Purification and characterization of the flavoenzyme glutathione reductase from rat liver. *J. Biol. Chem.* **1975**, *250*, 5475–5480. [PubMed]
47. Beutler, E.; Duron, O.; Kelly, B.M. Improved method for the determination of blood glutathione. *J. Lab. Clin. Med.* **1963**, *61*, 882–888. [PubMed]
48. Jiang, Z.Y.; Hunt, J.V.; Wolff, S.P. Ferrous ion oxidation in the presence of xylenol orange for detection of lipid hydroperoxide in low-density lipoprotein. *Anal. Biochem.* **1992**, *202*, 384–389. [CrossRef]
49. Reznick, A.Z.; Packer, L. Oxidative damage to proteins: Spectrophotometric method for carbonyl assay. *Methods Enzymol.* **1994**, *233*, 357–363. [CrossRef] [PubMed]
50. Bradford, M.M.A. A rapid and sensitive method for the quantification of microgram quantities of protein utilizing the principle of protein dye binding. *Anal. Biochem.* **1976**, *72*, 218–251. [CrossRef]
51. Kruger, N.J. The Bradford method for protein quantification. *Meth. Mol. Biol.* **1994**, *32*, 9–15. [CrossRef]

toxins

MDPI

Article

Anti-*Helicobacter pylori* Properties of the Ant-Venom Peptide Bicarinalin

Jesus Guzman [1], Nathan Téné [2], Axel Touchard [2], Denis Castillo [1], Haouaria Belkhelfa [3], Laila Haddioui-Hbabi [3], Michel Treilhou [2,*,†] and Michel Sauvain [1,4,†]

1 Laboratorios de Investigación y Desarrollo, Universidad Peruana Cayetano Heredia (UPCH), Lima 34, Peru; guzman_j29@hotmail.com (J.G.); denis.castillo.p@upch.pe (D.C.); michel.sauvain@ird.fr (M.S.)
2 EA7417-BTSB, Université Fédérale Toulouse Midi-Pyrénées, INU Champollion, 81012 Albi, France; nathan.tene@univ-jfc.fr (N.T.); T.Axel@hotmail.fr (A.T.)
3 Fonderephar, Université de Toulouse, Faculté des Sciences Pharmaceutiques, 31062 Toulouse, France; haouaria_belkhelfa@hotmail.com (H.B.); laila.haddioui@fonderephar.com (L.H.-H.)
4 UMR 152 PHARMADEV, Université de Toulouse, IRD, 31062 Toulouse, France
* Correspondence: michel.treilhou@univ-jfc.fr
† These authors contributed equally to this work.

Received: 6 December 2017; Accepted: 23 December 2017; Published: 29 December 2017

Abstract: The venom peptide bicarinalin, previously isolated from the ant *Tetramorium bicarinatum*, is an antimicrobial agent with a broad spectrum of activity. In this study, we investigate the potential of bicarinalin as a novel agent against *Helicobacter pylori*, which causes several gastric diseases. First, the effects of synthetic bicarinalin have been tested against *Helicobacter pylori*: one ATCC strain, and forty-four isolated from stomach ulcer biopsies of Peruvian patients. Then the cytoxicity of bicarinalin on human gastric cells and murine peritoneal macrophages was measured using XTT and MTT assays, respectively. Finally, the preventive effect of bicarinalin was evaluated by scanning electron microscopy using an adherence assay of *H. pylori* on human gastric cells treated with bicarinalin. This peptide has a potent antibacterial activity at the same magnitude as four antibiotics currently used in therapies against *H. pylori*. Bicarinalin also inhibited adherence of *H. pylori* to gastric cells with an IC_{50} of 0.12 $\mu g \cdot mL^{-1}$ and had low toxicity for human cells. Scanning electron microscopy confirmed that bicarinalin can significantly decrease the density of *H. pylori* on gastric cells. We conclude that Bicarinalin is a promising compound for the development of a novel and effective anti-*H. pylori* agent for both curative and preventive use.

Keywords: bicarinalin; antimicrobial peptide; *Helicobacter pylori*; gastric cells; bacterial adhesion; SEM

1. Introduction

Helicobacter pylori is a unique bacteria able to colonize human stomach mucosa [1,2]. This helix-shaped Gram-negative bacteria expresses outer membrane proteins which enable it to bind epithelial gastric cells, and secretes ureases which enable it to overcome stomach acidity. It is estimated that half of the world's population is infected with *H. pylori*, making this pathogen one of the most common bacterial infections globally [3,4]. The colonization of stomachs by *H. pylori* results in gastric inflammation (gastritis), and a persistent colonization is recognized as the leading factor in the development of gastric ulcers and cancers [5,6]. *H. pylori* can be eradicated by a proton pump inhibitor combined with two or three antibiotics (i.e., amoxicillin, clarithromycin, metronidazole, and levofloxacin) [7]. However, in recent years, the overuse of this therapeutic strategy has promoted the emergence of antibiotic resistant strains, and antibiotic resistance is now the main reason for treatment failure. Therefore, finding alternative anti-*H. pylori* therapies is of considerable interest [6,8]. Several natural products have already been proven to actively suppress *H. pylori*, contributing significantly to the therapeutic arsenal against gastrointestinal infections and diseases [9].

In this context, antimicrobial peptides (AMPs) may provide an alternative approach in the treatment of *H. pylori*. These peptides are naturally found in a variety of organisms and are an essential part of the innate immune system of both invertebrates and vertebrates [10–12]. AMPs can generally be defined as short (10 to 60 amino acids) with an overall positive charge (generally +2 to +9). They have a substantial proportion of hydrophobic residues (>30%), enabling them to disrupt bacterial membranes. They therefore could be used against a broad range of bacteria including some that are resistant to conventional antibiotics [13]. Consequently, they have great potential as new antibiotics against both human and animal pathogens, although, to date, clinical trials have mostly demonstrated their efficacy as topical agents [14].

Peptides are the predominant class of toxins in most arthropod venoms, and multiple AMPs have been reported in the venoms of scorpions [15], spiders [16], centipedes [17], wasps [18] and ants [19,20]. Our research group has previously isolated the antimicrobial polycationic and c-terminally amidated peptide bicarinalin in the venom of the myrmicine ant *Tetramorium bicarinatum*. Recent antimicrobial bioassay-based studies on several pathogens confirmed that bicarinalin is an effective and fast-acting molecule with a broad spectrum of antimicrobial activity and a moderate cytotoxicity against human lymphocytes [13,21]. Several studies argued that AMPs, including bicarinalin, are suitable for the development of novel preservatives in the food industry [22]. Given this, were the peptide used as a preservative it might also prevent some gastric diseases by acting against *H. pylori* once ingested.

There have been no studies to date of venom peptides as potential anti-*H. pylori* agents. Consequently, we embarked on an investigation of bicarinalin with a view towards the development of an antimicrobial agent to protect the human stomach against the colonization of *H. pylori*. In this study, we demonstrate that *H. pylori* strains isolated from Peruvian patients present antimicrobial resistance to the antibiotics clarithromycin and levofloxacin and sensitivity to metronidazole. Then, we show that bicarinalin peptide has a strong antimicrobial activity against both reference and Peruvian-patient strains of *H. pylori*. Finally, we show that bicarinalin has low cytotoxicity for both peritoneal macrophages and gastric cells, but efficiently limited the adherence of *H. pylori* to human gastric cells.

2. Results

Clarithromycin, levofloxacin, metronidazole and amoxicillin are the conventional drugs used in a triple therapy to treat stomach infection by the gram-negative bacteria *H. pylori* even though the eradication rate is currently less than 80% in most parts of the world. Antibiotic resistance is the main reason for treatment failure. In this study, we isolated 44 clinical strains of *H. pylori* from cultures of gastric tissues of 95 biopsies obtained from Peruvian patients with dyspeptic symptoms. In Table 1, we show the antimicrobial activities of conventional antibiotics against both clinical strains and the reference ATCC strain. The results show that conventional antibiotics are more efficient with clinical strains except for metronidazole, which has a MIC_{50} 16-fold higher against ATCC strain.

Table 1. Antibacterial activities of bicarinalin and reference antibiotics against *H. pylori* strains.

***H. pylori* Strain**	**MIC_{50} µmol·L^{-1} (µg·mL^{-1})**				
	Bicarinalin	**Clarithromycin**	**Levofloxacin**	**Metronidazole**	**Amoxicillin**
ATCC 43504	3.9 (8.6)	0.042 (0.03)	0.17 (0.06)	374.4 (64)	0.035 (0.014)
Peruvian patients	0.99 (2.2)	0.66 (0.5)	1.94 (0.7)	23.4 (4)	<0.082 (0.03)

A previous study conducted in Peru in 2008 highlighted that the antimicrobial resistance rates of *H. pylori* fluctuated between 6.7% and 27% for clarithromycin, was around 50% for metronidazole, 36.9% for levofloxacin, and 7% for amoxicillin [23]. In our study, the resistance rates to clarithromycin and levofloxacin in Peru increased to 52.3% and 45.5%, respectively (Figure 1). In contrast, the resistance rate to metronidazole decreased to 29.6%, while the resistance rate to amoxicillin was stable at 4.6%.

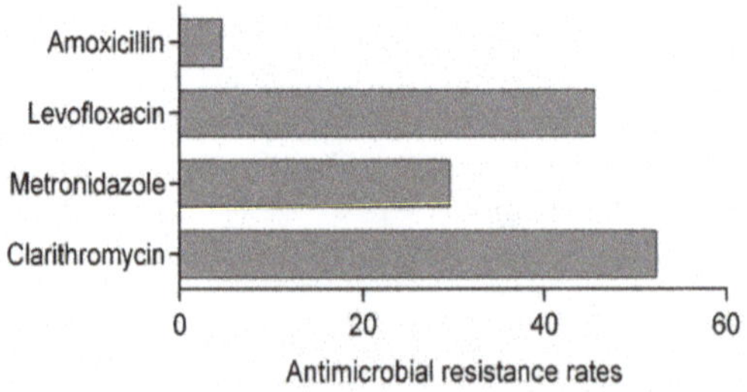

Figure 1. Antimicrobial resistance rates of Peruvian clinical *H. pylori* strains to conventional antibiotics according the EUCAST antimicrobial breakpoints for *H. pylori*: clarithromycin: S ≤ 0.7 μmol·L^{-1}, R > 1.4 μmol·L^{-1}; metronidazole: S ≤ 46.8 μmol·L^{-1}, R > 46.8 μmol·L^{-1}; levofloxacin: S ≤ 1.34 μmol·L^{-1}, R > 1.34 μmol·L^{-1}; amoxicillin: S ≤ 0.28 μmol·L^{-1}, R > 0.28 μmol·L^{-1}.

Cationic peptides have great potential in the development of novel antimicrobial agents, particularly for topical application. They are in general strongly antimicrobial and are efficient against a broad spectrum of pathogens, including those resistant to conventional antibiotics. The bicarinalin peptide displayed antimicrobial activity against all *H. pylori* and was 3.3 times more potent against clinical strains than the ATCC strain (Table 1).

We conducted a scanning electron microscopy analysis to directly evaluate the effect of bicarinalin against *Helicobacter pylori*. As shown in Figure 2, the microscopy revealed the membrane perturbation of *H. pylori* bacteria treated with 60 μg·mL^{-1} of bicarinalin, although membrane perturbation appeared slight even with 10 μg·mL^{-1}. The ability of antimicrobial peptides to effect membrane permeability is well known [24], including for bicarinalin [13].

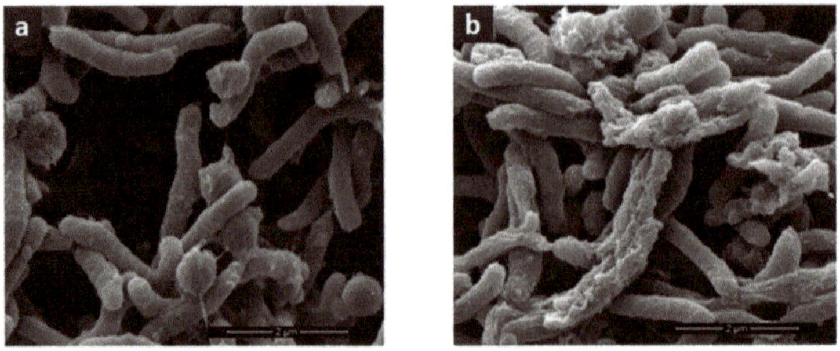

Figure 2. Scanning-electron microscopy analysis of *H. pilori*. (**a**) Without antimicrobial peptide; (**b**) with bicarinalin (60 μg·mL^{-1}). (SEM mag = 20,000).

The second aim of this study was to evaluate the effect of bicarinalin on the inhibition of adhesion of *H. pylori* (ATCC strain) on the gastric cell line (N87). The adhesion of *Helicobacter pylori* to gastric cells in the presence of bicarinalin was measured by radioactivity. We established that 50% of the adhesion (IC_{50}) is inhibited by a concentration of 0.12 μg·mL^{-1} (Log $IC_{50} = -0.92$) i.e., 0.054 μmol·L^{-1}. Above 0.56 μg·mL^{-1} (Log −0.25) i.e., 0.25 μmol·L^{-1} of bicarinalin, the inhibition of *H. pylori* adhesion on gastric cells reaches its maximum. On the other hand, at concentrations lower than 0.032 μg·mL^{-1} (log −1.5), bicarinalin no longer exhibits any significant anti-adhesive effect (Figure 3).

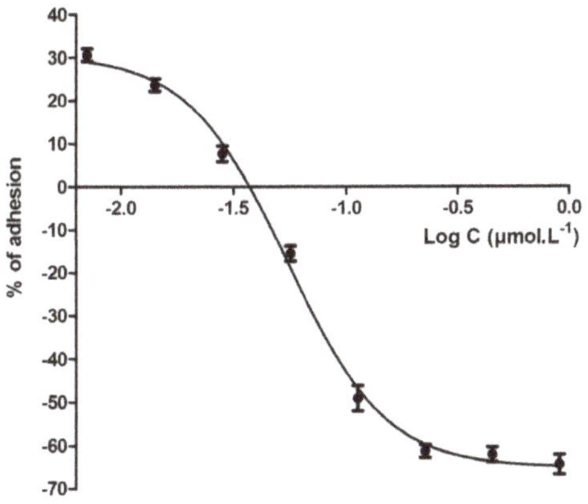

Figure 3. Anti-adhesive effect of *helicobacter pylori* on gastric cells.

On the basis of these results, electron microscopy images were performed in vitro on gastric cells (N87); with no *H. pylori*, with *H. pylori* but without bicarinalin treatment, and with *H. pylori* with bicarinalin treatment. With no *H. pylori*, Figure 4a shows a uniform cell carpet, while with *H. pylori* and without bicarinalin, isolated and aggregated bacteria adhering to the gastric cell carpet are observed (Figure 4b). The effect of bicarinalin on the adhesion of *H. pylori* to gastric cells was investigated at 0.015 and 0.25 µg·mL^{-1} (Figure 4c,d respectively). The similarity between Figure 4b,c does not reveal an effect of bicarinalin at 0.015 µg·mL^{-1}. In contrast, Figure 4d shows a significant decrease in the density of bacteria on the surface of the cellular carpet. This result is in accordance with those obtained by the radioactive counting of Figure 3, which shows that 0.25 µg·mL^{-1} provides the maximum inhibition of adhesion. However, electron microscopy shows that the maximum inhibition observed by radioactivity does not mean that no bacteria adhere, since some bacteria and aggregates are still present.

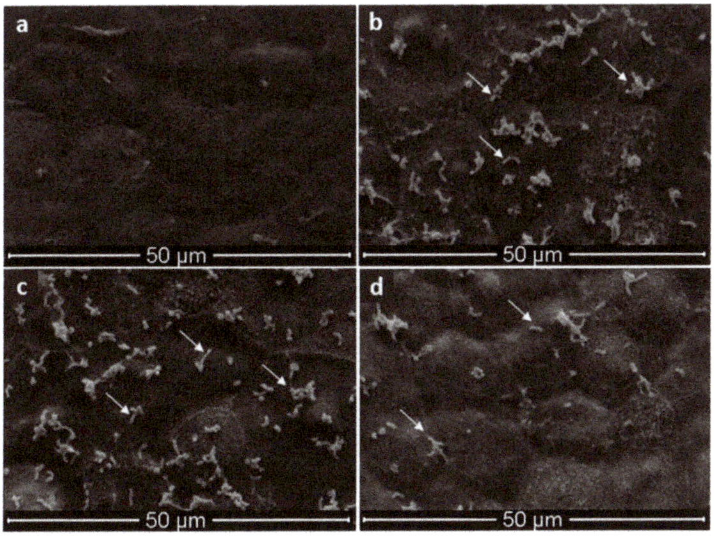

Figure 4. SEM images of cultured cellular carpet of human stomach: (**a**) no *H. pylori*; (**b**) *H. pylori* present; (**c**) *H. pylori* present with 0.015 µg·mL^{-1} of bicarinalin or (**d**) with 0.25 µg·mL^{-1} of bicarinalin (SEM mag = 1000; Arrows show single or aggregated bacteria).

To complete this study, the cytotoxicity of bicarinalin on both peritoneal macrophages and gastric cell lines was determined. Cytotoxic concentrations of 50% were measured at 39.2 $\mu mol \cdot L^{-1}$ and 1.7 $\mu mol \cdot L^{-1}$, respectively (Table 2). This resulted in a selectivity index (SI) of 39 between macrophages and *H. pylori*, and 17 between gastric cells and *H. pylori*. The selectivity index was determined as the ratio of the concentration of the bicarinalin that reduced *Helicobacter pylori* viability to 50% (MIC_{50}) to the concentration of the bicarinalin needed to inhibit the cytopathic effect to 50% of the control cells (CC_{50} of gastric cells and peritoneal macrophages).

Table 2. Cytotoxicity of bicarinalin.

	CC_{50} ($\mu mol \cdot L^{-1}$)	SI
Peritoneal macrophages (Balb/C)	39.2	>39 [a]
Gastric cells (N87)	1.7 *	>17 [b]

a: SI = CC_{50} */MIC_{50}; b: SI = CC_{50} **/IC_{50}

3. Discussion

Antimicrobial peptides (AMPs) have promise as antibacterial agents to overcome multi-drug resistant bacteria, however, systemic therapies have yet to be launched. Currently, topical application of AMPs is preferred. However, this raises the question of whether AMPs could be used to treat human pathogens colonizing mucosal surfaces, such as *Helicobacter pylori*. Previous studies have highlighted the remarkable antimicrobial activity of the ant venom peptide bicarinalin on a broad range of human pathogens and have suggested that it could be developed as a food preservative. Continuing with this idea, we investigated the effect of bicarinalin on the stomach bacteria, *H. pylori*.

Bicarinalin has a direct cytotoxic effect on *H. pylori* (ATCC 43504 strain), having a MIC_{50} of 3.9 $\mu mol \cdot L^{-1}$ that is comparable to that of anti-*H. pylori* peptides isolated from the frog *Odorrana grahami* (Odorranaina, MIC_{50} of 8.1 $\mu mol \cdot L^{-1}$) [25], those isolated from the fishes *Epinephelus coioides* and *Pardachirus marmoratus*, Epi-1 (MIC_{50} = 8.1 $\mu mol \cdot L^{-1}$) and pardaxin (>7.5 $\mu mol \cdot L^{-1}$) [26]. Nevertheless, bicarinalin was also active against forty-four clinical strains of *H. pylori* and requires a lower molar concentration to inhibit the growth of 50% of clinical isolates (MIC_{50} = 0.99 $\mu mol \cdot L^{-1}$), suggesting a better activity profile than clarithromycin, levofloxacin and metronidazole (Table 1). The cytotoxicity of bicarinalin on the gastric cells is quite similar to the MIC_{50} for the ATCC strain, which would indicate that bicarinalin is not ideal for use as a curative treatment. However, the cytotoxicity of bicarinalin on the gastric cell line NCI-N87 (IC_{50} > 1.7 $\mu mol \cdot L^{-1}$) compared to the MIC_{50} for clinical strains, led to a selectivity index higher than 17 (Table 2) which could makes it an interesting lead molecule to overcome *H. pylori* even though extending works should be carried out to try to decrease the cytotoxicity on gastric cells.

Anti-adhesion therapy is an attractive novel approach to fight drug-resistant bacteria [27]. This approach has been validated by several studies, which include *H. pylori* [28–30]. Bicarinalin inhibits the adhesion of *H. pylori* to the gastric cell model with an IC_{50} < 0.098 $\mu mol \cdot L^{-1}$ (<0.25 $\mu g \cdot mL^{-1}$), which is about forty times lower than the MIC_{50} obtained in the antimicrobial assay of the ATCC *H. pylori* strain (43504) and around ten times lower than the bicarinalin MIC_{50} tested on the *H. pylori* strains isolated from patients. These results suggest that bicarinalin can inhibit a key step in the establishment of infection of the gastric epithelial cells by *H. pylori* in addition to the direct cytotoxic effect observed at higher concentrations.

Electron microscopy confirms a significant reduction in the adhesion of bacteria to gastric cells from 0.25 $\mu g \cdot mL^{-1}$ of bicarinalin, whereas visible effects on the plasma membrane of bacteria do not appear until 10 $\mu g \cdot mL^{-1}$. This suggests that the integrity of the bacterial membrane and its ability to adhere to gastric cells is impacted at lower concentrations than those needed to observe membrane perturbations by SEM. Therefore bicarinalin can be considered effective against *H. pylori* at relatively low concentrations: less than 1 $\mu g \cdot mL^{-1}$ with an SI always greater than 10.

4. Conclusions

In summary, our data show that bicarinalin has important direct antimicrobial action on different strains of *H. pylori* isolated from dyspeptic patients as well as the reference strain. Furthermore, bicarinalin has an indirect action on *H. pylori* by inhibiting bacterial adhesion on the surface of gastric epithelial cells. Therefore, we conclude that bicarinalin could be considered as a novel alternative compound for curative and preventive therapies against *H. pylori* and contribute to controlling this emerging global health problem and the issues associated with antimicrobial resistance. However, future investigations should be conducted to study the activity of bicarinalin as well as its stability in vivo.

5. Materials and Methods

5.1. Bicarinalin Synthesis

Bicarinalin is a C-terminally amidated peptide of twenty residues (KIKIPWGKVKDFLVGGM KAV-NH_2) that was synthesized on a Liberty microwave assisted automated peptide synthesizer (CEM, Saclay, France) at a higher than 99% purity grade, as previously described in Rifflet et al. [21]. The purity and the molecular identity of the synthetic peptide were controlled using MALDI-TOF mass spectrometry.

5.2. Microorganism Strains and Growth Conditions

The *H. pylori* clinical strains were obtained from patients recruited at the Gastroenterology Service of the Cayetano Heredia Medical Clinic in Peru who presented symptoms of dyspepsia. Gastric tissues from dyspeptic patients were extracted via endoscopic gastric biopsy and were transported in 1 mL of BHI broth/FBS/glycerol (*v*/*v*/*v*; 80/10/10) at 4 °C. Gastric tissues were homogenized using a 40 μm diameter cell disintegration mesh incorporated into a BDFalcon® tube. The resulting homogenate was subjected to serial dilutions of 10^{-1} and 10^{-2}; and cultivated on blood agar plates composed of a BHI agar supplemented with: 10% *v*/*v* defibrinated sheep blood/water, Amphotericin B, and Skirrow Campylobacter selective supplement. The plates were incubated at 37 °C in an atmosphere of 5% O_2 and 10% of CO_2 for five to seven days [31]. Small transparent colonies were grown and were then re-cultured on fresh blood agar plates. The isolated *H. pylori* strains were characterized by microbiological screening according to culture characteristics (small, slightly hemolytic), morphological features (curved bacillary or spiral Gram negative bacteria), biochemical tests (catalase, oxidase and positive urease), and conventional PCR (23S rRNA gene) [32,33].The characterized *H. pylori* strains were collected and resuspended in 5 mL of BHI/FBS/glycerol (*v*/*v*/*v*; 80/10/10). The suspensions were homogenized by vortexing and stored at −70 °C.

Brain heart infusion (BHI) broth, fetal bovine serum (FBS) amphotericin B were supplied by Sigma Aldrich France. Glycerol was supplied by HiMedia USA. Defibrinated sheep blood and Campylobacter selective supplement containing Vancomycin, Trimetropin and Polymyxin B was supplied by OXOID France. The *H. pylori* reference strain (ATCC 43504) used was purchased from the ATCC®. The BHI broth, FBS and the reference antibiotics; clarithromycin, metronidazole, amoxicillin and levofloxacin were supplied by Sigma Aldrich USA. IsoVitalex was supplied by BD BBL USA.

5.3. Antimicrobial Assays

Minimal inhibitory concentrations (MIC) of the four reference antibiotics plus Bicarinalin were determined by a standard broth microdilution assay following the guidelines of the Clinical and Laboratory Standards Institute (CLSI) [34]. Bacterial innocula of *H. pylori* from both clinical strains and the reference strain (ATCC 43504) were suspended at 10^7 to 10^8 CFU·mL^{-1} in BHI broth medium/FBS/IsoVitalex (*v*/*v*/*v*; 89/10/1) [35,36]. In addition, we evaluated the activity of four antimicrobials used in eradication therapy of *H. pylori*: clarithromycin, metronidazole, amoxicillin and levofloxacin. These were added to the medium at different concentrations: from 0.25 μg·mL^{-1}

to 2 μg·mL^{-1}, from 2 μg·mL^{-1} to 16 μg·mL^{-1}, from 0.03 μg·mL^{-1} to 0.24 μg·mL^{-1} and from 0.25 μg·mL^{-1} to 2 μg·mL^{-1}. The serial dilutions for each antibiotic were calculated based on the cut-off points recommended by the European Committee on Antimicrobial Susceptibility Testing (EUCAST) [37]. The synthetic peptide bicarinalin was added to the medium at several concentrations between 0.1 and 10 μg·mL^{-1}. The cultures were incubated at 37 °C in an atmosphere of 5% O_2 and 10% CO_2 for 72 h. The MIC values were visually determined and were defined as the lowest concentration where antibiotics or bicarinalin induced a complete inhibition of visible growth in the culture. The MIC of both antibiotics and bicarinalin were calculated by a Probit logistic regression analysis of percentages of inhibition accumulated versus the distribution of MICs observed in the isolated strains of *H. pylori* for each antibiotic and bicarinalin. Strains were categorized as sensitive or resistant according to cut-off points recommended by EUCAST [37]. The MIC assays were performed in triplicate.

5.4. Anti-Adherence Effect

The NCI-N87 gastric cell line was cultured in RPMI 1640 medium supplemented with 10% FBS, 1% IsoVitalex, 1% penicillin and streptomycin at 37 °C in a 5% CO_2 humidified atmosphere. After 48 h of incubation, the single-cell layer obtained was removed with 5 mL of 0.05% trypsin-EDTA solution for 5 min at 37 °C. Trypsin was inactivated by the addition of 10 mL of RPMI 1640 medium supplemented with 10% FBS. The cells were harvested by centrifugation at 3500 *g* for 5 min at 20 °C [38]. The cell viability was checked by trypan blue assay and the cell suspension was adjusted to 1×10^6 viable cells·mL^{-1}.

H. pylori strain ATCC 43504 cultures (2×10^8 bacteria·mL^{-1}) were inoculated in 10 mL BHI broth with 30 μL of tritium adenine solution (1 μCi) and incubated for 48 h at 37 °C in an atmosphere of 5% O_2 and 10% CO_2. To eliminate non-incorporated cells in the cultures, the bacteria were washed three time using PBS buffer (centrifugation 2500 *g*/10 min at 5 °C). The inhibition of adhesion was evaluated on a 96-well plate previously prepared by placing 500 μL/well of 2×10^8 bacteria·mL^{-1} suspension of treated bacteria and the gastric cells (N87). The cells were treated with serial dilutions of bicarinalin concentrations between 0.9 and 0.007 μmol·L^{-1} (2 and 0.015 μg·mL^{-1}) at 37 °C for 24 h. Then, non-adherent bacteria were eliminated by PBS washing (three times). At the end of the treatment, the gastric cells received 500 μL of a lysis solution (SDS 0.1% (*w*/*v* in NaOH 5 mol·L^{-1}) and were then incubated at 37 °C for 12 h [39]. Radioactivity was measured with a beta-liquid scintillation system (Perkin Elmer, San Diego, CA, USA). The percentages of inhibition of adherence was calculated as follows:

$$\% \text{ Adherence inhibition} = \frac{(\text{CPM control} - \text{CPM treatment}) \times 100}{\text{CPM control}} \; [\text{CPM} = \text{counts per minute}]$$

The required concentration of bicarinalin to inhibit the adherence of 50% of bacteria was expressed as IC_{50}, which was calculated by a logistic regression analysis of probit.

5.5. Cytotoxicity of Bicarinalin

The cytotoxic effect of bicarinalin on human gastric cells (NCI-N87) and murine peritoneal macrophages (RAW 264) were determined with XTT and MTT assays, respectively [40].

Suspensions of trypsinized gastric cells (10^6 cells·mL^{-1}) were incubated with phenol at 0.5% in the medium (as a positive control) or serial dilutions of bicarinalin concentrations between 0.9 and 0.007 μmol·L^{-1} (2 and 0.015 μg·mL^{-1}) for 24 h at 37 °C. Then, 50 μL of XTT was added to each well and incubated at 37 °C for 2 h. Absorbance was read at 450 nm in the Chameleon-Hidex® plate reader.

Murine macrophages were cultured in RPMI 1640 medium and incubated at 37 °C in a 5% CO_2 atmosphere. Then, 0.05% of a trypsin-EDTA solution (Invitrogen®) was added and incubated for 2 min. Subsequently, 100 μL of suspensions of the macrophages (1×10^5 macrophages·mL^{-1}) were distributed in each well and incubated for 24 h. Then, the microdilution assay was prepared in a system of three serial dilutions of 50 μmol·L^{-1} maximum concentration of bicarinalin and incubated for 48 h at 37 °C in 5% CO_2. MTT reagent was added for 4 h and the reaction was stopped by adding

100 μL of a solution of isopropanol/SDS/water (*v*/*v*/*v*; 50/10/40) over 30 min. Finally, absorbance was read at 570 nm in the Chameleon-Hidex® plate reader.

All experiments were conducted in triplicate. The CC_{50} values for both gastric cells and macrophages were obtained by a logistic regression analysis Probit based on the calculation of the percentage of viability calculated as follows:

$$\% \text{ Viability} = \frac{\text{Abs control} - \text{Abs treatment}}{\text{Abs control}} \times 100$$

5.6. Scanning Electron Microscopy

5.6.1. Helicobacter SEM

Helicobacter pylori ATCC 43504 cultured on blood agar was used to prepare five tubes of inoculum of 1×10^8 CFU·mL^{-1} using brucella broth. Then, bicarinalin was added to each tube to achieve final concentrations of 15, 30, 60, 120, and 240 μg·mL^{-1} which were incubated for 1 h. The bacteria were washed three times using PBS buffer and centrifuged at 3000 rpm during 10 min at 5 °C. The PBS was replaced by 2% glutaraldehyde in 0.1 mol·L^{-1} Sorensen phosphate buffer (pH 7.4).

5.6.2. Gastric cells SEM

H. pylori ATCC 43504 cultured on blood agar was used to prepare an initial inoculum of 1×10^7 to 1×10^8 CFU·mL^{-1}. After washing the inoculum, the optical density was adjusted at 2×10^8 CFU·mL^{-1} using the cell culture medium RPMI 1640. This last inoculum was then added to a microwell plate with previously adhered gastric cells, and two concentrations of bicarinalin (0.25 and 0.016 μg·mL^{-1}). After two hours of incubation under microaerophilic conditions, cell culture medium was removed and replaced by 2% glutaraldehyde in 0.1 mol·L^{-1} Sorensen's phosphate buffer (pH 7.4).

5.6.3. Scanning Electron Microscopy

The bacterial cells (alone or with gastric cells) were fixed in 2% glutaraldehyde in 0.1 mol·L^{-1} Sorensen phosphate buffer (pH 7.4) for at least 4 h at 4 °C. After sedimentation, the pellets were resuspended in water and adhered to poly-lysine coated glass coverslips. The bacteria were then dehydrated in a graded ethanol series and dried by critical point drying with a Leica EM CPD 300. The samples were coated with 6 nm platinum on a Leica EM Med 020 before being examined on a FEI Quanta 250 FEG scanning electron microscope, at an accelerating voltage of 5 kV.

Acknowledgments: The authors acknowledge the CONCYTEC from Peru for the attribution of a master grant to one of us to realize part of the studies.

Author Contributions: L.H.-H., M.S. and M.T. conceived and designed the experiments; J.G., H.B., D.C., N.T. performed the experiments and data analysis; J.G., A.T., M.S., and M.T. contributed to writing and theoretical discussions; M.S. and M.T. coordinated the study.

Conflicts of Interest: The authors declare no conflict of interest.

References

1. Keilberg, D.; Ottemann, K.M. How *Helicobacter pylori* senses, targets and interacts with the gastric epithelium. *Environ. Microbiol.* **2016**, *18*, 791–806. [CrossRef] [PubMed]
2. Valenzuela, M.; Cerda, O. Overview on chemotaxis and acid resistance in *Helicobacter pylori*. *Biol. Res.* **2003**, *36*, 429–436. [CrossRef] [PubMed]
3. Eusebi, L.H.; Zagari, R.M.; Bazzoli, F. Epidemiology of *Helicobacter pylori* infection. *Helicobacter* **2014**, *19*, 1–5. [CrossRef] [PubMed]
4. National Cancer Institute. *Helicobacter pylori* and Cancer. 2013. Available online: https://www.cancer.gov/about-cancer/causes-prevention/risk/infectious-agents/h-pylori-fact-sheet#q1 (accessed on 13 September 2017).

5. Dubois, A. Spiral bacteria in the human stomach: The gastric helicobacters. *Emerg. Infect. Dis.* **1995**, *1*, 79–85. [CrossRef] [PubMed]
6. International Agency for Research on Cancer (IARC). Schistosomes, Liver flukes and *Helicobacter pylori*: Monographs on the evaluation of Carcinogenic Risks to Human. IARC: Lyon, Francia, France; 1994. *IARC Sci. Publ.* **1994**, *6*, 177–241.
7. European Helicobacter Pylori Study Group (EHPSG). Current European concepts in the management of *Helicobacter pylori* infection. The Maastricht Consensus Report. *Gut* **1997**, *41*, 8–16.
8. Espino, A. Infección por *Helicobacter pylori*. *Gastroenterol. Latinoam.* **2010**, *21*, 323–327.
9. Cogo, L.L.; Monteiro, C.L.B.; Miguel, M.D.; Miguel, O.G.; Cunico, M.M.; Ribeiro, M.L.; de Carmago, E.R.; Kussen, G.M.B.; da Silva Nogueira, K.; Dalla Costa, L.M. Anti-*Helicobacter pylori* activity of plant extracts traditionally used for the treatment of gastrointestinal disorders. *Brazilian J. Microbiol.* **2010**, *41*, 304–309. [CrossRef] [PubMed]
10. Touchard, A.; Aili, S.R.; Fox, E.G.P.; Escoubas, P.; Orivel, J.; Nicholson, G.M.; Dejean, A. The biochemical toxin arsenal from ant venoms. *Toxins* **2016**, *8*, 30. [CrossRef] [PubMed]
11. Epand, R.M.; Vogel, H.J. Diversity of antimicrobial peptides and their mechanisms of action. *Biochim. Biophys. Acta* **1999**, *1462*, 11–28. [CrossRef]
12. Li, Y.; Xiang, Q.; Zhang, Q.; Huang, Y.; Su, Z. Overview on the recent study of antimicrobial peptides: Origins, functions, relative mechanisms and application. *Peptides* **2012**, *37*, 207–215. [CrossRef] [PubMed]
13. Téné, N.; Bonnafé, E.; Berger, F.; Rifflet, A.; Guilhaudis, L.; Ségalas-Milazzo, I.; Pipy, B.; Coste, A.; Leprince, J.; Treilhou, M. Biochemical and biophysical combined study of bicarinalin, an ant venom antimicrobial peptide. *Peptides* **2016**, *79*, 103–113. [CrossRef] [PubMed]
14. Kang, S.-J.; Park, S.J.; Mishig-Ochir, T.; Lee, B.-J. Antimicrobial peptides: Therapeutic potentials. *Expert Rev. Anti-Infect. Ther.* **2014**, *12*, 1477–1486. [CrossRef] [PubMed]
15. Cao, L.; Dai, C.; Li, Z.; Fan, Z.; Song, Y.; Wu, Y.; Cao, Z.; Li, W. Antibacterial activity and mechanism of a scorpion venom peptide derivative in vitro and in vivo. *PLoS ONE* **2012**, *7*, e40135. [CrossRef] [PubMed]
16. Abreu, T.F.; Sumitomo, B.N.; Nishiyama, M.Y.; Oliveira, U.C.; Souza, G.H.M.F.; Kitano, E.S.; Zelanis, A.; Serrano, S.M.T.; Junqueira, I.; Azevedo, D.; et al. Peptidomics of *Acanthoscurria gomesiana* spider venom reveals new toxins with potential antimicrobial activity. *J. Proteom.* **2017**, *151*, 232–242. [CrossRef] [PubMed]
17. Peng, K.; Kong, Y.; Zhai, L.; Wu, X.; Jia, P.; Liu, J.; Yu, H. Two novel antimicrobial peptides from centipede venoms. *Toxicon* **2010**, *55*, 274–279. [CrossRef] [PubMed]
18. Perez-Riverol, A.; Roberto, J.; Musacchio, A.; Sergio, M.; Brochetto-Braga, M.R. Wasp venomic: Unravelling the toxins arsenal of *Polybia paulista* venom and its potential pharmaceutical applications. *J. Proteom.* **2017**, *161*, 88–103. [CrossRef] [PubMed]
19. Pluzhnikov, K.A.; Kozlov, S.A.; Vassilevski, A.A.; Vorontsova, O.V.; Feofanov, A.V.; Grishin, E.V. Linear antimicrobial peptides from *Ectatomma quadridens* ant venom. *Biochimie* **2014**, *107*, 211–215. [CrossRef] [PubMed]
20. Wanandy, T.; Gueven, N.; Davies, N.W.; Brown, S.G.A.; Wiese, M.D. Pilosulins: A review of the structure and mode of action of venom peptides from an Australian ant *Myrmecia pilosula*. *Toxicon* **2015**, *98*, 54–61. [CrossRef] [PubMed]
21. Rifflet, A.; Gavalda, S.; Téné, N.; Orivel, J.; Leprince, J.; Guilhaudis, L.; Génin, E.; Treilhou, M. Identification and characterization of a novel antimicrobial peptide from the venom of the ant *Tetramorium bicarinatum*. *Peptides* **2012**, 1–8. [CrossRef] [PubMed]
22. Téné, N.; Roche-Chatain, V.; Rifflet, A.; Bonnafé, E.; Lefranc, B.; Leprince, J.; Treilhou, M. Potent bactericidal effects of bicarinalin against strains of the *Enterobacter* and *Cronobacter* genera. *Food Control* **2014**, *42*, 202–206. [CrossRef]
23. Ramos, A.R.; Sánchez, R.S. Helicobacter pylori 25 anos después (1983–2008): Epidemiologia, microbiologia, patogenia, diagnóstico y tratamiento. *Rev. Gastroenterol. Perú* **2009**, *29*, 158–170.
24. Raghuraman, H.; Chattopadhyay, A. Melittin: A membrane-active peptide with diverse functions. *Biosci. Rep.* **2007**, *27*, 189–223. [CrossRef] [PubMed]
25. Chen, L.; Li, Y.; Li, J.; Xu, X.; Lai, R.; Zou, Q. An antimicrobial peptide with antimicrobial activity against *Helicobacter pylori*. *Peptides* **2007**, *28*, 1527–1531. [CrossRef] [PubMed]

26. Narayana, J.L.; Huang, H.; Wu, C.; Chen, J. Epinecidin-1 antimicrobial activity: In vitro membrane lysis and In vivo efficacy against *Helicobacter pylori* infection in a mouse model. *Biomaterials* **2015**, *61*, 41–51. [CrossRef] [PubMed]
27. Ofek, I.; Hasty, D.L.; Sharon, N. Anti-adhesion therapy of bacterial diseases: Prospects and problems. *FEMS Immunol. Med. Microbiol.* **2003**, *38*, 181–191. [CrossRef]
28. Wittschier, N.; Lengsfeld, C.; Vorthems, S.; Stratmann, U.; Ernst, J.F.; Verspohl, E.J.; Hensel, A. Large molecules as anti-adhesive compounds against pathogens. *J. Pharm. Pharmacol.* **2007**, *59*, 777–786. [CrossRef] [PubMed]
29. Wittschier, N.; Faller, G.; Hensel, A. Aqueous extracts and polysaccharides from liquorice roots (*Glycyrrhiza glabra* L.) inhibit adhesion of *Helicobacter pylori* to human gastric mucosa. *J. Ethnopharmacol.* **2009**, *125*, 218–223. [CrossRef] [PubMed]
30. O'Mahony, R.; Al-Khtheeri, H.; Weerasekera, D.; Fernando, N.; Vaira, D.; Holton, J.; Basset, C. Bactericidal and anti-adhesive properties of culinary and medicinal plants against *Helicobacter pylori*. *World J. Gastroenterol.* **2005**, *11*, 7499–7507. [CrossRef] [PubMed]
31. Ndip, R.N.; MacKay, W.G.; Farthing, M.J.G.; Weaver, L.T. Culturing *Helicobacter pylori* from Clinical Specimens: Review of Microbiologic Methods. *J. Pedriatr. Gasteroenterol. Nutr.* **2003**, *36*, 616–622. [CrossRef]
32. Rimbar, E.; Sasatsu, M.; Graham, D.Y. PCR detection of *Helicobacter pylori* in clinical samples. *Methods Mol. Biol.* **2013**, *943*, 279–287. [CrossRef]
33. NHS. NHS UK Standards for Microbiology Investigations. In *Identification of Helicobacter Species*; Service NH: London, UK, 2015.
34. CLSI. CLSI Methods for antimicrobial dilution and disk susceptibility testing of infrequently isolated or fastidious bacteria. In *Approved Guideline*, 2nd ed.; PA Clinical and Laboratory Standards Institute: Wayne, NJ, USA, 2010.
35. Piccolomini, R.; Di Bonaventura, G.; Festi, D.; Catamo, G.; Laterza, F.; Neri, M. Optimal combination of media for primary isolation of *Helicobacter pylori* from gastric biopsy specimens. *J. Clin. Microbiol.* **1997**, *35*, 1541–1544. [PubMed]
36. Hachem, C.Y.; Clarridge, J.E.; Reddy, R.; Flamm, R.; Evans, D.G.; Tanaka, S.K.; Graham, D.Y. Antimicrobial susceptibility testing of *Helicobacter pylori*: Comparison of E-test, broth microdilution and disk diffusion for ampicillin, clarithromycin and metronidazole. *Diagn. Microbiol. Infect. Dis.* **1996**, *24*, 37–41. [CrossRef]
37. EUCAST. *EUCAST Breakpoint Tables for Interpretation of MICs and Zone Diameters*; Contract No.: Version 5.0.; European Committee on Antimicrobial Susceptibility Testing: Växjö, Sweden, 2015.
38. Diesing, A.; Nossol, C.; Faber-Zuschratter, H.; Zuschratter, W.; Renner, L.; Sokolova, O.; Naumann, M.; Rothkotter, H.-J. Rapid interaction of *Helicobacter pylori* with microvilli of the polar human gastric epithelial cell line NCI-N87. *Anat. Rec.* **2013**, *296*, 1800–1805. [CrossRef] [PubMed]
39. Jung, Y.J.; Lee, K.L.; Kim, B.K.; Kim, J.W.; Jeong, J.B.; Kim, S.G.; Kim, J.S.; Jung, H.C.; Song, I.S. Usefulness of NCI-N87 cell lines in *Helicobacter pylori* infected gastric mucosa model. *Korean J. Gastroenterol.* **2006**, *47*, 357–362. [PubMed]
40. Horemans, T.; Kerstens, M.; Clais, S.; Struijs, K.; van den Abbeele, P.; Van Assche, T.; Maes, L.; Cos, P. Evaluation of the anti-adhesive effect of milk fat globule membrane glycoproteins on *Helicobacter pylori* in the human NCI-N87 cell line and C57BL/6 mouse model. *Helicobacter* **2012**, *17*, 312–318. [CrossRef] [PubMed]

toxins

MDPI

Review

Snake Venoms in Cancer Therapy: Past, Present and Future

Li Li [1], Jianzhong Huang [1,*] and Yao Lin [2,*]

[1] Engineering Research Center of Industrial Microbiology, College of Life Sciences, Fujian Normal University, Fuzhou 350117, China; lili@fjnu.edu.cn

[2] Provincial University Key Laboratory of Cellular Stress Response and Metabolic Regulation, College of Life Sciences, Fujian Normal University, Fuzhou 350117, China

* Correspondence: hjz@fjnu.edu.cn (J.H.); yaolin@fjnu.edu.cn (Y.L.)

Received: 1 August 2018; Accepted: 26 August 2018; Published: 29 August 2018

Abstract: Cancer is one of the leading causes of morbidity and mortality worldwide, and the discovery of new drugs for cancer therapy is one of the most important objectives for the pharmaceutical industry. Snake venoms are complex mixtures containing different peptides, proteins, enzymes, carbohydrates and other bioactive molecules, which are secreted by the snake in the predation or defending against threats. Understanding the snake venoms may turn the toxins into a valuable source of new lead compounds in drug discovery. Captopril, the first angiotensin-converting enzyme inhibitor approved in 1981 by FDA, was designed based on the structure of a peptide isolated from the snake venom. The earliest reports about snake venoms used in cancer treatments appeared in the 1930s. Since then, numerous studies on the activities, isolations, purifications and structure elucidations of the components from snake venoms were published. The comprehensive structural and functional investigations of snake venoms would contribute to the development of novel anti-cancer drugs. Our review will focus on the past, present and the future of the studies on snake venoms in cancer target therapy.

Keywords: snake venom; cancer; target therapy

Key Contribution: In this review, the application of snake venoms and their potential future combinational technologies for cancer therapy were thoroughly discussed.

---

1. Introduction

Cancer is one of the leading causes of morbidity and mortality worldwide. According to GLOBOCAN, there were approximately 14.1 million new cases diagnosed and 8.2 million deaths from cancer in 2012 globally [1]. Surgery and chemotherapy are still the main strategies for cancer therapy [2]. Target therapy, which interferes with a specific molecular target and usually causes fewer toxicities, is becoming more and more popular in chemotherapy [3,4]. Recently, the compounds purified and characterized from snake venom displayed a tremendous potential as agents targeting specific molecular pathways in cancer cells [5,6].

Snake venoms are complex mixtures of proteins, peptides and other bioactive molecules secreted by the venom gland of snakes and injected by unique fangs of snakes to debilitate and digest their prey. World Health Organization has placed snakebite envenoming on its list of top 20 priority neglected tropical diseases, which kills more than 100,000 people and maims 400,000 people annually [7]. The various clinical manifestations of snakebite victims are caused by the highly complex and diverse compositions of snake venoms, which can selectively recognize their different biological targets [8]. Although snakebite envenoming is a life-threatening public health problem, snake venoms are recognized as a potential resource of biologically active compounds. In China, snake wine or snake venom liquor is supplied as traditional Chinese medicine [9]. Snake venoms are also discovered and developed as drug leads in the modern drug industry. For example, captopril was the first angiotensin-converting enzyme inhibitor approved in 1981 by FDA for the treatment of hypertension and some types of congestive heart failure. Captopril was actually designed based on BPP_{5a}, a bradykinin-potentiating pentapeptide isolated from venoms of *Bothrops jararaca* [10].

Common venom components could be classified as enzyme and non-enzyme components. Enzymatic snake venoms include phospholipase A_2 (PLA_2), L-amino acid oxidases (LAAO), metalloproteases (SVMP), serine proteases (SVSP), 5′-nucleotidases, acetylcholinesterases and hyaluronidases. Non-enzymatic components include disintegrins (DIS), three-finger toxins (3FTx), Kunitz peptides, cysteine-rich secretory proteins (CRiSP), C-type lectins (CTL) and natriuretic peptides (NP) [11]. There has been a long history of research on exploring the therapeutic potential of snake venoms for cancer.

2. Early-Stage Study on Snake Venoms in Cancer Therapy

The effect of snake venoms was first investigated in the 1930s. For example, Essex et al. treated 15 tumor-bearing white rats with intravenous injections of different doses of venoms from rattlesnakes. However, after 6 successive weeks, cancer progresses between the experimental and control groups were similar [12]. Kurotchkin et al. discovered that cobra venom could destroy cells of the Fujinami rat sarcoma, which seemed to require direct contact between the venom and tumor cells [13]. Ligneris et al. showed that African snake venoms had no effect on the great majority of tumors in humans [14]. Notwithstanding, there were no encouraging results on cancer suppression, the pain relief effects of snake venoms were shown in some cases, whose advantages were long-acting and no morphine dependence [14]. In 1936, Macht reported the experimental and clinical study of cobra venoms as pain-relieving agents. In total, 105 cancer patients were injected with a dose of 2–5 mouse units, which was defined as the quantity of venom solution enough to kill a 22 g white mouse within 18 h after intraperitoneal injection [15]. Among the patients, 30 cases showed definite relief and 38 cases showed marked relief. Only 13.3% of the patients showed doubtful results or no relief [16].

Meanwhile, enzymes of snake venoms attracted attention of investigators for their potent biological significances. In 1938, Jynegar et al. found the activity of cholinesterase in cobra venom [17], and Zeller found that cholinesterase exists in many types of snake venom [18]. The activity of hyaluronidase in snake venom was noted by Duran-Reynals in 1936 [19]. The first nonhydrolytic enzyme, L-amino acid oxidase (LAAO), was reported by Zeller in 1944 [20].

In summary, the studies on the inhibitory effect of crude snake venoms towards tumor cells showed doubtful results at the early stages. The snake venom was used as the mixture and their main clinical effect for cancer therapy was pain relief for the patients with hopelessly malignant tumors.

3. Development of Snake Venoms for Cancer Target Therapy

The isolation and characterization of the components from snake venoms began in the 1940s. After that, numerous components including enzymes, non-enzyme proteins and peptides were purified, sequenced and structurally elucidated. L-amino acid oxidase (LAAO) was first isolated and characterized by Singer et al. from moccasin snake venom in the early 1950s [21–24]. LAAO is a FAD (Flavin Adenine Dinucleotide)—containing an enzyme that converts L-amino acid stereospecific into the corresponding α-keto acid with hydrogen peroxide and ammonia as byproducts [25].

Phospholipase A_2 (PLA_2) from *Crotalus adamanteus* was purified and partially characterized by Saito and Hanahan in 1962 [26]. In 1969, Wu and Tinker studied PLA_2 from *Crotalus atrox* [27]. The PLA_2 enzyme hydrolyzes glycerophospholipid to form lysophopholipid and fatty acid. Snake venoms often contain multiple types of PLA_2 isoenzymes, resulting in extra difficulty for purification. For example, eight PLA_2 (Pa-1G, Pa-3, Pa-5, Pa-9C, Pa-10A, Pa-12A, Pa-12C and Pa-15) have been isolated and sequenced from the venom of Australian king brown snake (*Pseudechis australis*) [28].

The lectin-related proteins in snake venom have been classified into true C-type lectins (containing the CRD domain) and C-type lectin-like proteins (containing the CRD-related non-carbohydrate-binding domains) [29,30]. A slice of C-type lectins or C-type lectin-like proteins were isolated from snake venoms in 1970s. Batroxobin, a lectin from *Bothrops atrox* venom, was isolated by Stocker and Barlow in 1976 [31]. Kirby et al. purified and characterized thrombocytin from *B. atrox* venom in 1979 [32]. It was found that C-type lectin binds to a sugar moiety at the presence of Ca^{2+} and contains the carbohydrate recognition domain (CRD).

Snake venom metalloproteinase (SVMP) is a major component in most viperidvenoms, and one of the import enzymes contributing to the toxicity of snake venom [33]. In 1978, Bjarnason and Tu purified SVMPs from western diamondback rattlesnake (*Crotalus atrox*) venom and showed that the zinc in each of these SVMPs was at approximate 1:1 ratio to the relevant protein. Removal of zinc from SVMPs abolished both proteolytic and hemorrhagic activities of the SVMPs [34].

Disintegrins are a family of integrin inhibitory proteins with low molecular weight, tripeptide sequence arginine-glycine-aspartic acid (RGD), and cysteine-rich peptides isolated from various snake venoms [35]. The integrin binding function usually depends upon the RGD motif. However, some disintegrins lacking this RGD motif can also bind and block integrins. In the late 1980s, Huang et al. purified and determined the primary structure of trigramin, a disintegrin from *Trimeresurus gramineus* snake venom, which kicked off a promising research field of the inhibition of integrin function by snake venom [36–38]. The isolation and characterization of components of snake venoms paved the way for cancer targeted therapy in modern medicine. Here, we discuss the army of snake venoms with different mechanisms of actions in cancer therapy (Table 1).

Table 1. Compounds with antitumor activities isolated from snake venoms.

Target/Mechanism	Protein Names	Compounds	Snakes	Reference
antiangiogenesis	Leucurogin	disintegrin	*Bothrops leucurus*	[39]
	Contortrostatin	disintegrin	*Agkistrodon contortrix contortrix*	[40]
	Obtustatin	disintegrin	*Vipera lebetina obtusa*	[41]
	Adinbitor	disintegrin	*A. halys brevicaudus stejneger*	[42,43]
	Salmosin	disintegrin	*A. halys brevicaudus*	[44]
apoptosis induction	LAAO	LAAO	*A. halys*	[45,46]
	AHP-LAAO	LAAO	*A. halys pallas*	[47]
	LAAO	LAAO	*V. berus berus*	[48]
	disintegrin	disintegrin	*Naja naja*	[49]
	VAP and VAP2	metalloprotease/disintegrin	*Crotalus atrox*	[50,51]
	stejnitin	SVMP	*Trimeresurus stejnegeri*	[52]

3.1. Antiangiogenesis

Human tumor growth is accompanied by neovascularization to provide essential nutrition and oxygen. Angiogenesis supports tumor cell extension and invasion into nearby normal tissue and is required to distant metastasis. Antiangiogenesis is a propitious strategy for cancer targeted therapy. Quite a few angiogenesis inhibitors for the treatment of cancer have been approved by FDA including Bevacizumab (targeting vascular endothelial growth factor, VEGF), Sorafenib (tyrosine kinase inhibitror, TKI), Sunitinib (TKI) et al. [53].

Disintegrins purified from snake venoms showed antiangiogenesis effects. Leucurogin, a disintegrin cloned from *Bothrops leucurus* (white-tailed-jararaca), showed significant anticancer activities against Ehrlich tumor implanted in mice with the administration of 10 μg/day. Antiangiogenesis effect of leucurogin was assessed and confirmed by the sponge implant model in mice [39]. Contortrostatin is a homodimeric peptide isolated from the venom of *Agkistrodon contortrix contortrix*, a subspecies of the southern copperhead snake, and contains a RGD sequence [40]. Contortrostatin showed anti-angiogenic activity against the primary tumor of human breast cancer MDA-MB-435 carried in mice. Obtustatin, a disintegrin isolated from *Vipera lebetina obtusa* venom has no classical RGD sequence [54]. Obtustatin reduced tumor size in the Lewis lung syngeneic mouse model and showed

84% inhibition of angiogenesis activities in the experiments of chick Chorioallantoic Membrane (CAM) assay [41]. Adinbitor is a disintegrin cloned from *Agkistrodon halys brevicaudus stejneger* with 73 amino acid residues including 12 cysteines and a RGD motif. Adinbitor can inhibit bFGF-induced proliferation of ECV304 cells with IC_{50} of 0.89 μM. In the Chick CAM angiogenesis assay, adinbitor showed the activities against bFGF-induced angiogenesis both in vivo and in vitro [42,43]. Salmosin was purified from the snake venom of *Agkistrodon halys brevicaudus* in 1998 [55]. Salmosin can prevent the bFGF induced bovine capillary endothelial cell proliferation. Treatment with salmosin significantly suppressed the growth of both the metastatic and solid tumor in mouse xenografts of Lewis lung carcinoma cells, and the tumor specific antiangiogenic activity of salmosin was considered related to the blockade of $\alpha_v\beta_3$ integrin [44].

3.2. Apoptosis Induction

Apoptosis is a process of programmed cell death to delete unnecessary cells in normal tissues and to keep cellular homeostasis. Any critical defect in the apoptotic process may lead to uncontrolled growth of cells and result in cancer [56,57]. A subset of snake venom proteins have demonstrated antitumoral activities by inducing apoptosis. In 1993, Araki et al. found that some hemorrhagic snake venoms induced apoptosis of vascular endothelial cells. However, the active component was unknown [58]. After the report of Araki et al., increasing LAAOs from snake venoms have increasingly been shown to induce apoptosis. Suhr and Kim purified and characterized a LAAO from the venom of *Agkistrodon halys*, and exposure to this LAAO resulted in the apoptosis of cultured L1210 cells [45]. Later, the research of Suhr et al. suggested that the activity of apoptosis induction of LAAO was not solely due to the production of H_2O_2 in the reaction of LAAO [46].

AHP-LAAO, a novel snake-venom LAAO, was isolated from *A. halys pallas* venom in 2004. The AHP-LAAO inhibited the proliferation of HeLa cells at 0.5 μg/mL and induced DNA fragmentation and nuclear morphological changes [47]. Samel et al. purified and characterized a homodimer LAAO from the venom of the common viper *Vipera berus berus*. The DNA fragmentation gel pattern indicated that the LAAO from *V. berus berus* induced apoptosis in cultured K562 and HeLa cells, and the inhibition of apoptosis by catalase suggested the role of hydrogen peroxide in the process [48].

Not only LAAO but also some disintegrins showed the activities of apoptosis induction. Thangam et al. purified the disintegrin from the venom of the Indian cobra snake (*Naja naja*), whose anticancer activity was at IC_{50} of 2.5 ± 0.5 μg/mL, 3.5 ± 0.5 μg/mL, and 3 ± 0.5 μg/mL for the MCF-7, A549 and HepG2 cell lines respectively. The DNA fragment analysis and AO/EtBr staining assay suggested that this disintegrin induced the apoptosis of the cancer cell lines [49]. Two metalloprotease/disintegrin family proteins, VAP and VAP2, were purified from the venom of the rattlesnake *Crotalus atrox*. The apoptosis-inducing activities seemed to be specific towards endothelial cells [50,51]. Han et al. characterized stejnitin, a SVMP from the venom of *Trimeresurus stejnegeri*. Stejnitin comprises metalloproteinase and disintegrin, and the DNA fragmentation and flow cytometry analysis suggested stejnitin induces apoptosis in ECV304 cells [52].

4. Future Directions

Though there was an ample evidence about the therapeutic potentials of snake venoms in the treatment of cancer, more research is needed. Most components of snake venoms, including PLA_2s, LAAOs, metalloproteases, disintegrins and other peptides show cytotoxicity to cancer cells. However, the discrimination between normal and cancer cells is the main problem in cancer treatment [59]. From our perspective, future research could pour more attention into these actions (Figure 1).

4.1. Isolation and Characterization of New Active Molecules from Snake Venoms by Snake Venomics

One of the major barriers in exploring snake venom is the low amount isolated from the venom glands, especially of rare snakes. The snake venomics will increase the discovery of the snake venom proteins and peptides for the development of new drugs for potential use [60].

4.2. New Drug Delivery System/Coupled with Monoclonal Antibody

One of the plausible strategies to develop clinical anti-cancer versions of cytotoxins is to conjugate the drugs with monoclonal antibodies that recognize and bind to specific epitopes on malignant cancer cells. As an example, Zhao et al. used an anti-nasopharyngeal carcinoma monoclonal antibody BAC5 conjugated with the venom of the Chinese cobra, which showed strong effects against nasopharyngeal carcinoma cells in vitro [61].

The other method for targeted cancer therapy by cytotoxin from snake venom is to combine the snake venom with silica nanoparticles. Al-Sadoon et al. demonstrated that the snake venom extracted from *Walterinnesia aegyptia* (WEV) combined with silica nanoparticles (NP) can inhibit the proliferation of human breast carcinoma cell lines and strongly induced apoptosis without significant effects on normal breast epithelial MCF-10A cells [62]. Al-Sadoon et al. also evaluated the effects of WEV+NP in the therapy of multiple myeloma in the nude mouse model. WEV+NP showed greater activities than WEV alone in decreasing the surface expression of the chemokine receptors CXCR3, CXCR4 and CXCR6 and decreased migration of the cancer cells [63], suggesting this approach possesses the promising therapeutic potential for clinical application of snake venoms.

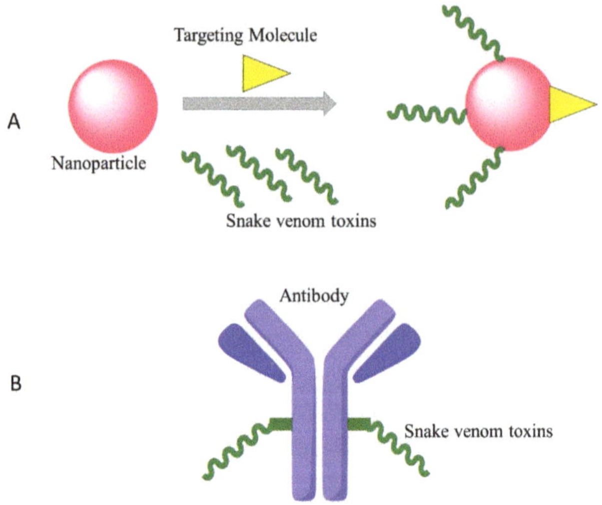

Figure 1. Effective targeting therapy by cytotoxins of snake venoms with new approaches. (**A**): Nanoparticles transport snake venoms to specific locations in the body; (**B**): Snake venoms conjugated with monoclonal antibodies for targeted therapy.

5. Conclusions

In conclusion, the application of the snake venoms in cancer therapy has evolved from the usage of the crude mixtures in the 1930s into the isolation of certain biologically active components targeting specific molecular pathways. Currently, the combination of snake venoms with other technologies such as nanoparticles is still at its early stage for cancer therapy and it can be expected that more combinational treatment will emerge. Snake venoms are no doubt valuable resources for cancer drug development.

Funding: This research is funded by the Natural Science Foundation of Fujian Province (2018J01727) and the scientific research innovation team construction program of Fujian Normal University (IRTL1702).

Conflicts of Interest: The authors declare no conflicts of interest.

References

1. Torre, L.A.; Bray, F.; Siegel, R.L.; Ferlay, J.; Lortet-Tieulent, J.; Jemal, A. Global cancer statistics, 2012. *CA Cancer J. Clin.* **2015**, *65*, 87–108. [CrossRef] [PubMed]
2. Miller, K.D.; Siegel, R.L.; Lin, C.C.; Mariotto, A.B.; Kramer, J.L.; Rowland, J.H.; Stein, K.D.; Alteri, R.; Jemal, A. Cancer treatment and survivorship statistics, 2016. *CA Cancer. J. Clin.* **2016**, *66*, 271–289. [CrossRef] [PubMed]
3. Gibbs, J.B. Mechanism-based target identification and drug discovery in cancer research. *Science* **2000**, *287*, 1969–1973. [CrossRef] [PubMed]
4. Sawyers, C. Targeted cancer therapy. *Nature* **2004**, *432*, 294–297. [CrossRef] [PubMed]
5. Aarti, C.; Khusro, A. Snake venom as anticancer agent-current perspective. *Int. J. Pure Appl. Biosci.* **2013**, *1*, 24–29.
6. Koh, D.; Armugam, A.; Jeyaseelan, K. Snake venom components and their applications in biomedicine. *Cell. Mol. Life Sci.* **2006**, *63*, 3030–3041. [CrossRef] [PubMed]
7. Gutiérrez, J.M.; Calvete, J.J.; Habib, A.G.; Harrison, R.A.; Williams, D.J.; Warrell, D.A. Snakebite envenoming. *Nat. Rev. Dis. Prim.* **2017**, *3*, 17063. [CrossRef] [PubMed]
8. Chan, Y.S.; Cheung, R.C.F.; Xia, L.; Wong, J.H.; Ng, T.B.; Chan, W.Y. Snake venom toxins: Toxicity and medicinal applications. *Appl. Microbiol. Biotechnol.* **2016**, *100*, 6165–6181. [CrossRef] [PubMed]
9. Ding, B.; Xu, Z.; Qian, C.; Jiang, F.; Ding, X.; Ruan, Y.; Ding, Z.; Fan, Y. Antiplatelet aggregation and antithrombosis efficiency of peptides in the snake venom of deinagkistrodon acutus: Isolation, identification, and evaluation. *Evid. Based Complement. Altern. Med.* **2015**, *2015*, 412841. [CrossRef] [PubMed]
10. Cushman, D.W.; Ondetti, M.A. History of the design of captopril and related inhibitors of angiotensin converting enzyme. *Hypertension* **1991**, *17*, 589–592. [CrossRef] [PubMed]
11. Sanhajariya, S.; Duffull, S.; Isbister, G. Pharmacokinetics of snake venom. *Toxins* **2018**, *10*, 73. [CrossRef] [PubMed]
12. Essex, H.E.; Priestley, J.T. Effect of rattlesnake venom on flexner-jobling's carcinoma in the white rat (*mus norvegicus albinus.*). *Proc. Soc. Exp. Biol. Med.* **1931**, *28*, 550–551. [CrossRef]
13. Kurotchkin, T.; Spies, J. Effects of cobra venom on the Fujinami rat sarcoma. *Proc. Soc. Exp. Biol. Med.* **1935**, *32*, 1408–1410. [CrossRef]
14. Des Ligneris, M.; Grasset, E. Clinical experiments on the effect of African snake venoms on human cancer cases with or without concomitant deep therapy. *Am. J. Cancer* **1936**, *26*, 512–520. [CrossRef]
15. Drueck, C.J. Cdbra venom and opiates in the pain of cancer of the rectum. *Anesth. Analg.* **1942**, *21*, 41–45.
16. Macht, D.I. Experimental and clinical study of cobra venom as an analgesic. *Proc. Natl. Acad. Sci. USA* **1936**, *22*, 61–71. [CrossRef] [PubMed]
17. Iyengar, N.; Sehra, K.; Mukerji, B.; Chopra, R. Choline esterase in cobra venom. *Curr. Sci.* **1938**, *7*, 51–53.
18. Zeller, E. Occurrence and nature of the cholinesterase of snake venoms. *Experientia* **1947**, *3*, 375–379. [CrossRef]
19. Duran-Reynals, F. The invasion of the body by animal poisons. *Science* **1936**, *83*, 286–287. [CrossRef] [PubMed]
20. Zeller, E.; Maritz, A. A new L-amino acid oxidase. *Helv. Chim. Acta* **1944**, *27*, 1888–1902. [CrossRef]
21. Singer, T.P.; Kearney, E.B. The L-amino acid oxidases of snake venom. II. Isolation and characterization of homogeneous L-amino acid oxidase. *Arch. Biochem.* **1950**, *29*, 190–209. [PubMed]
22. Kearney, E.; Singer, T.P. The L-amino acid oxidases of snake venom. III. Reversible inactivation of L-amino acid oxidases. *Arch. Biochem. Biophys.* **1951**, *33*, 377–396. [CrossRef]
23. Kearney, E.; Singer, T.P. The L-amino acid oxidases of snake venom. IV. The effect of anions on the reversible inactivation. *Arch. Biochem. Biophys.* **1951**, *33*, 397–413. [CrossRef]
24. Kearney, E.; Singer, T.P. The L-amino acid oxidases of snake venom. V. Mechanism of the reversible inactivation. *Arch. Biochem. Biophys.* **1951**, *33*, 414–426. [CrossRef]

25. Moustafa, I.M.; Foster, S.; Lyubimov, A.Y.; Vrielink, A. Crystal structure of LAAO from *Calloselasma rhodostoma* with an L-phenylalanine substrate: Insights into structure and mechanism. *J. Mol. Biol.* **2006**, *364*, 991–1002. [CrossRef] [PubMed]
26. Saito, K.; Hanahan, D.J. A study of the purification and properties of the phospholipase a of *Crotalus adamanteus* venom. *Biochemistry* **1962**, *1*, 521–532. [CrossRef] [PubMed]
27. Wu, T.-W.; Tinker, D.O. Phospholipase A_2 from *Crotalus atrox* venom. I. Purification and some properties. *Biochemistry* **1969**, *8*, 1558–1568. [CrossRef] [PubMed]
28. Takasaki, C.; Yutani, F.; Kajiyashiki, T. Amino acid sequences of eight phospholipases A_2 from the venom of australian king brown snake, *Pseudechis australis*. *Toxicon* **1990**, *28*, 329–339. [CrossRef]
29. Drickamer, K. C-type lectin-like domains. *Curr. Opin. Struct. Biol.* **1999**, *9*, 585–590. [CrossRef]
30. Ogawa, T.; Chijiwa, T.; Oda-Ueda, N.; Ohno, M. Molecular diversity and accelerated evolution of C-type lectin-like proteins from snake venom. *Toxicon* **2005**, *45*, 1–14. [CrossRef] [PubMed]
31. Stocker, K.; Barlow, G.H. The coagulant enzyme from *Bothrops atrox* venom (batroxobin). *Methods Enzymol.* **1976**, *45*, 214–223. [PubMed]
32. Kirby, E.P.; Niewiarowski, S.; Stocker, K.; Kettner, C.; Shaw, E.; Brudzynski, T.M. Thrombocytin, a serine protease from *Bothrops atrox* venom. 1. Purification and characterization of the enzyme. *Biochemistry* **1979**, *18*, 3564–3570. [CrossRef] [PubMed]
33. Markland, F.S., Jr.; Swenson, S. Snake venom metalloproteinases. *Toxicon* **2013**, *62*, 3–18. [CrossRef] [PubMed]
34. Bjarnason, J.B.; Tu, A.T. Hemorrhagic toxins from western diamondback rattlesnake (*Crotalus atrox*) venom: Isolation and characterization of five toxins and the role of zinc in hemorrhagic toxin e. *Biochemistry* **1978**, *17*, 3395–3404. [CrossRef] [PubMed]
35. Gould, R.J.; Polokoff, M.A.; Friedman, P.A.; Huang, T.-F.; Holt, J.C.; Cook, J.J.; Niewiarowski, S. Disintegrins: A family of integrin inhibitory proteins from viper venoms. *Proc. Soc. Exp. Biol. Med.* **1990**, *195*, 168–171. [CrossRef] [PubMed]
36. Ouyang, C.; Huang, T.-F. Potent platelet aggregation inhibitor from *Trimeresurus gramineus* snake venom. *Biochim. Biophys. Acta Gen. Subj.* **1983**, *757*, 332–341. [CrossRef]
37. Huang, T.F.; Holt, J.C.; Lukasiewicz, H.; Niewiarowski, S. Trigramin. A low molecular weight peptide inhibiting fibrinogen interaction with platelet receptors expressed on glycoprotein IIb-IIIa complex. *J. Biol. Chem.* **1987**, *262*, 16157–16163. [PubMed]
38. Huang, T.F.; Holt, J.C.; Kirby, E.P.; Niewiarowski, S. Trigramin: Primary structure and its inhibition of von willebrand factor binding to glycoprotein IIb/IIIa complex on human platelets. *Biochemistry* **1989**, *28*, 661–666. [CrossRef] [PubMed]
39. Higuchi, D.A.; Almeida, M.C.; Barros, C.C.; Sanchez, E.F.; Pesquero, P.R.; Lang, E.A.S.; Samaan, M.; Araujo, R.C.; Pesquero, J.B.; Pesquero, J.L. Leucurogin, a new recombinant disintegrin cloned from *Bothrops leucurus* (white-tailed-jararaca) with potent activity upon platelet aggregation and tumor growth. *Toxicon* **2011**, *58*, 123–129. [CrossRef] [PubMed]
40. Zhou, Q.; Nakada, M.T.; Arnold, C.; Shieh, K.Y.; Markland, F.S., Jr. Contortrostatin, a dimeric disintegrin from *Agkistrodon contortrix contortrix*, inhibits angiogenesis. *Angiogenesis* **1999**, *3*, 259–269. [CrossRef] [PubMed]
41. Marcinkiewicz, C.; Weinreb, P.H.; Calvete, J.J.; Kisiel, D.G.; Mousa, S.A.; Tuszynski, G.P.; Lobb, R.R. Obtustatin: A potent selective inhibitor of $\alpha1\beta1$ integrin in vitro and angiogenesis in vivo. *Cancer Res.* **2003**, *63*, 2020–2023. [PubMed]
42. Wang, J.-H.; Wu, Y.; Ren, F.; Lü, L.; Zhao, B.-C. Cloning and characterization of adinbitor, a novel disintegrin from the snake venom of *Agkistrodon halys brevicaudus stejneger*. *Acta Biochim. Biophys. Sin.* **2004**, *36*, 425–429. [CrossRef] [PubMed]
43. Wang, J.; Ren, F.; Wu, Y.; Tian, X.; Wu, Y.; Zhao, B. Cloning, expression and some biological functions of adinbitor, a disintegrin from *Agkistrodon halys brevicaudus stejneger*. *Chin. J. Biochem. Mol. Biol.* **2004**, *20*, 745–749.
44. Kang, I.-C.; Chung, K.-H.; Lee, S.-J.; Yun, Y.; Moon, H.-M.; Kim, D.-S. Purification and molecular cloning of a platelet aggregation inhibitor from the snake (*Agkistrodon halys brevicaudus*) venom. *Thromb. Res.* **1998**, *91*, 65–73. [CrossRef]
45. Suhr, S.-M.; Kim, D.-S. Identification of the snake venom substance that induces apoptosis. *Biochem. Biophys. Res. Commun.* **1996**, *224*, 134–139. [CrossRef] [PubMed]

46. Suhr, S.-M.; Kim, D.-S. Comparison of the apoptotic pathways induced by L-amino acid oxidase and hydrogen peroxide. *J. Biochem.* **1999**, *125*, 305–309. [CrossRef] [PubMed]
47. Zhang, H.; Teng, M.; Niu, L.; Wang, Y.; Liu, Q.; Huang, Q.; Hao, Q.; Dong, Y.; Liu, P. Purification, partial characterization, crystallization and structural determination of AHP-LAAO, a novel L-amino-acid oxidase with cell apoptosis-inducing activity from *Agkistrodon halys pallas* venom. *Acta Crystallogr. Sect. D Biol. Crystallogr.* **2004**, *60*, 974–977. [CrossRef] [PubMed]
48. Samel, M.; Vija, H.; Rönnholm, G.; Siigur, J.; Kalkkinen, N.; Siigur, E. Isolation and characterization of an apoptotic and platelet aggregation inhibiting L-amino acid oxidase from *Vipera berus berus* (common viper) venom. *Biochim. Biophys. Acta Proteins Proteom.* **2006**, *1764*, 707–714. [CrossRef] [PubMed]
49. Thangam, R.; Gunasekaran, P.; Kaveri, K.; Sridevi, G.; Sundarraj, S.; Paulpandi, M.; Kannan, S. A novel disintegrin protein from *Naja naja* venom induces cytotoxicity and apoptosis in human cancer cell lines in vitro. *Process Biochem.* **2012**, *47*, 1243–1249. [CrossRef]
50. Shinako, M.; Hiroshi, H.; Satohiko, A. Two vascular apoptosis-inducing proteins from snake venom are members of the metalloprotease/disintegrin family. *Eur. J. Biochem.* **1998**, *253*, 36–41.
51. Masuda, S.; Araki, S.; Yamamoto, T.; Kaji, K.; Hayashi, H. Purification of a vascular apoptosis-inducing factor from hemorrhagic snake venom. *Biochem. Biophys. Res. Commun.* **1997**, *235*, 59–63. [CrossRef] [PubMed]
52. Han, Y.-P.; Lu, X.-Y.; Wang, X.-F.; Xu, J. Isolation and characterization of a novel P-II class snake venom metalloproteinase from *Trimeresurus stejnegeri*. *Toxicon* **2007**, *49*, 889–898. [CrossRef] [PubMed]
53. Jayson, G.C.; Kerbel, R.; Ellis, L.M.; Harris, A.L. Antiangiogenic therapy in oncology: Current status and future directions. *Lancet* **2016**, *388*, 518–529. [CrossRef]
54. Moreno-Murciano, M.P.; Monleón, D.; Calvete, J.J.; Celda, B.; Marcinkiewicz, C. Amino acid sequence and homology modeling of obtustatin, a novel non-RGD-containing short disintegrin isolated from the venom of *Vipera lebetina obtusa*. *Protein Sci.* **2003**, *12*, 366–371. [CrossRef] [PubMed]
55. Kang, I.-C.; Lee, Y.-D.; Kim, D.-S. A novel disintegrin salmosin inhibits tumor angiogenesis. *Cancer Res.* **1999**, *59*, 3754–3760. [PubMed]
56. Kerr, J.F.; Winterford, C.M.; Harmon, B.V. Apoptosis. Its significance in cancer and cancer therapy. *Cancer* **1994**, *73*, 2013–2026. [CrossRef]
57. Ghobrial, I.M.; Witzig, T.E.; Adjei, A.A. Targeting apoptosis pathways in cancer therapy. *CA Cancer J. Clin.* **2005**, *55*, 178–194. [CrossRef] [PubMed]
58. Araki, S.; Ishida, T.; Yamamoto, T.; Kaji, K.; Hayashi, H. Induction of apoptosis by hemorrhagic snake venom in vascular endothelial cells. *Biochem. Biophys. Res. Commun.* **1993**, *190*, 148–153. [CrossRef] [PubMed]
59. Gasanov, S.E.; Alsarraj, M.A.; Gasanov, N.E.; Rael, E.D. Cobra venom cytotoxin free of phospholipase A_2 and its effect on model membranes and T leukemia cells. *J. Membr. Biol.* **1997**, *155*, 133–142. [CrossRef] [PubMed]
60. Calvete, J.J.; Juárez, P.; Sanz, L. Snake venomics. Strategy and applications. *J. Mass Spectrom.* **2007**, *42*, 1405–1414. [CrossRef] [PubMed]
61. Zhao, Y.S.; Yang, H.L.; Liu, C.Z. Inhibitory effects of immunotargeting of Chinese cobra cytotoxin and iodine-131 against nasopharyngeal carcinoma cells in vitro. *J. South. Med. Univ.* **2008**, *28*, 1235–1236. (In Chinese)
62. Al-Sadoon, M.K.; Abdel-Maksoud, M.A.; Rabah, D.M.; Badr, G. Induction of apoptosis and growth arrest in human breast carcinoma cells by a snake (*Walterinnesia aegyptia*) venom combined with silica nanoparticles: Crosstalk between Bcl2 and caspase 3. *Cell. Physiol. Biochem.* **2012**, *30*, 653–665. [CrossRef] [PubMed]
63. Al-Sadoon, M.K.; Rabah, D.M.; Badr, G. Enhanced anticancer efficacy of snake venom combined with silica nanoparticles in a murine model of human multiple myeloma: Molecular targets for cell cycle arrest and apoptosis induction. *Cell. Immunol.* **2013**, *284*, 129–138. [CrossRef] [PubMed]

toxins

MDPI

Review

The Development of Toad Toxins as Potential Therapeutic Agents

Ji Qi [1], Abu Hasanat Md Zulfiker [2], Chun Li [1], David Good [1,3] and Ming Q. Wei [1,*]

1 Menzies Health Institute Queensland and School of Medical Science, Griffith University, Gold Coast, QLD 4222, Australia; ji.qi2@griffithuni.edu.au (J.Q.); ivy_sampras@hotmail.com (C.L.); david.good@acu.edu.au (D.G.)

2 Department of Biomedical Sciences, Marshall University, Huntington, WV 25701, USA; zulfikermsgu@outlook.com

3 School of Physiotherapy, Australian Catholic University, Banyo, QLD 4014, Australia

* Correspondence: m.wei@griffith.edu.au; Tel.: +61-07-567-80745

Received: 3 July 2018; Accepted: 15 August 2018; Published: 20 August 2018

Abstract: Toxins from toads have long been known to contain rich chemicals with great pharmaceutical potential. Recent studies have shown more than 100 such chemical components, including peptides, steroids, indole alkaloids, bufogargarizanines, organic acids, and others, in the parotoid and skins gland secretions from different species of toads. In traditional Chinese medicine (TCM), processed toad toxins have been used for treating various diseases for hundreds of years. Modern studies, including both experimental and clinical trials, have also revealed the molecular mechanisms that support the development of these components into medicines for the treatment of inflammatory diseases and cancers. More recently, there have been studies that demonstrated the therapeutic potential of toxins from other species of toads, such as Australian cane toads. Previous reviews mostly focused on the pharmaceutical effects of the whole extracts from parotoid glands or skins of toads. However, to fully understand the molecular basis of toad toxins in their use for therapy, a comprehensive understanding of the individual compound contained in toad toxins is necessary; thus, this paper seeks to review the recent studies of some typical compounds frequently identified in toad secretions.

Keywords: toad toxins; Chansu; Huachansu; cane toad; bufadienolides; indolealkylamines; inflammation; cancer; obsessive–compulsive disorder (OCD)

Key Contribution: This paper has reviewed the recent progress in the chemical and biological studies of toad toxins with emphasis on single compounds.

---

1. Introduction

Toad toxins from parotoid or skin glands have significant therapeutic value for a plethora of diseases [1]. In China and other East and Southeast Asian countries, toad toxins traditionally refer to the processed and dried venom from parotoid glands of the toad *Bufo bufo gargarizans* [2]. In traditional Chinese medicine (TCM) it is known as Chansu, while in Japan it is known as Senso, which has been recorded since the Tang Dynasty (618–907 B.C.) [3]. These products have been used for treating pain and inflammatory diseases with more than a dozen remedies on the market [4]. Similarly, the water extracts from the skins of *B. b. gargarizans* is known as Huachansu (Cinobufacini), which was developed in China about 20 years ago, and had been successfully used to treat various types of cancers with low toxicity and few side effects [5,6]. Both molecular and clinical data have revealed the chemical constituents, as well as the mechanisms of action from their use [7,8]. Although different groups of constituents may have diverse functions, it is well known now that bufadienolides, such as bufalin and cinobufagin, are considered as the main bioactive compounds in toad toxins. These groups of compounds are C-24 steroids with similar properties as cardiac glycosides medications such as digoxin. The pharmaceutical use of bufadienolide is primarily considered as a Na^+/K^+-ATPase inhibitor for treating congestive heart failure and arterial hypertension, due to its property of high binding affinity to phosphoenzyme [9,10]. However, there have been reports indicating that an overdose of cardiac glycosides may cause prolonged blockage of Na^+/K^+-ATPase in these cells, resulting in cardiac arrest [11]. Recent studies have also revealed the therapeutic potential of bufadienolides in immunomodulation, anti-inflammation, and anti-neoplastic activity [12,13]

It has also been found that ancient people of Mesoamerica had used toads, *B. marinus* or *B. alvarius*, as a hallucinogen via licking toad skins directly, or smoking the prepared powder [14]. Studies have shown that indolealkylamines (IAAs) in toad skin, primarily bufotenine, are responsible for these hallucinogenic effects [15]. IAAs are biogenic amines and derivatives of 5-hydroxytryptamine, producing their effects through binding of serotonin receptors [3]. Due to the hallucinogenic effect, the use of bufotenine has increased in New York, USA in last century, and has drawn the attention of scientists to study the potential of bufotenine for the treatment of neuropsychiatric disorders [16].

Past significant studies have primarily focused on Chansu and Huachansu, due to their likely effect on cancer treatment. Recent studies have increasingly examined the therapeutic potential of other species of toads. An example of this is the studies of Australian cane toads (*B. marinus*), which originated from North American, but were introduced into Australia in 1935 to control cane beetles. The cane toads have become a biological and environmental disaster in northern Australia, due to their fast reproduction speed and lack of natural predators [17]. There are numerous scientists who have now started to consider the pharmaceutical potential of these cane toads [18–20]. In a recent study, the umbilical arteries isolated from human fetal placentas have been used as a model in studies comparing the cardiac glycoside-like activity of cane toad skin extracts prepared in different extraction procedures [18]. The inhibitory effect of cane toad skin aqueous extracts (CTSAE) on Na^+/K^+-ATPase was also demonstrated in other experimental models [18]. In our laboratory, we have recently shown the anti-inflammatory effect of CTSAE via inhibiting the release and expression of TNF-α and IL-6, and the suppression of nuclear factor (NF)-kappa (κ)B in vitro [19]. A further study from us has also indicated that CTSAE enhanced the expression of 5-HT2AR and D2R, with the modulation of Gq/11-PLCβ signaling pathway and c-FOS transcription factor, which may improve the therapeutic effect on certain diseases, such as obsessive–compulsive disorder (OCD) [20].

Previous reviews are mostly focused on the therapeutic effects of whole extracts, such as Chansu and Huachansu. Thus, we believe that it is important to understand the effect of individual compounds, which would enable us to explore the development of toad toxins as medicines.

2. Chemicals Components in Different Species of Toads

Several classes of compounds have been identified from the parotoid or skins glands of toads, including peptides, steroids, indole alkaloids, bufogargarizanines, organic acid, and others [2,21–25]. Bufadienolides and indolealkylamines are considered as the two main groups of compounds with therapeutic potential (Table 1) [26,27].

Table 1. The identification of significant bioactive compounds in different species of toads.

Name	Classification	Formula	Species of Toad			
			B. b. gargarizans	*B. marinus*	*B. alvarius*	*B. melanosticus*
Bufalin	Bufadienolides	$C_{24}H_{34}O_4$	+	+	+	+
Cinobufagin	Bufadienolides	$C_{26}H_{34}O_6$	+	−	−	−
Arenobufagin	Bufadienolides	$C_{24}H_{32}O_6$	+	+	+	+
Gamabufotalin	Bufadienolides	$C_{24}H_{34}O_5$	+	−	+	+
Telocinobufagin	Bufadienolides	$C_{24}H_{34}O_5$	+	+	+	+
Marinobufagin	Bufadienolides	$C_{24}H_{32}O_5$	+	+	+	+

Table 1. *Cont.*

Name	Classification	Formula	Species of Toad			
			B. b. gargarizans	*B. marinus*	*B. alvarius*	*B. melanosticus*
Bufotenine	Indolealkylamine	$C_{12}H_{16}N_2O$	+	+	+	+
Bufotenidine	Indolealkylamine	$C_{13}H_{18}N_2O$	+	−	−	+
Dehydrobufotenine	Indolealkylamine	$C_{12}H_{14}N_2O$	+	+	−	+
Bufothionine	Indolealkylamine	$C_{12}H_{15}N_2O_3S$	+	+	+	−
5-methoxytryptamine	Indolealkylamine	$C_{11}H_{14}N_2O$	−	+	+	−
Indole-3-acetic acid	Indolealkylamine	$C_{10}H_9NO_2$	−	−	+	−

+: Present; −: Not present.

A previous study has investigated the toad venoms from different *Bufo* species, in which 43 compounds were identified in the methanolic extracts of the different samples. Gamabufotalin, arenobufagin, telocinobufagin, bufotalin, cinobufotalin, bufalin, cinobufagin, and resibufogenin, were identified as major constituents of Chansu. Low levels of resibufogenin, but no cinobufagin was observed in the samples from *B. melanosticus*, *B. marinus*, and *B. viridis*. Three compounds, telocinobufagin, marinobufagin, and bufalin, were found in all samples [2]. These results have been confirmed by other studies using different analytical methods [28,29]. The indolealkylamines in Chansu have been analyzed in another study, including bufotenine, bufotenidine, bufobutanoic acid, serotonin, bufotenine *N*-oxide and *N*-methyl serotonin were also identified [30].

The chemical constituents of bufadienolides and indolealkylamines have also been identified in Huachansu. There were eight bufadienolide compounds, including bufalin, cinobufagin, recinobufagin, cinobufotalin, telocinobufagin, gamabufotalin, arenobufagin, and bufotalin, which were detected in an injected preparation of Huachansu from a previous study [31]. Additionally, the indolylalkylamines—including bufotenine, bufotenidine, cinobufotenine, and serotonin—were found in Huachansu, as mentioned in previous literature [32,33].

Several studies have evaluated the chemical compounds in other species of toads, such as cane toads collected from sites in Australia. They found that cane toad parotoid gland secretion contains bufadienolides, including high levels of marinobufagin; medium levels of bufalin, telocinobufagin, arenobufagin, and marinobufotoxin; low levels of resibufogenin, hellebrigenin, marinobufagin-3-pimeloyl-L-arginine ester, bufalin-3-pimeloyl-L-arginine ester, and bufalitoxin; and detectable levels of many other biotransformed bufadienolides [34]. A recent study performed by us, using high-performance liquid chromatography coupled with a hybrid quadrupole-time of flight mass spectrometer (HPLC/MS-Q-TOF), examined the chemicals in secretions of the cane toad parotoid glands. We found the presence of twelve key chemicals in the secretion, including several major bufadienolides, which was further confirmed by calculating the exact differences between the theoretical and measured mass of each assumed compound [35]. Following this study, a similar analytical method was used in our laboratory to assess the chemicals in extracts from cane toad skin, and up to 42 constituents, including both bufadienolides and indolylalkylamines, were identified [22].

Other numerous studies have been carried out in different species of toads, such as *B. melanosticus* from different regions [24,25,36,37]. Taken together, these data have provided us with the chemical profiles of toad toxins, which are essential for the study of their pharmaceutical effects.

3. The Bioactivity Studies of Bufadienolides

The potential pharmaceutical effects of bufadienolides contained in toad toxins have been studied in recent years. Several in vitro studies have demonstrated that they have predominant effects on the inhibition of different tumor cell growth, inducing cell cycle arrest, apoptosis, and in regulating the expression of malignant related genes/proteins in human cancer cells [38–41] (Table 2). Here, we reviewed and listed the major compounds from some of the major studies (Figure 1).

3.1. Bufalin

Bufalin is a major compound in Chansu, Huachansu, as well as the toxins of other toad species, such as *B. marinus*. Several studies have demonstrated its anti-inflammatory and anticancer effects through inhibiting NF-κB pathway, which is a crucial pathway in both anti-inflammation and cancer [42,43]. The effect of bufalin on the treatment of the asthmatic response has been studied in a murine model. The mouse asthma model was developed by ovalbumin (OVA)-induced BALB/c mice. The results demonstrated that bufalin reduces hyperresponsiveness, and inhibits the OVA-induced activation of inflammatory cells, including macrophages, eosinophils, lymphocytes, and neutrophils and cytokines, including IL-4, IL-5, and IL-13. Histological staining examined the reduction of inflammatory cell infiltration and goblet cell hyperplasia, while the blockage of NF-κB was evaluated by Western blot [44]. The anti-inflammatory and analgesic effects of bufalin have also been studied in a carrageenan-induced paw oedema model. Bufalin downregulated the expression of nitric oxide synthase (iNOS), cyclooxygenase-2 (COX-2), interleukin-1β (IL-1β), interleukin-6 (IL-6), and tumor necrosis factor-α (TNF-α), to which the inhibitory effect on the master switch of NF-κB signaling is attributed [45].

The antimetastatic effect of bufalin was studied in human hepatocellular carcinoma SK-Hep1 cells to determine if bufalin plays an important role in mortality of cancer patients, in which the expression of matrix metalloproteinases (MMPs), such as MMP-2 and -9 are inhibited, while phosphoinisitide-3-kinase (PI3K) and phosphorylation of AKT are reduced with the suppression of NF-κB [46]. Another study has shown the antimetastasis effects of bufalin on NCI-H460 lung cancer cells, with similar mechanisms [47].

The anticancer property of bufalin has been validated in a wide range of cancer cells, including leukemia, prostate, gastric, liver, and breast. Studies have indicated that bufalin inhibits tumor growth through the induction of programed cell death via multiple pathways [48].

In a study using an animal model, bufalin has been shown to suppress the growth of BEL-7402 cells, human hepatocellular carcinoma (HHC) cells, in an orthotopic transplantation tumor model in nude mice [49]. This study has also shown bufalin-induced apoptosis in a tumor model by activating Bax without causing apparent toxicity [49]. In another study, nude mice injected with HCCLM3-R cells were studied after treatment with bufalin. Significant antitumor activities, and the reduction of the metastatic growth with the inhibition of AKT/GSK3β/β-catenin/E-cadherin signaling pathways, were found [50]. A study has also investigated anticolorectal cancer (CRC) effects of bufalin in HCT116 orthotopic xenograft model in mice. The results have indicated that bufalin inhibits tumor growth by inducing cell apoptosis through the intrinsic apoptotic pathway [51]. A human lung cancer cell line, NCI-H460, injected into a BALB/C nu/nu mouse model, was also studied after bufalin treatment, confirming a reduction in tumor size without significant drug-related toxicity [52].

3.2. Cinobufagin

Cinobufagin from toad *B. b. gargarizans* is known as the second major compound in Chansu and Huachansu; however, it is not detected in some other species of toads, such as Australian cane toad.

In a previous study, we have demonstrated that cinobufagin inhibited the growth of colon, prostate, skin, and lung cancers, in vitro. Specifically, cinobufagin induced apoptosis of HCT116 and HT29 via the caspase-3-dependent and -independent pathway, respectively. The inhibition of hypoxia-inducing factor-1 alpha subunit 75 has been demonstrated both in vitro and in vivo [53]. Further study has shown that cinobufagin inhibited the expression of cortactin in HCT116 cells, and HCT116 xenograft tumors in nude mice in vivo [54].

A study has also investigated the potential anti-osteosarcoma (OS) effect and the mechanisms of action of cinobufagin. The in vitro studies have indicated that cinobufagin induced the cell cycle arrest and apoptosis in OS cells with the involvement of Notch pathway suppression. Moreover, in the in vivo xenograft OS mouse model, cinobufagin inhibited OS cell growth with a suitable drug tolerance [55].

3.3. Arenobufagin

Arenobufagin has been shown to act against the growth of esophageal squamous cell carcinoma (ESCC) by triggering the activation of p53 through its phosphorylation, and caspase through intrinsic and extrinsic pathways both in vitro and in vivo. This study has also shown the selective effect in killing tumor cells and low toxicity toward Het-1A human normal esophageal squamous cells. Transfection of cells with p53 small interfering RNA can reverse this effect. Moreover, in vivo studies have confirmed the anticancer effect of arenobufagin by inhibiting the tumor growth through activation of the p53 pathway [56].

Arenobufagin has also shown anti-neoplastic activity against HCC HepG2 cells, as well as the corresponding multidrug-resistant HepG2/ADM cells, increasing Bax/Bcl-2 expression ratio, and inhibiting the phosphatidylinositol 3-kinase (PI3K)/Akt/mammalian target of rapamycin (mTOR) pathway. Arenobufagin inhibited the growth of HepG2/ADM xenograft tumors, which were associated with poly (ADP-ribose) polymerase cleavage, light chain 3-II activation, and mTOR inhibition [57].

Another study has demonstrated an antimetastasis and epithelial–mesenchymal transition (EMT) inhibitory effect of arenobufagin in PC3 cells by suppressing β-catenin. These results are also verified in a xenograft tumor mouse model [58].

Arenobufagin has also been shown anti-angiogenic activity through inhibiting vascular endothelial growth factor (VEGF)-induced viability, migration, invasion, and tube formation in human umbilical vein endothelial cells (HUVECs). Additionally, this effect has been confirmed via an in vivo model. Computer simulations suggested that arenobufagin interacted with the ATP-binding sites of VEGFR-2 by docking. Furthermore, arenobufagin inhibited VEGF-induced VEGFR-2 autophosphorylation, and suppressed the activity of VEGFR-2-mediated signaling cascades [59].

3.4. Gamabufotalin

There has been a study showing that gamabufotalin plays a role in angiogenesis inhibition through the blockage of VEGF-induced HUVEC proliferation, migration, invasion, and tubulogenesis. This study also demonstrated the effect of gamabufotalin in decreasing vessel density in human lung tumor xenograft implanted in nude mice, while inhibiting vascularization in matrigel plugs impregnated in C57/BL6 mice. Further studies, including computer simulations and Western blot analysis, have revealed that gamabufotalin interacted with the ATP-binding sites of VEGFR-2 using molecular docking. Furthermore, Western blot analysis indicated that the inhibitory effect of gamabufotalin for angiogenesis was due to the suppression of the VEGFR-2 signaling pathway [60].

The therapeutic potential of gamabufotalin in human multiple myeloma (MM) cells has also been studied. Results have shown that gamabufotalin inhibited cell growth and induced apoptosis via the activation of the ubiquitination process of c-Myc. The anticancer effect and inhibition of MM-induced osteolysis of gamabufotalin were further validated in a xenograft mouse model and SCID-hu model, separately [61].

Gamabufotalin has also shown effect in blocking the NF-kB pathway. A study has shown gamabufotalin strongly suppressed COX-2 expression by inhibiting the phosphorylation of IKKβ via targeting the ATP-binding site, which in turn, prevents NF-κB binding and p300 recruitment to COX-2 promoter in a range of human NSCLC, H1299, A549, H322, and H460 cell lines. In in vivo studies, gamabufotalin suppressed the tumor weight and size with the decreasing protein levels of COX-2 and phosphorylated p65 NF-κB in the tumor tissues of xenograft mice [62].

3.5. Other Key Bufadienolides

The immunoregulatory effect of telocinobufagin, another major compound in Chansu, was studied in vitro. The activation of several cytokines and immunocytes was observed [63]. Telocinobufagin and marinobufagin isolated from skin secretions of the Brazilian toad *B. rubescens* have been shown to exhibit antimicrobial activity inhibitory action over *Staphylococcus aureus* and *Escherichia coli* [23].

Differently from telocinobufagin, marinobufagin is a minor constituent in Chansu and Huachansu. However, it has been identified as the main component in the toxins of cane toads. Currently, there are still very few functional studies of marinobufagin.

Table 2. Molecular targets of bufadienolides found in a wide range of preclinal models.

Compound	Experimental Models	Molecular Targets	References
Bufalin	In vitro/In vivo	Macrophages, eosinophils, lymphocytes, and neutrophils and cytokines including IL-4, IL-5, and IL-13, NF-κB	[44]
	In vivo	iNOS, COX-2, IL-1β, IL-6, TNF-α, NF-κB	[45]
	In vitro	MMP-2, MMP-9, PI3K, AKT, NF-κB	[46,47]
	In vivo	Bax	[49]
	In vivo	AKT/GSK3β/β-catenin/E-cadherin	[50]
	In vivo	PTEN/phosphate-PTEN, AKT/phosphate-AKT, Bad, Bcl-xl, Bax, or Caspase-3	[51]
Cinobufagin	In vitro/In vivo	Caspase-3, hypoxia-inducing factor-1 alpha	[53]
	In vivo	Cortactin	[54]
	In vitro	Notch pathway	[55]
Arenobufagin	In vitro/In vivo	p53 pathway	[56]
	In vitro/In vivo	Bax/Bcl-2, PI3K/Akt/ mTOR pathway. ADP-ribose polymerase, light chain 3-II	[57]
	In vivo	β-catenin	[58]
	In vitro/In vivo	VEGFR-2 pathway	[59]
Gamabufotalin	In vivo	VEGFR-2 pathway	[60]
	In vitro/In vivo	c-Myc	[61]
	In vitro/In vivo	IKKβ, NF-κB, COX-2, p65	[62]
Telocinobufagin	In vitro	CD4, CD8, IL-2, IL-12, IFN-γ, TNF-α, IL-4	[63]

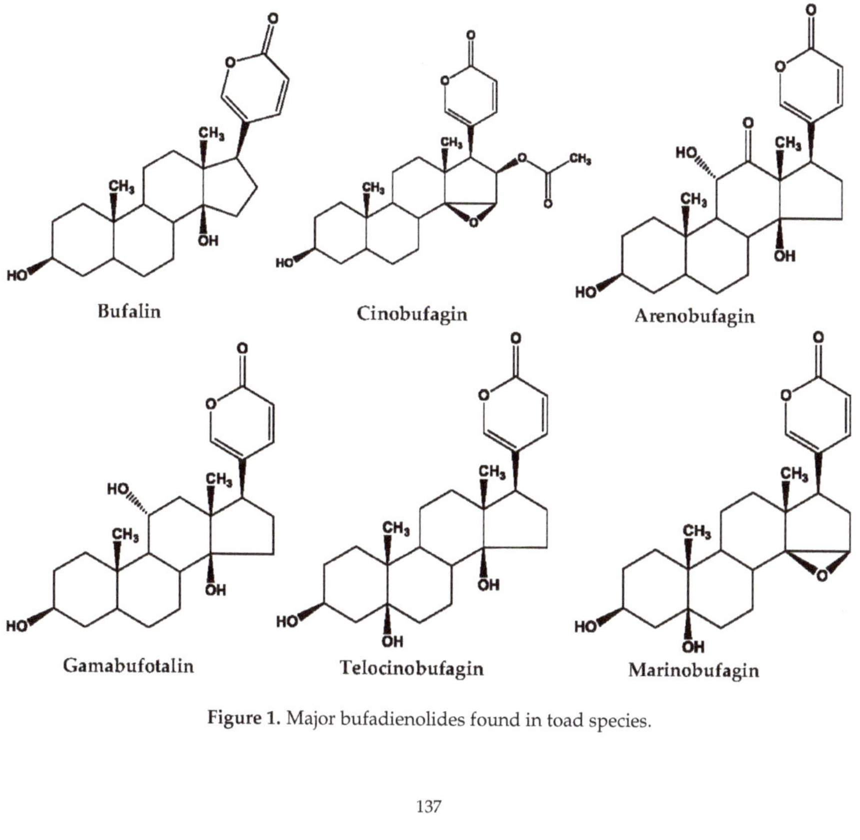

Figure 1. Major bufadienolides found in toad species.

4. Indolealkylamines

Indolealkylamines (IAAs) are known as derivatives of 5-hydroxytryptamine (5-HT), which primarily affect the central nervous system (CNS). To date, at least fourteen IAAs, including 5-methoxy-*N*,*N*-dimethyltryptamine (5-MeO-DMT) have been characterized; among these, bufotenine, bufotenidine, and cinobufotenine have been identified in the skins of toad species. Some IAAs are clinically used as antimigraine therapies, whereas the misuse of these chemicals may cause drug abuse. Recently, IAAs in toad toxins are considered as potential therapeutic compounds in developing new agents for treating several neurologic disorders, such as schizophrenia, depression, anxiety, obsessive–compulsive disorders, and chronic pain conditions, due to their potential 5-HT2A receptor selectivity in the CNS [49].

Some of the major IAAs found in toad toxins having pharmaceutical values, which are summarized below (Figure 2).

4.1. *Bufotenine*

Bufotenine was first identified from Senso in Japan, and Chansu in China [64]. Bufotenine binds to the 5-HT2A receptor in vitro, with a similar affinity to that of 4-bromo-2,5-dimethoxy-phenylisopropylamine (DOB) [65,66]. For many years, the activity of bufotenine remained a controversy, as to whether it was a hallucinogen or psychotomimetic. Though there are few reports about the significant pharmaceutical value of bufotenine, it was found to have potent psychotropic properties, and other psychotic symptoms, due to the similar physiological and structural features to lysergic acid diethylamide (LSD) in the 5HT2 receptor [67,68]. Bufotenine was also reported to be used as a biomarker in the diagnosis of various psychiatric disorders, such as schizophrenia and autism [69]. Recently, bufotenine isolated from the parotoid gland secretions of *Bufo bufo* was also reported to have cholinergic properties in α7 nicotinic acetylcholine receptors [70].

4.2. *Bufotenidine*

Like bufotenine, bufotenidine was also isolated from Senso in Japan and Chansu in China [65]. Bufotenidine was obtained from the skin of *Leptodactylus vilarsi melin*, which was found to have a hypertensive effect [71]. Bufotenidine showed marked neuromuscular blocking activity by producing the characteristic head drop in rabbits in doses of 5.2 ± 0.9 mg/kg iv. It also showed potent ganglionic stimulation and significant cholinergic-like action [72]. Recently, bufotenidine was isolated from the parotid gland secretions of *Bufo bufo* and reported to have cholinergic properties in α7 nicotinic acetylcholine receptors [71].

4.3. *Dehydrobufotenine*

Dehydrobufotenine was isolated from parotoid glands and skins of many species of toads, such as *B. marinus*, *B. arenarum*, and *B. b. gargarizans*, as the principal indolealkylamine [25,73]. Dehydrobufotenine was reported to show potent in vitro cytotoxicity against human tumor cell lines that were thought to act as DNA topoisomerase II inhibitors [74,75]. Additionally, dehydrobufotenine was used as a dry powder inhaler (DPI) in preparation of an antitumor drug for treating lung tumor [76].

4.3.1. Bufothionine

Bufothionine was found in the skin of various toad species [25]. Recently, it was identified in Cinobufacini injection [77] and the skin of *B. b. gargarizans* [73,78]. Bufothionine was reported to inhibit the proliferation of human hepatocellular carcinoma cell lines [77]. Bufothionine was also reported to have cytotoxic activity against the murine leukemia cell line P388, and human hepatocellular carcinoma cell lines SMMC-7721 and BEL-7402 [79]. A new formulation of this alkaloid after isolation from toad skin has already been used for cancer therapy [80]. Bufothionine powder for inhalation was found in medicine for the treatment of pulmonary neoplasm [81].

4.3.2. Other Indolealkylamines

Though there are at least 14 IAA identified in various toad species, very few of them were found to have significant pharmaceutical value, except the above four described here. Among others, 5-methoxytryptamine was reported to have antioxidant and radioprotective effects in various biological systems [82]. Indole-3-acetic acid is another IAA which was found in the skin of *Bufo alvarius* [25]. Several pharmacological activities of indole-3-acetic acid was reported, including anti-inflammatory, antipyretic [83], antifungal [84], hypoglycemic [85], and anticancer [86].

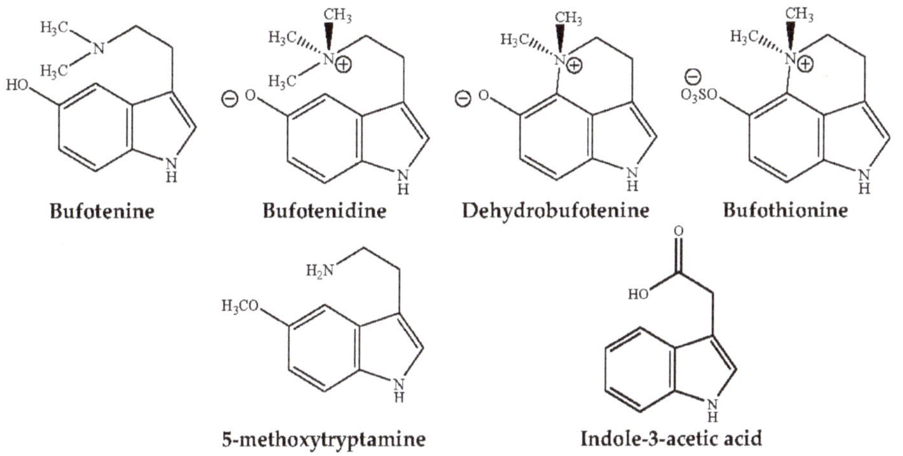

Figure 2. Major indolealkylamines found in toad species.

5. Conclusions

The studies of toad toxins in the past years have demonstrated new perspectives for their pharmaceutical use, not only for treating cardiac failure, but also for other therapeutic purposes, for example, as anti-inflammatory, immunoregulatory, and anticancer compounds. The understanding of the chemical basis of toad toxins has provided the basis to develop new therapeutic agents from different species of toads. Primarily, due to the environmental pollution in China, there is currently a shortage of toad resources in the pharmaceutical industry, while toads in some countries are becoming natural disasters that need to be managed, such as the cane toad in Australia. Therefore, the development of toad medicines from different resources is acutely needed.

Currently, several fundamental questions remain to be resolved to fully reveal the potential use of toad secretions. Although a number of in vitro studies have been done by researchers on Chansu, Huachansu, and single compounds, regarding their effects and mechanisms, the in vivo and clinical studies are still very limited. Thus, it is important to perform more animal studies to decode the potential of toad toxins in treating various diseases, such as cancer. Additionally, digitalis toxicity has always been a main concern for scientists in using toad medicines in patients [87]. Several reports have shown cardiotoxicity caused by bufadienolides [88]. Therefore, there is an urgent need to study the toxicities and the maximum tolerated dose of toad toxins. Beyond that, how to reduce the side effects is the next step to be considered. The knowledge of TCM formulas may provide us with some good points for resolving this problem. Chansu is generally used as a recipe with other herbs, to prepare formulations such as She Xiang Bao Xin Wan for the treatment of cardiovascular diseases, or Mei Hua Dian She Wan and Liu Shen Wan for the treatment of inflammatory diseases. Another herb, by the name of Bezoar Bovis, was frequently used in preparations of Chansu (Table 3). There has been a study showing that the use of Bezoar Bovis protects Chansu-induced acute toxicity in mice. Further study has shown that the taurine derived from Bezoar Bovis also prevented Chansu- or bufalin-caused cardiotoxicity, and reduced the mortality in animal models [75,76]. Other studies have also indicated that the use of nanoparticles may help improve antitumor activity while reducing the side effects of toad medicines [89].

Table 3. Some classic recipes contain Chansu in traditional Chinese medicine (TCM).

Recipe Name	Treatment Purpose	Main Ingredients
Liu Shen Wan	Inflammatory and infectious diseases, etc.	Chansu, Pearl Powder, Bezoar Bovis, Musk, Realgar, and Bornel
She Xiang Bao Xin Wan	Congestive heart failure	Chansu, Ginseng, Bezoar Bovis, Musk, Cinnamon, Liquidambar, and Borneol
Mei Hua Dian She Wan	Relieves swelling and pain	Chansu, Borneol, Cinnabar, Myrrh, Bezoar Bovis, Borax, Frankincense, Musk, Draco Seed, Realgar, Bear Gall, Blood Dracon, Pearl Powder, and Cinnabar

Moreover, the quality control of using natural products as therapeutic agents has always been a high concern for researchers. Some chemistry studies have indicated that the same species of toads obtained from different geographical regions, and under different conditions of weather, time, and other environmental factors, will result in an impact on their chemical compositions. Therefore, chemical analysis to quantify the various compounds present, and quality control to ensure the consistency of preparations in the study, are crucial issues that need to be considered [2,32].

Taken together, toad toxins from different species have a promising role in treating various diseases. However, the molecular mechanisms, drug safety, and the demand for quality control need to be resolved in future studies. No doubt though, the application of toad toxins as novel therapeutic agents will contribute to the world in many different aspects in terms of scientific research, pharmaceutical industry, environmental protection, and economic growth.

Author Contributions: J.Q. conceived and drafted the manuscript. A.H.M.Z. contributed the part of writing of indolealkylamines in the paper and drew the figures of the chemical compounds. C.L. contributed the part of writing of bufadienolides in the paper. D.G. and M.Q.W. proofread the paper and gave the scientific instructions.

Funding: This research received no external funding.

Acknowledgments: J.Q. was a receipt of Griffith University Postgraduate Research Scholarship; The authors appreciate the support from Menzies Health Institute Queensland and School of Medical Science Griffith University, Australia.

Conflicts of Interest: The authors declare that they have no conflict of interests.

References

1. Yang, Q.; Zhou, X.; Zhang, M.; Bi, L.; Miao, S.; Cao, W.; Xie, Y.; Sun, J.; Tang, H.; Li, Y. Angel of human health: Current research updates in toad medicine. *Am. J. Transl. Res.* **2015**, *7*, 1. [PubMed]
2. Gao, H.; Zehl, M.; Leitner, A.; Wu, X.; Wang, Z.; Kopp, B. Comparison of toad venoms from different Bufo species by HPLC LC-DAD-MS/MS. *J. Ethnopharmacol.* **2010**, *131*, 368–376. [CrossRef] [PubMed]
3. Meng, Q.; Yau, L.-F.; Lu, J.-G.; Wu, Z.-Z.; Zhang, B.-X.; Wang, J.-R.; Jiang, Z.-H. Chemical profiling and cytotoxicity assay of bufadienolides in toad venom and toad skin. *J. Ethnopharmacol.* **2016**, *187*, 74–82. [CrossRef] [PubMed]
4. Qi, J.; Tan, C.; Hashimi, S.M.; Zulfiker, A.H.M.; Good, D.; Wei, M.Q. Toad glandular secretions and skin extractions as anti-inflammatory and anticancer agents. *Evid.-Based Complement. Altern. Med.* **2014**, *2014*. [CrossRef] [PubMed]
5. Meng, Z.; Yang, P.; Shen, Y.; Bei, W.; Zhang, Y.; Ge, Y.; Newman, R.A.; Cohen, L.; Liu, L.; Thornton, B. Pilot study of huachansu in patients with hepatocellular carcinoma, nonsmall-cell lung cancer, or pancreatic cancer. *Cancer* **2009**, *115*, 5309–5318. [CrossRef] [PubMed]

6. Meng, Z.; Shen, Y.; Yang, P.; Robert, N.; Bei, W.; Zhang, Y.; Yongqian, G.; Lorenzo, C.; Razelle, K.; Liu, L. Phase I study of huachansu in hepatocellular carcinoma, non-small cell lung cancer, and pancreatic cancer: A preliminary report. *Chin. Oncol.* **2001**, *5*, 006.
7. Wang, L.; Raju, U.; Milas, L.; Molkentine, D.; Zhang, Z.; Yang, P.; Cohen, L.; Meng, Z.; Liao, Z. Huachansu, containing cardiac glycosides, enhances radiosensitivity of human lung cancer cells. *Anticancer Res.* **2011**, *31*, 2141–2148. [PubMed]
8. Efuet, E.T.; Ding, X.-P.; Cartwright, C.; Pan, Y.; Cohen, L.; Yang, P. Huachansu mediates cell death in non-Hodgkin's lymphoma by induction of caspase-3 and inhibition of MAP kinase. *Int. J. Oncol.* **2015**, *47*, 592–600. [CrossRef] [PubMed]
9. Laursen, M.; Gregersen, J.L.; Yatime, L.; Nissen, P.; Fedosova, N. Structures and characterization of digoxin-and bufalin-bound Na^+, K^+-ATPase compared with the ouabain-bound complex. *Proc. Natl. Acad. Sci. USA* **2015**, *112*, 1755–1760. [CrossRef] [PubMed]
10. Prassas, I.; Diamandis, E.P. Novel therapeutic applications of cardiac glycosides. *Nat. Rev. Drug Discov.* **2008**, *7*, 926. [CrossRef] [PubMed]
11. Kelly, R.A.; Smith, T.W. Recognition and management of digitalis toxicity. *Am. J. Cardiol.* **1992**, *69*, 108–119. [CrossRef]
12. Kamboj, A.; Rathour, A.; Kaur, M. Bufadienolides and their medicinal utility: A review. *Int. J. Pharm. Pharm. Sci.* **2013**, *5*, 20–27.
13. Baldo, E.C.F.; Anjolette, F.A.P.; Arantes, E.C.; Baldo, M.A. Toad Poison and Drug Discovery. *Toxicon* **2015**, 1–22. [CrossRef]
14. Davis, W.; Weil, A.T. Identity of a New World psychoactive toad. *Anc. Mesoam.* **1992**, *3*, 51–59. [CrossRef]
15. Weil, A.T.; Davis, W. Bufo alvarius: A potent hallucinogen of animal origin. *J. Ethnopharmacol.* **1994**, *41*, 1–8. [CrossRef]
16. Chamakura, R.P. Bufotenine-A Hallucinogen in Ancient Snuff Powders of South America and a Drug of Abuse on the Streets of New York City. *Forensic Sci. Rev.* **1994**, *6*, 1–18. [PubMed]
17. Phillips, B.L.; Brown, G.P.; Greenlees, M.; Webb, J.K.; Shine, R. Rapid expansion of the cane toad (*Bufo marinus*) invasion front in tropical Australia. *Aust. Ecol.* **2007**, *32*, 169–176. [CrossRef]
18. Leitch, I.; Lim, T.; Boura, A. Novel drugs from toad skins. *RIDC Publ. Aust.* **2000**, *17*, 1–65.
19. Zulfiker, A.H.M.; Hashimi, S.M.; Qi, J.; Grice, I.D.; Wei, M.Q. Aqueous and Ethanol Extracts of Australian Cane Toad Skins Suppress Pro-Inflammatory Cytokine Secretion in U937 Cells via NF-κB Signaling Pathway. *J. Cell. Biochem.* **2016**, *117*, 2769–2780. [CrossRef] [PubMed]
20. Zulfiker, A.H.M.; Hashimi, S.M.; Good, D.A.; Grice, I.D.; Wei, M.Q. Cane Toad Skin Extract—Induced Upregulation and Increased Interaction of Serotonin 2A and D2 Receptors via Gq/11 Signaling Pathway in CLU213 Cells. *J. Cell. Biochem.* **2017**, *118*, 979–993. [CrossRef] [PubMed]
21. Wang, D.L.; Qi, F.H.; Tang, W.; Wang, F.S. Chemical constituents and bioactivities of the skin of Bufo bufo gargarizans Cantor. *Chem. Biodivers.* **2011**, *8*, 559–567. [CrossRef] [PubMed]
22. Zulfiker, A.H.M.; Sohrabi, M.; Qi, J.; Matthews, B.; Wei, M.Q.; Grice, I.D. Multi-constituent identification in Australian cane toad skin extracts using high-performance liquid chromatography high-resolution tandem mass spectrometry. *J. Pharm. Biomed. Anal.* **2016**, *129*, 260–272. [CrossRef] [PubMed]
23. Cunha Filho, G.A.; Schwartz, C.A.; Resck, I.S.; Murta, M.M.; Lemos, S.S.; Castro, M.S.; Kyaw, C.; Pires, O.R., Jr.; Leite, J.R.S.; Bloch, C., Jr. Antimicrobial activity of the bufadienolides marinobufagin and telocinobufagin isolated as major components from skin secretion of the toad Bufo rubescens. *Toxicon* **2005**, *45*, 777–782. [CrossRef] [PubMed]
24. Cunha-Filho, G.A.; Resck, I.S.; Cavalcanti, B.C.; Pessoa, C.Ó.; Moraes, M.O.; Ferreira, J.R.; Rodrigues, F.A.; dos Santos, M.L. Cytotoxic profile of natural and some modified bufadienolides from toad Rhinella schneideri parotoid gland secretion. *Toxicon* **2010**, *56*, 339–348. [CrossRef] [PubMed]
25. Zulfiker, M.; Hasanat, A.; Mariottini, G.L.; Qi, J.; Grice, I.D.; Wei, M.Q. Indolealkylamines from toad vertebrates and sea invertebrates-their identification and potential activities on the central nervous system. *Cent. Nerv. Syst. Agents Med. Chem.* **2016**, *16*, 197–207. [CrossRef] [PubMed]
26. Garg, A.D.; Hippargi, R.V.; Gandhare, A.N. Toad skin-secretions: Potent source of pharmacologically and therapeutically significant compounds. *Int. J. Pharmacol.* **2008**, *5*, 17.
27. Liu, M.; Feng, L.-X.; Hu, L.-H.; Liu, X.; Guo, D.-A. Advancement in research of anti-cancer effects of toad venom (ChanSu) and perspectives. *World J. Tradit. Chin. Med.* **2015**, *1*, 12–23. [CrossRef]

28. Li, J.; Zhang, Y.; Lin, Y.; Wang, X.; Fang, L.; Geng, Y.; Zhang, Q. Preparative separation and purification of bufadienolides from ChanSu by high-speed counter-current chromatography combined with preparative HPLC. *Química Nova* **2013**, *36*, 686–690. [CrossRef]
29. Ye, M.; Guo, H.; Guo, H.; Han, J.; Guo, D. Simultaneous determination of cytotoxic bufadienolides in the Chinese medicine ChanSu by high-performance liquid chromatography coupled with photodiode array and mass spectrometry detections. *J. Chromatogr. B* **2006**, *838*, 86–95. [CrossRef] [PubMed]
30. Zhang, P.; Cui, Z.; Liu, Y.-S.; Sheng, Y. Isolation and identification of the indolealkylamines from the traditional Chinese medicine Toad Venom. *J. Shenyang Pharm. Univ.* **2006**, *4*, 005.
31. Wu, X.; Zhao, H.; Wang, H.; Gao, B.; Yang, J.; Si, N.; Bian, B. Simultaneous determination of eight bufadienolides in cinobufacini injection by HPLC coupled with triple quadrupole mass spectrometry. *J. Sep. Sci.* **2012**, *35*, 1893–1898. [CrossRef] [PubMed]
32. Liu, C.; Cao, W.; Chen, Y.; Qu, D.; Zhou, J. Comparison of toad skins Bufo bufo gargarizans Cantor from different regions for their active constituents content and cytotoxic activity on lung carcinoma cell lines. *Pharmacogn. Mag.* **2014**, *10*, 207. [CrossRef] [PubMed]
33. Meng, Z.; Garrett, C.; Shen, Y.; Liu, L.; Yang, P.; Huo, Y.; Zhao, Q.; Spelman, A.; Ng, C.; Chang, D. Prospective randomised evaluation of traditional Chinese medicine combined with chemotherapy: A randomised phase II study of wild toad extract plus gemcitabine in patients with advanced pancreatic adenocarcinomas. *Br. J. Cancer* **2012**, *107*, 411. [CrossRef] [PubMed]
34. Shine, R. The ecological impact of invasive cane toads (*Bufo marinus*) in Australia. *Q. Rev. Biol.* **2010**, *85*, 253–291. [CrossRef] [PubMed]
35. Jing, J.; Ren, W.C.; Li, C.; Bose, U.; Parekh, H.S.; Wei, M.Q. Rapid identification of primary constituents in parotoid gland secretions of the Australian cane toad using HPLC/MS-Q-TOF. *Biomed. Chromatogr.* **2013**, *27*, 685–687. [CrossRef] [PubMed]
36. Maciel, N.M.; Schwartz, C.A.; Junior, O.R.P.; Sebben, A.; Castro, M.S.; Sousa, M.V.; Fontes, W.; Schwartz, E.N.F. Composition of indolealkylamines of Bufo rubescens cutaneous secretions compared to six other Brazilian bufonids with phylogenetic implications. *Comp. Biochem. Physiol. Part B Biochem. Mol. Biol.* **2003**, *134*, 641–649. [CrossRef]
37. Barry, T.L.; Petzinger, G.; Zito, S.W. GC/MS comparison of the West Indian aphrodisiac "Love Stone" to the Chinese medication "chan su": Bufotenine and related bufadienolides. *J. Forensic. Sci.* **1996**, *41*, 1068–1073. [CrossRef] [PubMed]
38. Moreno, Y.; Banuls, L.; Urban, E.; Gelbcke, M.; Dufrasne, F.o.; Kopp, B.; Kiss, R.; Zehl, M. Structure—Activity relationship analysis of bufadienolide-induced in vitro growth inhibitory effects on mouse and human cancer cells. *J. Natl. Prod.* **2013**, *76*, 1078–1084. [CrossRef] [PubMed]
39. Dong, Y.; Yin, S.; Li, J.; Jiang, C.; Ye, M.; Hu, H. Bufadienolide compounds sensitize human breast cancer cells to TRAIL-induced apoptosis via inhibition of STAT3/Mcl-1 pathway. *Apoptosis* **2011**, *16*, 394–403. [CrossRef] [PubMed]
40. Yu, C.H.; Kan, S.F.; Pu, H.F.; Chien, E.J.; Wang, P.S. Apoptotic signaling in bufalin-and cinobufagin-treated androgen-dependent and-independent human prostate cancer cells. *Cancer Sci.* **2008**, *99*, 2467–2476. [CrossRef] [PubMed]
41. Yuan, B.; He, J.; Kisoh, K.; Hayashi, H.; Tanaka, S.; Si, N.; Zhao, H.-Y.; Hirano, T.; Bian, B.; Takagi, N. Effects of active bufadienolide compounds on human cancer cells and $CD4^{+}$ $CD25^{+}$ $Foxp3^{+}$ regulatory T cells in mitogen-activated human peripheral blood mononuclear cells. *Oncol. Rep.* **2016**, *36*, 1377–1384. [CrossRef] [PubMed]
42. Yin, P.-H.; Liu, X.; Qiu, Y.-Y.; Cai, J.-F.; Qin, J.-M.; Zhu, H.-R.; Li, Q. Anti-tumor activity and apoptosis-regulation mechanisms of bufalin in various cancers: New hope for cancer patients. *Asian Pac. J. Cancer Prev.* **2012**, *13*, 5339–5343. [CrossRef] [PubMed]
43. Karin, M. NF-κB as a critical link between inflammation and cancer. *Cold Spring Harbor Perspect. Biol.* **2009**. [CrossRef]
44. Zhakeer, Z.; Hadeer, M.; Tuerxun, Z.; Tuerxun, K. Bufalin inhibits the inflammatory effects in asthmatic mice through the suppression of nuclear factor-kappa B activity. *Pharmacology* **2017**, *99*, 179–187. [CrossRef] [PubMed]
45. Wen, L.; Huang, Y.; Xie, X.; Huang, W.; Yin, J.; Lin, W.; Jia, Q.; Zeng, W. Anti-Inflammatory and Antinociceptive Activities of Bufalin in Rodents. *Mediat. Inflamm.* **2014**, *2014*, 171839. [CrossRef] [PubMed]

46. Chen, Y.Y.; Lu, H.F.; Hsu, S.C.; Kuo, C.L.; Chang, S.J.; Lin, J.J.; Wu, P.P.; Liu, J.Y.; Lee, C.H.; Chung, J.G. Bufalin inhibits migration and invasion in human hepatocellular carcinoma SK-Hep1 cells through the inhibitions of NF-kB and matrix metalloproteinase-2/-9-signaling pathways. *Environ. Toxicol.* **2015**, *30*, 74–82. [CrossRef] [PubMed]
47. Wu, S.-H.; Hsiao, Y.-T.; Kuo, C.-L.; Yu, F.-S.; Hsu, S.-C.; Wu, P.-P.; Chen, J.-C.; Hsia, T.-C.; Liu, H.-C.; Hsu, W.-H. Bufalin inhibits NCI-H460 human lung cancer cell metastasis in vitro by inhibiting MAPKs, MMPs, and NF-κB pathways. *Am J. Chin. Med.* **2015**, *43*, 1247–1264. [CrossRef] [PubMed]
48. Takai, N.; Kira, N.; Ishii, T.; Yoshida, T.; Nishida, M.; Nishida, Y.; Nasu, K.; Narahara, H. Bufalin, a traditional oriental medicine, induces apoptosis in human cancer cells. *Asian Pac. J. Cancer Prev.* **2012**, *13*, 399–402. [CrossRef] [PubMed]
49. Han, K.Q.; Huang, G.; Gu, W.; Su, Y.H.; Huang, X.Q.; Ling, C.Q. Anti-tumor activities and apoptosis-regulated mechanisms of bufalin on the orthotopic transplantation tumor model of human hepatocellular carcinoma in nude mice. *World J. Gastroenterol.* **2007**, *13*, 3374–3379. [CrossRef] [PubMed]
50. Zhang, Z.J.; Yang, Y.K.; Wu, W.Z. Bufalin attenuates the stage and metastatic potential of hepatocellular carcinoma in nude mice. *J. Transl. Med.* **2014**, *12*, 57. [CrossRef] [PubMed]
51. Wang, J.; Chen, C.; Wang, S.; Zhang, Y.; Yin, P.; Gao, Z.; Xu, J.; Feng, D.; Zuo, Q.; Zhao, R.; et al. Bufalin Inhibits HCT116 Colon Cancer Cells and Its Orthotopic Xenograft Tumor in Mice Model through Genes Related to Apoptotic and PTEN/AKT Pathways. *Gastroenterol. Res. Pract.* **2015**, *2015*. [CrossRef] [PubMed]
52. Wu, S.H.; Bau, D.T.; Hsiao, Y.T.; Lu, K.W.; Hsia, T.C.; Lien, J.C.; Ko, Y.C.; Hsu, W.H.; Yang, S.T.; Huang, Y.P. Bufalin induces apoptosis in vitro and has Antitumor activity against human lung cancer xenografts in vivo. *Environ. Toxicol.* **2017**, *32*, 1305–1317. [CrossRef] [PubMed]
53. Li, C.; Hashimi, S.M.; Cao, S.; Mellick, A.S.; Duan, W.; Good, D.; Wei, M.Q. The mechanisms of chansu in inducing efficient apoptosis in colon cancer cells. *Evid.-Based Complement. Altern. Med.* **2013**, *2013*. [CrossRef] [PubMed]
54. Li, C.; Hashimi, S.M.; Cao, S.; Qi, J.; Good, D.; Duan, W.; Wei, M.Q. Chansu inhibits the expression of cortactin in colon cancer cell lines in vitro and in vivo. *BMC Complement. Altern. Med.* **2015**, *15*, 207. [CrossRef] [PubMed]
55. Cao, Y.; Yu, L.; Dai, G.; Zhang, S.; Zhang, Z.; Gao, T.; Guo, W. Cinobufagin induces apoptosis of osteosarcoma cells through inactivation of Notch signaling. *Eur. J. Pharmacol.* **2017**, *794*, 77–84. [CrossRef] [PubMed]
56. Lv, J.; Lin, S.; Peng, P.; Cai, C.; Deng, J.; Wang, M.; Li, X.; Lin, R.; Lin, Y.; Fang, A.; et al. Arenobufagin activates p53 to trigger esophageal squamous cell carcinoma cell apoptosis in vitro and in vivo. *OncoTargets Ther.* **2017**, *10*, 1261–1267. [CrossRef] [PubMed]
57. Zhang, D.-M.; Liu, J.-S.; Deng, L.-J.; Chen, M.-F.; Yiu, A.; Cao, H.-H.; Tian, H.-Y.; Fung, K.-P.; Kurihara, H.; Pan, J.-X. Arenobufagin, a natural bufadienolide from toad venom, induces apoptosis and autophagy in human hepatocellular carcinoma cells through inhibition of PI3K/Akt/mTOR pathway. *Carcinogenesis* **2013**, *34*, 1331–1342. [CrossRef] [PubMed]
58. Chen, L.; Mai, W.; Chen, M.; Hu, J.; Zhuo, Z.; Lei, X.; Deng, L.; Liu, J.; Yao, N.; Huang, M. Arenobufagin inhibits prostate cancer epithelial-mesenchymal transition and metastasis by down-regulating β-catenin. *Pharmacol. Res.* **2017**, *123*, 130–142. [CrossRef] [PubMed]
59. Li, M.; Wu, S.; Liu, Z.; Zhang, W.; Xu, J.; Wang, Y.; Liu, J.; Zhang, D.; Tian, H.; Li, Y. Arenobufagin, a bufadienolide compound from toad venom, inhibits VEGF-mediated angiogenesis through suppression of VEGFR-2 signaling pathway. *Biochem. Pharmacol.* **2012**, *83*, 1251–1260. [CrossRef] [PubMed]
60. Tang, N.; Shi, L.; Yu, Z.; Dong, P.; Wang, C.; Huo, X.; Zhang, B.; Huang, S.; Deng, S.; Liu, K. Gamabufotalin, a major derivative of bufadienolide, inhibits VEGF-induced angiogenesis by suppressing VEGFR-2 signaling pathway. *Oncotarget* **2016**, *7*, 3533. [CrossRef] [PubMed]
61. Yu, Z.; Li, T.; Wang, C.; Deng, S.; Zhang, B.; Huo, X.; Zhang, B.; Wang, X.; Zhong, Y.; Ma, X. Gamabufotalin triggers c-Myc degradation via induction of WWP2 in multiple myeloma cells. *Oncotarget* **2016**, *7*, 15725. [CrossRef] [PubMed]
62. Yu, Z.; Guo, W.; Ma, X.; Zhang, B.; Dong, P.; Huang, L.; Wang, X.; Wang, C.; Huo, X.; Yu, W. Gamabufotalin, a bufadienolide compound from toad venom, suppresses COX-2 expression through targeting IKKβ/NF-κB signaling pathway in lung cancer cells. *Mol. Cancer* **2014**, *13*, 203. [CrossRef] [PubMed]

63. Cao, Y.; Song, Y.; An, N.; Zeng, S.; Wang, D.; Yu, L.; Zhu, T.; Zhang, T.; Cui, J.; Zhou, C. The effects of telocinobufagin isolated from Chan Su on the activation and cytokine secretion of immunocytes in vitro. *Fundam. Clin. Pharmacol.* **2009**, *23*, 457–464. [CrossRef] [PubMed]
64. Sementsov, A. Poisons of toads. *Farm. Zh. (Kharkov)* **1939**, *12*, 19–22.
65. Shen, H.W.; Jiang, X.L.; Winter, J.C.; Yu, A.M. Psychedelic 5-methoxy-*N*,*N*-dimethyltryptamine: Metabolism, pharmacokinetics, drug interactions, and pharmacological actions. *Curr. Drug Metab.* **2010**, *11*, 659–666. [CrossRef] [PubMed]
66. Roth, B.; Choudhary, M.; Khan, N.; Uluer, A. High-affinity agonist binding is not sufficient for agonist efficacy at 5-hydroxytryptamine2A receptors: Evidence in favor of a modified ternary complex model. *J. Pharmacol. Exp. Ther.* **1997**, *280*, 576–583. [PubMed]
67. Raisanen, M.; Karkkainen, J. Mass fragmentographic quantification of urinary *N*,*N*-dimethyltryptamine and bufotenine. *J. Chromatogr.* **1979**, *162*, 579–584. [CrossRef]
68. Takeda, N.; Ikeda, R.; Ohba, K.; Kondo, M. Bufotenine reconsidered as a diagnostic indicator of psychiatric disorders. *Neuroreport* **1995**, *6*, 2378–2380. [CrossRef] [PubMed]
69. Emanuele, E.; Colombo, R.; Martinelli, V.; Brondino, N.; Marini, M.; Boso, M.; Barale, F.; Politi, P. Elevated urine levels of bufotenine in patients with autistic spectrum disorders and schizophrenia. *Neuro Endocrinol. Lett.* **2010**, *31*, 117–121. [PubMed]
70. Kryukova, E.V.; Lebedev, D.S.; Ivanov, I.A.; Ivanov, D.A.; Starkov, V.G.; Tsetlin, V.I.; Utkin, Y.N. *N*-methyl serotonin analogues from the *Bufo bufo* toad venom interact efficiently with the alpha7 nicotinic acetylcholine receptors. *Dokl. Biochem. Biophys.* **2017**, *472*, 52–55. [CrossRef] [PubMed]
71. Erspamer, G.F.; Cei, J.M. Biogenic amines and active polypeptides in the skin of *Leptodactylus vilarsi* melin. *Biochem. Pharmacol.* **1970**, *19*, 321–325. [CrossRef]
72. Gyermek, L.; Bindler, E. Action of indole alkylamines and amidines on the inferior mesenteric ganglion of the cat. *J. Pharmacol. Exp. Ther.* **1962**, *138*, 159–164. [PubMed]
73. Dai, L.P.; Gao, H.M.; Wang, Z.M.; Wang, W.H. Isolation and structure identification of chemical constituents from the skin of Bufo bufo gargarizans. *Yao Xue Xue Bao* **2007**, *42*, 858–861. [PubMed]
74. Radisky, D.C.; Radisky, E.S.; Barrows, L.R.; Copp, B.R.; Kramer, R.A.; Ireland, C.M. Novel cytotoxic topoisomerase II inhibiting pyrroloiminoquinones from Fijian sponges of the genus Zyzzya. *J. Am. Chem. Soc.* **1993**, *115*, 1632–1638. [CrossRef]
75. Barrows, L.R.; Radisky, D.C.; Copp, B.R.; Swaffar, D.S.; Kramer, R.A.; Warters, R.L.; Ireland, C.M. Makaluvamines, marine natural-products, are active anticancer agents and DNA topo-ii inhibitors. *Anti-Cancer Drug Des.* **1993**, *8*, 333–347.
76. Zhang, Z.; Chen, Y.; Jia, X.; He, J.; Wang, J. Dehydrobufotenine dry powder inhaler (DPI), its preparation method and application in preparation of antitumor drug for treating lung tumor. *Am. Chem. Soc.* **2012**, *8*, 55–66.
77. Xie, R.-F.; Li, Z.-C.; Gao, B.; Shi, Z.-N.; Zhou, X. Bufothionine, a possible effective component in cinobufocini injection for hepatocellular carcinoma. *J. Ethnopharmacol.* **2012**, *141*, 692–700. [CrossRef] [PubMed]
78. Gao, B.; Luo, C. *Method for Extracting Bufothionine from Skin of Bufo Bufo Gargarizans Cantor*; Anhui Jinchan Biochemical Co., Ltd.: Huaibei, China, 2012; 10p.
79. Dai, Y.H.; Shen, B.; Xia, M.Y.; Wang, A.D.; Chen, Y.L.; Liu, D.C.; Wang, D. A New Indole Alkaloid from the Toad Venom of Bufo bufo gargarizans. *Molecules* **2016**, *21*, 349. [CrossRef] [PubMed]
80. Zhou, Y.; Liu, J. *New Formulations of Alkaloid from Toad Skin for Cancer Therapy*, China, 2017; 14p.
81. Chen, Y.; Zhang, Z.; Jia, X.; Zhou, L.; Wu, Q. *Bufothionine Powder for Inhalation and its Preparation and Application for Treating Pulmonary Neoplasm*, China, 2011; 8p.
82. Ujváry, I. Psychoactive natural products: Overview of recent developments. *Annali dell'Istituto Superiore Sanita* **2014**, *50*, 12–27.
83. Winter, C.A.; Risley, E.A.; Nuss, G.W. Anti-inflammatory and antipyretic activities of indo-methacin, 1-(p-chlorobenzoyl)-5-methoxy-2-methyl-indole-3-acetic acid. *J. Pharmacol. Exp. Ther.* **1963**, *141*, 369–376. [PubMed]
84. Khamna, S.; Yokota, A.; Lumyong, S. Actinomycetes isolated from medicinal plant rhizosphere soils: Diversity and screening of antifungal compounds, indole-3-acetic acid and siderophore production. *World J. Microbiol. Biotechnol.* **2009**, *25*, 649–655. [CrossRef]

85. Seltzer, H.S. Quantitative effects of glucose, sulfonylureas, salicylate, and indole-3-acetic acid on the secretion of insulin activity into pancreatic venous blood. *J. Clin. Investig.* **1962**, *41*, 289–300. [CrossRef] [PubMed]
86. Folkes, L.K.; Wardman, P. Oxidative activation of indole-3-acetic acids to cytotoxic species—A potential new role for plant auxins in cancer therapy. *Biochem. Pharmacol.* **2001**, *61*, 129–136. [CrossRef]
87. Gowda, R.M.; Cohen, R.A.; Khan, I.A. Toad venom poisoning: Resemblance to digoxin toxicity and therapeutic implications. *Heart* **2003**, *89*, e14. [CrossRef] [PubMed]
88. Gao, H.; Popescu, R.; Kopp, B.; Wang, Z. Bufadienolides and their antitumor activity. *Nat. Prod. Rep.* **2011**, *28*, 953–969. [CrossRef] [PubMed]
89. Hu, K.; Zhu, L.; Liang, H.; Hu, F.; Feng, J. Improved antitumor efficacy and reduced toxicity of liposomes containing bufadienolides. *Arch. Pharm. Res.* **2011**, *34*, 1487. [CrossRef] [PubMed]

toxins

MDPI

Review

Malaysian Cobra Venom: A Potential Source of Anti-Cancer Therapeutic Agents

Syafiq Asnawi Zainal Abidin [1,2], Yee Qian Lee [2], Iekhsan Othman [1,2] and Rakesh Naidu [2,*]

1 Liquid Chromatography Mass Spectrometry (LCMS) Platform, Monash University Malaysia, Jalan Lagoon Selatan, Bandar Sunway 47500, Selangor Darul Ehsan, Malaysia; syafiq.asnawi@monash.edu (S.A.Z.A.); iekhsan.othman@monash.edu (I.O.)

2 Jeffrey Cheah School of Medicine and Health Sciences, Monash University Malaysia, Jalan Lagoon Selatan, Bandar Sunway 47500, Selangor Darul Ehsan, Malaysia; Yee.Lee@monash.edu

* Correspondence: rakesh.naidu@monash.edu

Received: 17 December 2018; Accepted: 24 January 2019; Published: 1 February 2019

Abstract: Cancer is a deadly disease and there is an urgent need for the development of effective and safe therapeutic agents to treat it. Snake venom is a complex mixture of bioactive proteins that represents an attractive source of novel and naturally-derived anticancer agents. Malaysia is one of the world's most biodiverse countries and is home to various venomous snake species, including cobras. *Naja kaouthia*, *Naja sumatrana*, and *Ophiophagus hannah* are three of the most common cobra species in Malaysia and are of medical importance. Over the past decades, snake venom has been identified as a potential source of therapeutic agents, including anti-cancer agents. This present review highlights the potential anticancer activity of the venom and purified venom protein of *N. kaouthia*, *N. sumatrana*, and *O. hannah*. In conclusion, this review highlights the important role of the venom from Malaysian cobras as an important resource that researchers can exploit to further investigate its potential in cancer treatment.

Keywords: snake venom; Malaysian cobras; *N. kaouthia*; *N. sumatrana*; *O. hannah*; anticancer

Key Contribution: In this review, the anticancer activity of the venom and purified venom proteins of Malaysian common cobra species are discussed.

1. Introduction

Cancer is a major health problem that affects people all over the world. Globally, 25% of human mortality is due to cancer. In the United States, cancer was reported as the second leading cause of death in 2015. Approximately 1,735,350 new cancer cases and 609,640 deaths due to cancer were expected in 2018 [1]. In Malaysia, cancer remains one of the leading causes of death, and a total of 64,725 deaths were reported from 2007 to 2011 [2]. Cancer is a group of diseases that arises from the uncontrollable proliferation of malignant cells. It is a multigenic and multistage disease due to multifactorial etiology [3]. Briefly, high levels of exposure to carcinogens such as radiation, tobacco, and oncogenic viruses increase the risk of DNA damage in the cells. The DNA-repair mechanism will be initiated at this stage. However, when the damage is too extensive, the repair of lesions fails. This gives rise to changes in the expression of genes (such as tumor-suppressor genes) in the cells, which further alter the signalling pathways resulting in unrestricted cell growth [4].

Even though the incidence of cancer is increasing globally, the mortality rate of cancer is reported to be in decline for the past 20 years [5]. This may be due to the advancement of therapeutic regimes over the past few decades. Various therapeutic options such as surgery, chemotherapy, radiotherapy, and immunotherapy are employed in treating localized cancer. Surgery in combination with chemotherapy is still the main treatment option. Unfortunately, owing to its cytotoxic activity via the inhibition of nucleic acid synthesis, chemotherapy often results in the death of fast growing cells such as white blood cells, hair follicles, and cells lining the gastrointestinal tract in addition to cancer cells [6]. Therefore, patients often suffer from side effects such as nausea and hair loss. A weakened immune system due to the reduction of white blood cell levels during chemotherapy increases patients' susceptibility to infection. The development of drug resistance during chemotherapy further complicates the treatment of cancer. Hence, there is an urgent need for effective cancer therapeutics with lesser side effects.

Malaysia, as the 12th most biodiverse country in the world, is the home to approximately 170,000 species of flora and fauna. Since ancient times, natural resources have been exploited to treat diseases and improve human health. For instance, the use of *Orthosiphon aristatus* (Misai Kuching) as a natural remedy against diabetes is common among indigenous communities in Malaysia [7,8]. Poisonous animals also play a vital role in the discovery of novel therapeutic candidates. For example, bee venom therapy is used to relieve pain symptoms and treat diseases such as rheumatoid arthritis [9,10] and other neurological diseases [11,12]. Venom from the Indian black scorpion was found to induce DNA fragmentation and reduce the proliferation of human leukemic cells [13]. Additionally, a novel peptide named Gonearrestide from scorpion venom showed the inhibition of primary colon cancer cells and solid tumor growth [14]. Commercialized drugs such as Captopril® and Enalpril® are two successful antihypertensive drugs developed based on bradykinin peptides derived from the venom of the snake *Bothrops jararaca* [15,16]. Ziconotide is another FDA-approved analgesic medication derived from ω-conotoxin that was found in the venom of *Conus magus*, a marine snail [17]. By reviewing the pharmaceutical potential of animal venoms, we conclude that the complex mixture of proteins may have the potential to be an important source of therapeutic agents.

Snake venom has been associated with various therapeutic applications—as a thrombolytic agent in cardiovascular disorders [18], anti-microbial activities [19], as an anti-viral agent [20] and in antiparasitic, and antifungal activities [21,22]. Undeniably, the anticancer activities of snake venom represent one of its most attractive therapeutic features and they have been actively researched and reviewed over the past decade [23–25]. The venom from Malaysian common cobras has been characterized, and proteins with anticancer potential have been described. However, while there are numerous reviews focusing on the anticancer activities of snake venom in general, none have focused on the Malaysian common cobra species, i.e., *Naja kaouthia*, *Naja sumatrana*, and *Ophiophagus hannah*. The abundance of these cobra species provides valuable access for researchers to further investigate the venom activity. Therefore, the present review highlights the anticancer activity of the venom components of Malaysian cobra species.

2. Malaysian Common Cobras

Malaysian venomous snake species can be divided into two families, Viperidae and Elapidae [26,27]. Viperidae can be further divided into three families, Azemiopinae (Fea's viper), Crotalinae (pit vipers), and Viperinae (true vipers). Malaysian vipers belong to the subfamily Crotalinae, which can be distinguished by the loreal pit on either side of the eyes [26]. Additional characteristics of pit vipers include hollow and retractile fangs on a moveable maxillary bone; a stocky, keel-scaled body with elliptical pupils; and that they are ovoviparous [26]. Elapidae is represented by cobras, kraits, and coral snakes that produce neurotoxic venom. It is characterized as a family of snakes with short and sharp fangs located anteriorly on the maxillary bone, with smooth-scaled body with rounded pupils, and that are oviparous [26]. Three cobra species, namely *Naja kaouthia*, *Naja sumatrana* and *Ophiophagus hannah* are the most common cobras in Malaysia. *N. kaouthia* or monocled cobra was formerly known as *Naja naja siamensis*, a subspecies of the Indian cobra (*Naja naja*) [26]. *N. sumatrana* is a spitting cobra and it is the most common Elapid of the ten species in the family. Both *N. kaouthia* and *N. sumatrana* can inhabit a wide range of environments, ranging from natural to anthropogenic landscapes. Members of the *Naja* genus are well known to be aggressive and envenomation is common for both species as humans infringe on their niche during the progress of urbanization. The third Malaysian cobra species is *O. hannah* or king cobra. *Ophiophagus*, meaning snake eater in Greek, is a monotypic genus, where the king cobra is the only species in this genus. It is the longest venomous snake species and is a dreadful assailant that is famous for its agility. The fatality rate incurred by king cobra envenomation is relatively high, although bites are rarely reported [26]. Table 1 provides a summary of comparisons between the cobras. In spite of their toxicity, the venom of Malaysian cobras demonstrates a wide range of therapeutic potential through antibacterial [28,29], anticancer [30–33], anticonvulsant [34], and antithrombotic [35] activities.

Table 1. Comparison of the cobra species in Malaysia.

	Naja kaouthia	*Naja sumatrana*	*Ophiophagus hannah*
Common name	Monocled cobra	Equatorial spitting cobra	King cobra
Characteristics	Absence of occipitals, brown to greyish-brown body, with white circle hood mark	Absence of occipitals, black body, without hood mark, white marking on throat	Large head; small hood; adult has yellow, green, brown, or black body; presence of a pair of occipitals behind parietals
Length	Usually 4–5 feet, occasionally can reach up to 7.5 feet	Usually 3–3.9 feet, occasionally can reach up to 4.9 feet	Usually 8–18 feet
Distribution in Malaysia	Peninsular Malaysia, mainly in the northern part of peninsular Malaysia	Peninsular Malaysia, Sabah, and Sarawak	Peninsular Malaysia, Sabah, and Sarawak
Habitat	Not habitat-specific, can adapt to a wide range of habitats such as grassland and paddy fields	Not habitat-specific, can adapt to a wide range of habitats such as primary and secondary forests and human-surrounding environments	Habitat-specific, mainly inhabits forests
Proteomic composition of the venom	3FTx, PLA_2, ohanin, CRVP, SVMP, vNGF, cardiotoxin, CVF, cytotoxin, and neurotoxin [36,37]	PLA_2, neurotoxins, cardiotoxin, cytotoxin, 3FTX, CVF, SVMP, CRVP, natriuretic peptide, aminopeptidase, thaicobrin, complement-depleting factor, vNGF, and cobra serum albumin [38]	Natriuretic peptides, 3FTx, Kunitz-type inhibitor, PLA2, ohanin, CRVP, cystatin, insulin-like growth factor, SVMP, LAAO, SVSP, vNGF, vPDE, PLB, AChE, 5′NUC, and neprilysins [31,39]

Abbreviations: 3FTx—three-finger toxin, PLA_2—phospholipase A_2, CRVP—cysteine-rich venom protein, SVMP—snake venom metalloproteinase, vNGF—venom nerve-growth factor, CVF—cobra venom factor, LAAO—L-amino acid oxidase, vPDE—venom phosphodiesterase, SVSP—snake venom serine protease, PLB—phospholipase B, AChE—acetylcholinesterase, 5′NUC—5′-nucleotidase.

3. Proteomic Composition of the Venom from *N. kaouthia*, *N. sumatrana*, and *O. hannah*

Snake venom is a natural resource that can be readily obtained, especially from Malaysian *N. kaouthia*, *N. sumatrana*, and *O. hannah*. The evolutionary arms race has driven the diversification of toxins in snake venom. It is a complex mixture comprising: 1) proteins such as phospholipase A_2 (PLA_2), L-amino acid oxidase (LAAO), acetylcholinesterase, and protease; 2) peptides such as disintegrins; 3) low-molecular-weight organic compounds such as carbohydrates and histamines; and 4) inorganic ions such as magnesium, cobalt, iron, and potassium [40]. The cocktail of proteins in snake venom aids the snakes in capturing and digesting their prey. These proteins can be categorized as cytotoxins, hemotoxins, neurotoxins, and cardiotoxins [4]. However, the composition of snake venoms may have inter- and intraspecies variation, depending on habitat, diet, gender, and ontogenetic development [4,41].

The advancement of mass spectrometry techniques has allowed for the proteomic characterization of the venom from *N. kouthia*, *N. sumatrana*, and *O. hannah*. A combination of transcriptomic and proteomic analyses of *N. kaouthia* venom has identified proteins such as the three-finger toxin (3FTx), phospholipase A_2 (PLA_2), ohanin, cysteine-rich venom protein (CRVP), snake venom metalloproteinase (SVMP), venom nerve-growth factor (vNGF), cobra venom factor (CVF), cardiotoxin, cytotoxin, and neurotoxin [36,37]. The proteomic characterization of *N. sumatrana* venom identified proteins including PLA_2, neurotoxins, cardiotoxin, cytotoxin, 3FTx, CVF, SVMP, CRVP, natriuretic peptide, aminopeptidase, thaicobrin, complement-depleting factor, kaouthin-1, vNGF, and cobra serum albumin [38]. Similar proteins, such as 3FTx, SVMP, PLA_2, and LAAO, were also identified from the venom of *O. hannah* in addition to acetylcholinesterase (AChE), phospholipase B (PLB), 5′-nucleotidase (5′NUC), neprilysins, and cystatins [31,39].

The common and unique venom proteins from *N. kaouthia*, *N. sumatrana*, and *O. hannah* are summarized in Figure 1. Five proteins were found to be common to all three cobra species, namely, 3FTx, PLA_2, CRVP, SVMP, and vNGF. Between *N. kaouthia* and *N. sumatrana*, four shared proteins were identified, including cardiotoxin, cytotoxin, neurotoxin, and CVF. Ohanin was found in both *N. kaouthia* and *O. hannah* and natriuretic peptides were identified in both *N. sumatrana* and *O. hannah*. Cobra serum albumin, aminopeptidase, thaicobrin, and complement-depleting factor were unique in *N. sumatrana* venom. Nine proteins in *O. hannah* venom were identified to be unique when compared with *N. kaouthia* and *N. sumatrana*, such as, LAAO, Kunitz-type inhibitor, cystatin, insulin-like growth factor, venom phosphodiesterase (vPDE), 5′NUC, snake venom serine protease (SVSP), AChE, and neprilysins.

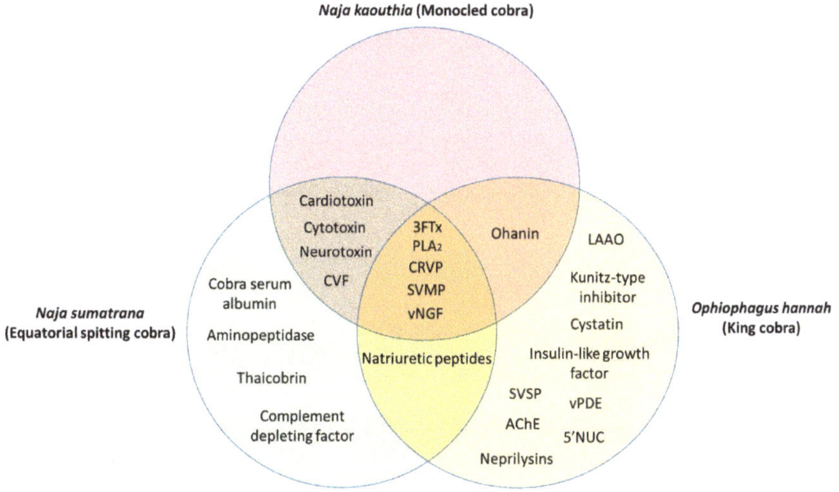

Figure 1. Common and unique proteins identified from the venom of *Naja kaouthia*, *Naja sumatrana*, and *Ophiophagus hannah*. Abbreviations: 3FTx—three-finger toxin, PLA_2—phospholipase A_2, CRVP—cysteine-rich venom protein, SVMP—snake venom metalloproteinase, vNGF—venom nerve-growth factor, CVF—cobra venom factor, LAAO—L-amino acid oxidase, vPDE—venom phosphodiesterase, SVSP—snake venom serine protease, AChE—acetylcholinesterase, 5′NUC—5′-nucleotidase.

4. Potential AntiCancer Activity of Malaysian Cobra Venom

The idea of utilizing snake venom as an important source of therapeutic agents and focusing on its anticancer properties has been extensively reviewed [23,24,42]. The investigation of snake venom's effects on cancers can be traced back as early as the 1930s [43,44]. Since then, various snake venom proteins—most notably, LAAO, PLA_2, SVMP/disintegrins, and snake venom C-type lectins (SNACLEC)—have been isolated and characterized for their activity as potential anticancer agents. The large amount of venom that can be obtained from the Malaysian common cobras renders them valuable for further investigation into potential therapeutic uses, especially as anticancer agents. The anticancer activity of the venom from the cobras is summarized in Table 2.

Table 2. Anticancer activity of Malaysian common cobra crude venom and protein components.

Species	Venom/Protein Component	Mechanism	Cancer Cell Type/Tissue	Reference
Ophiophagus hannah	Crude venom	Cytotoxic activity on pancreatic cancer cells (EC_{50}; 1.39 ng/mL), reduced migration activity, and induction of apoptosis in PaTu 8988t cells	Patu 8988t	[45]
	Crude venom	Reduced tumor-cell-induced angiogenesis in vivo	Zebrafish embryos	[45]
	L-amino acid oxidase (OH-LAAO)	Antiproliferative activity on murine melanoma, human fibrosarcoma, and murine epithelial cells	B16/F10, HT1080, and Balb/3T3	[30]
	OH-LAAO	Cytotoxic activity on human breast adenocarcinoma cells (EC_{50} 0.05 µg/mL) and apoptosis induction	MCF-7	[33]
	OH-LAAO	Apoptosis induction and inhibition of prostate tumor growth	PC-3 xenograft in nude mice	[32]
	OH-LAAO	Induced alteration of gene expression involved in cytotoxicity and apoptotic effects	MCF-7	[31]
	OH-LAAO	Modulation of proteins involved in stress response, ubiquitination, proteolysis, cell proliferation, and apoptosis	MCF-7	[33]
Naja kaouthia	Crude venom	Cytotoxic activity on pancreatic cancer cells (EC_{50} 1.42 ng/mL)	PaTu 8988t	[45]
	Crude venom	Venom at a nonlethal dose inhibited tumor-cell proliferation and showed cytotoxic activity and apoptosis induction in human lung cancer cells and leukemic cells	Ehrlich-ascites cells (EAC), U937, K562	[47]
	Cytotoxin CT3	Histopathological changes in leukemia cells treated with CT3	A549 and HL60	[48]
	Cardiotoxic–cytotoxic protein	Antiproliferative activity and apoptosis induction in human leukemic cells	U947 and K562	[49]
	kaotree (*N. kouthia*) and atroporin (*Crotalus atrox*) combination	Elevated cytotoxic activity in various human cancer cells	HBL-100, BT-20, ZR-75-1, HT-29, and Diji	[50]

4.1. Ophiophagus Hannah

In a recent study by Kerkkamp et.al [45], the cytotoxic activity of crude *O. hannah* and *N. kaouthia* venom was demonstrated on the human pancreatic cancer cell line (PaTu 8988t) at an EC_{50} value of 1.39 ng/mL and 1.42 ng/mL, respectively. Selective cytotoxic activity was demonstrated by the crude venoms with EC_{50} values of approximately 20 ng/mL on the control cell lines (ZF4 cells; zebrafish cells) [45]. Furthermore, in-vitro migration and apoptosis assays demonstrated the ability of crude *O. hannah* venom to reduce cell migration activity and induce apoptosis, respectively [45]. Using an in-vivo zebrafish model, PaTu 8988t cells were injected post fertilization of the zebrafish to induce an angiogenic response. Treatment with *O. hannah* venom successfully inhibited the angiogenesis induction of the cancer cells [45].

LAAO is one of the major enzymatic protein components in *O. hannah* venom. The enzyme is categorized under flavoenzymes, which convert L-amino acid into alpha-keto acids with hydrogen peroxide (H_2O_2) as a byproduct [40]. The cytotoxic activity of LAAO from different snake species, including pit vipers and cobras, has been demonstrated in various human cancer cells [51–53]. The production of H_2O_2 was noted to be the main cause of cytotoxicity in several experiments [30,54–56]. The apoptosis-inducing activities of snake venom LAAO were also associated with the secondary production of H_2O_2 [57,58], but the specific mechanism remains unknown [40]. Interestingly, a study has demonstrated that LAAO was able to trigger the apoptotic mechanism even without the presence of H_2O_2 [59]. In 2014, Lee et.al [46] demonstrated the antiproliferative activity of LAAO purified from *O. hannah* (OH-LAAO) on human breast adenocarcinoma (MCF-7) and human lung adenocarcinoma (A549) with EC_{50} values of 0.04 µg/mL and 0.05 µg/mL, respectively. The cytotoxic activity was demonstrated to be selective on the cancer cells with greater potency when compared with doxorubicin, an established chemotherapeutic agent [33]. The induction of apoptosis partly contributed to the cytotoxic mechanism, as demonstrated by the increased level of caspase-3/7 and DNA fragmentation [33]. Similarly, OH-LAAO demonstrated cytotoxic activity against human prostate adenocarcinoma (PC-3) with an EC_{50} value of 0.05 µg/mL [32]. The in-vivo activity of OH-LAAO on nude mice implanted with PC-3 cells showed a significant reduction of tumor size with no obvious tissue damage in their vital organs [32]. These findings support an earlier study of OH-LAAO cytotoxicity on several cancer cell lines by Ahn et al. [30]. The investigators demonstrated the antiproliferative activity of OH-LAAO in murine melanoma cells (B16/F10) and human fibrosarcoma cells (HT1080) with approximately 74% inhibition at a concentration of 2 µg/mL.

Fung and co-investigators [31] identified a total of 178 genes with significant alteration in MCF-7 cells treated with OH-LAAO. Amongst these were genes associated with the induction of apoptosis, such as, BMF (Bcl2 modifying factor), IGFBP3 (insulin-like growth-factor-binding protein 3), PLEKHF1 (Pleckstrin homology domain-containing, family F member 1), and PPARG (peroxisome proliferator-activated receptor gamma) [31]. Recent proteomic investigations by Fung et.al [46] further suggest that the use of OH-LAAO on MCF-7 induced 21 differentially expressed proteins with various biological functions including apoptosis, proteolysis, stress response, protein ubiquitination, and oxidoreduction. The authors concluded that the nonspecific oxidative modification of transcriptional factors caused by OH-LAAO is the key factor in the cell death and apoptosis induction.

4.2. *Naja Kaouthia*

Feofanov et.al [48] demonstrated that cytotoxin (CT3) from *N. kaouthia* induced strong cytotoxic activity in human lung adenocarcinoma (A549) and human promyelotic leukemia cells (HL60) at an EC_{50} value of 2.6 µM and 0.18 µM, respectively. The authors further suggest that the cytotoxic effects of CT3 in HL60 were noted by their ability to bind strongly to the plasma membrane followed by internalization of the protein [48]. Furthermore, lysosomes were identified as the primary target of the cytotoxin that triggered the cytolytic action on the cells. [48]. Permeabilization of the plasma membrane was noted as a downstream event following lysosome rupture [48]. Interestingly, cytotoxins from other cobra species such as *Naja oxiana* and *Naja haje* demonstrated weak internalization of the protein in the plasma membrane compared to CT3. In a separate study investigating the anticancer activity of *N. kaouthia* crude venom by Debnath et.al [47], nonlethal doses of the crude venom inhibited the proliferation of various cancer cell lines such as Ehrlich-ascites cells (EAC), human lung lymphoblasts (U937), and human myelogenous leukemia cells (K562). Morphological changes associated with apoptosis such as membrane blebbing, chromatin condensation, and fragmentation were common features in cells treated with the *N. kaouthia* venom [47]. Furthermore, the solid-tumor growth of sarcomas using a Balb/c mice model was significantly reduced when treated with *N. kaouthia* crude venom [47].

In a follow-up study by Debnath et al. [49] on *N. kaouthia*, a lethal protein named cardiotoxic–cytotoxic protein was purified from the crude venom. The protein was identified through sequence homology to cytotoxins and cardiotoxins from the venom of other cobra species and demonstrated significant antiproliferative activity on human leukemic cells (U937 and K562) in a dose-dependent manner [49]. The leukemic cells treated with the cardiotoxic–cytotoxic protein demonstrated an increase of caspase-3/-9 activity and an increase of the proapoptotic Bax level, which suggest the induction of apoptosis [49]. Additionally, a novel protein, named Kaotree, has been identified and characterized from the venom of *N. kaouthia* with reports of selective cytotoxic activity on transformed mammary epithelial cells (HBL-100), mammary gland carcinoma (BT-20), breast cancer cells (ZR-75-1), and colon adenocarcinoma (HT-29) [50]. The anticancer activity of Kaotree was demonstrated to have an enhanced killing effect when combined with Atroporin, a snake venom protein derived from *Crotalus atrox* [50]. A patent was filed for the anticancer activity of Kaotree and Atroporin (US Patent No: 5565431) with the claim of a novel method for treating cancer patients.

4.3. *Naja Sumatrana*

While the anticancer activity of the crude venom of *O. hannah* and *N. kaouthia* has been well documented, to date, there is no available literature on *N. sumatrana* venom. However, proteomic characterization of *N. sumatrana* venom has identified major proteins with well-reported anticancer activity, such as PLA_2, cardiotoxin, and neurotoxin [38]. The anticancer potential of snake venom PLA_2 has been well reviewed [60] and the protein has been isolated from various snake species such as vipers [61,62], sea snakes [63], and cobras [64]. The cytotoxicity and anticancer activity of PLA_2 has been demonstrated by MjTX-II, a PLA_2 isolated from *Bothrops moojeni*, on treated EAC cancer cells, human breast carcinoma (SK-BR-3), and human T-cell leukemia (Jurkat) [65]. Ammodytoxin C, PLA_2 purified from *Vipera ammodytes*, showed antitumoral activity against colon adenocarcinoma (Caco-2) [66]; and RVV-7, a cytotoxic PLA_2 from *Daboia russellii*, displayed significant inhibition in B16F10 tumors in C57BL/6 mice [67]. In addition to its cytotoxic activity, snake venom PLA_2 could also be employed as a model for drug development in humans. A study by Sales et al. [68] demonstrated similar interactions between PLA_2 from *Bothrops* species and human-secreted PLA_2 (HGIIA); which catalyzes the production of potent inflammatory molecules and is commonly associated with diseases [69]. Therefore, the anti-inflammatory activity of a novel drug can be studied using snake venom PLA_2 for therapy models in humans. Cardiotoxin III (CTX-III) isolated from *Naja atra* venom has demonstrated anticancer activities in oral squamous cell carcinoma (Ca9-22) [70], human breast cancer cells (MDA-MB-231) [71], and human neuroblastoma cells (SK-N-SH) [72]. Cell death by CTX-III in all cell lines was attributed to the induction of apoptosis through a significant increase of caspase-3/-9 activity and cell cycle arrest. Alpha-cobrotoxin (α-CbT), a neurotoxin purified from *N. kaouthia* venom, prolonged the survival of animals in a non-small cell lung cancer (NSCLC) mouse model [73].

5. Future Directions and Conclusions

Cancer remains one of the most critical health burden worldwide, including in Malaysia. The disease presents a major challenge in the discovery and production of therapeutic agents that are effective, nontoxic, and cause fewer side effects. Anticancer agents from naturally derived sources—especially animal venom—have demonstrated their potential in cancer therapy. Venom-based protein/peptide therapeutics has been a subject of interest over the past few decades. Venomous animals such as scorpions and snakes were widely studied for their therapeutic values with some advantages and some challenges coming to light [24,74,75]. One of the major promises of a venom-based therapeutic agent is the specificity and the selectivity of its interaction with the target molecule. This is crucial to developing an anticancer agent with the ability to differentiate between normal and cancerous cells, with significant impact on cell proliferation, migration, and angiogenesis [76]. Moreover, some venom peptides, such as cytotoxins from *N. kaouthia* [47] are small and are able to penetrate cancer cells to trigger the cytotoxic effect. However, because of its toxic nature and our incomplete understanding of the anticancer mechanism of the venom proteins/peptides, the number of venom-based therapeutic agents currently in the market are very low [77].

Technological advancement in proteomics and genomics approaches, such as mass spectrometry and sequencing techniques, have allowed multiple proteins from venom to be isolated and characterized for activity. Proteins such as LAAO, PLA_2, cytotoxin, and SVMP can be purified, characterized, and further investigated to determine the mechanisms by which they induce anticancer activity in vitro and in vivo. The majority of the studies mentioned in this review focused on the cytotoxicity of the venom/purified venom proteins on cancer cells, but further studies are needed to elucidate the potential anticancer mechanisms. Data obtained from the studies can serve as a template for further preclinical and clinical studies to demonstrate the safety and efficacy of these anticancer proteins. The delivery of the venom protein/peptide-based therapeutics represents another challenge in this field. However, continued advancements in the field of molecular biology, recombinant proteins, and drug delivery, such as nanoparticles, could possibly overcome the issues of bioavailability, pharmacokinetics, and efficacy of snake venom proteins as an anticancer drug. A study by Al-Sadoon et al. [77] demonstrated that the venom from *Walterinnesia aegyptia* combined with silica nanoparticles strongly induced apoptosis in human breast cancer cells with no significant impact on normal breast epithelial cells. In another study, the combination of venom and silica nanoparticles showed greater suppression of tumor growth in an in vivo model in nude mice compared to the venom alone [78].

In conclusion, this present review demonstrates that the venom from Malaysian common cobras could exert anticancer effects by modulating the cancer cell development mechanism and triggering apoptosis. The widespread availability of the cobras *N. kaouthia*, *N. sumatrana*, and *O. hannah* in the wild in Malaysia provides a valuable opportunity for researchers to further investigate their venom as a source of potential anticancer agents.

Author Contributions: S.A.Z.A and Y.Q.L. wrote the manuscript. I.O. and R.N. edited and reviewed the manuscript.

Funding: This research received no external funding.

Acknowledgments: We would like to thank Mr. Zainuddin Ismail (Bukit Bintang Enterprise, Perlis) for his expertise and advice in snake handling. Y.Q.L. is supported by Monash University Malaysia through a merit scholarship for their master's study.

Conflicts of Interest: The authors declare no conflict of interest.

References

1. Siegel, R.L.; Miller, K.D.; Jemal, A. Cancer statistics, 2018. *CA Cancer J. Clin.* **2018**, *68*, 7–30. [CrossRef] [PubMed]
2. Omar, Z.A.; Ibrahim Tamin, N. *National Cancer Registry Report: Malaysia Cancer Statistics-Data and Figure*; National Cancer Registry: Putrajaya, Malaysia, 2011; pp. 85–87.
3. Baskar, R.; Lee, K.A.; Yeo, R.; Yeoh, K.-W. Cancer and radiation therapy: Current advances and future directions. *Int. J. Med. Sci.* **2012**, *9*, 193. [CrossRef] [PubMed]
4. Jain, D.; Kumar, S. Snake venom: A potent anticancer agent. *Asian Pac. J. Cancer Prev.* **2012**, *13*, 4855–4860. [CrossRef] [PubMed]
5. Wang, L.; Dong, C.; Li, X.; Han, W.; Su, X. Anticancer potential of bioactive peptides from animal sources (review). *Oncol. Rep.* **2017**, *38*, 637–651. [CrossRef] [PubMed]
6. Liberio, M.S.; Joanitti, G.A.; Fontes, W.; Castro, M.S. Anticancer peptides and proteins: A panoramic view. *Protein Pept. Lett.* **2013**, *20*, 380–391. [PubMed]
7. Mustaffa, F.; Indurkar, J.; Ali, N.; Hanapi, A.; Shah, M.; Ismail, S.; Mansor, S. A review of malaysian medicinal plants with potential antidiabetic activity. *J. Pharm. Res.* **2011**, *4*, 4217–4224.

8. Tan, K.Y.; Tan, C.H.; Fung, S.Y.; Tan, N.H. Venomics, lethality and neutralization of *Naja kaouthia* (monocled cobra) venoms from three different geographical regions of Southeast Asia. *J. Proteome* **2015**, *120*, 105–125. [CrossRef] [PubMed]
9. Hong, S.-J.; Rim, G.S.; Yang, H.I.; Yin, C.S.; Koh, H.G.; Jang, M.-H.; Kim, C.-J.; Choe, B.-K.; Chung, J.-H. Bee venom induces apoptosis through caspase-3 activation in synovial fibroblasts of patients with rheumatoid arthritis. *Toxicon* **2005**, *46*, 39–45. [CrossRef] [PubMed]
10. Lee, J.-D.; Kim, S.-Y.; Kim, T.-W.; Lee, S.-H.; Yang, H.-I.; Lee, D.-I.; Lee, Y.-H. Anti-inflammatory effect of bee venom on type II collagen-induced arthritis. *Am. J. Chin. Med.* **2004**, *32*, 361–367. [CrossRef] [PubMed]
11. Cho, S.-Y.; Shim, S.-R.; Rhee, H.Y.; Park, H.-J.; Jung, W.-S.; Moon, S.-K.; Park, J.-M.; Ko, C.-N.; Cho, K.-H.; Park, S.-U. Effectiveness of acupuncture and bee venom acupuncture in idiopathic Parkinson's Disease. *Parkinsonism Relat. Disord.* **2012**, *18*, 948–952. [CrossRef] [PubMed]
12. Kim, J.-I.; Yang, E.J.; Lee, M.S.; Kim, Y.-S.; Huh, Y.; Cho, I.-H.; Kang, S.; Koh, H.-K. Bee venom reduces neuroinflammation in the MPTP-induced model of Parkinson's Disease. *Int. J. Neurosci.* **2011**, *121*, 209–217. [CrossRef] [PubMed]
13. Gupta, S.D.; Debnath, A.; Saha, A.; Giri, B.; Tripathi, G.; Vedasiromoni, J.R.; Gomes, A.; Gomes, A. Indian black scorpion (*Heterometrus bengalensis* Koch.) venom induced antiproliferative and apoptogenic activity against human leukemic cell lines U937 and K562. *Leuk. Res.* **2007**, *31*, 817–825. [CrossRef] [PubMed]
14. Li, B.; Lyu, P.; Xi, X.; Ge, L.; Mahadevappa, R.; Shaw, C.; Kwok, H.F. Triggering of cancer cell cycle arrest by a novel scorpion venom-derived peptide—Gonearrestide. *J. Cell Mol. Med.* **2018**, *22*, 4460–4473. [CrossRef] [PubMed]
15. Cushman, D.W.; Ondetti, M.A. History of the design of captopril and related inhibitors of angiotensin converting enzyme. *Hypertension* **1991**, *17*, 589–592. [CrossRef] [PubMed]
16. Ferreira, S. A bradykinin-potentiating factor (BPF) present in the venom of *Bothrops jararaca*. *Br. J. Pharmacol.* **1965**, *24*, 163–169. [CrossRef]
17. Pope, J.E.; Deer, T.R. Ziconotide: A clinical update and pharmacologic review. *Expert Opin. Pharmacother.* **2013**, *14*, 957–966. [CrossRef] [PubMed]
18. Koh, C.Y.; Kini, R.M. From snake venom toxins to therapeutics—Cardiovascular examples. *Toxicon* **2012**, *59*, 497–506. [CrossRef] [PubMed]
19. Wen, Y.L.; Wu, B.J.; Kao, P.H.; Fu, Y.S.; Chang, L.S. Antibacterial and membrane-damaging activities of β-bungarotoxin b chain. *J. Pept. Sci.* **2013**, *19*, 1–8. [CrossRef]
20. Muller, V.D.; Russo, R.R.; Cintra, A.C.; Sartim, M.A.; Alves-Paiva Rde, M.; Figueiredo, L.T.; Sampaio, S.V.; Aquino, V.H. Crotoxin and phospholipases A_2 from *Crotalus durissus terrificus* showed antiviral activity against dengue and yellow fever viruses. *Toxicon* **2012**, *59*, 507–515. [CrossRef]
21. Castillo, J.C.Q.; Vargas, L.J.; Segura, C.; Gutiérrez, J.M.; Pérez, J.C.A. In vitro antiplasmodial activity of phospholipases A_2 and a phospholipase homologue isolated from the venom of the snake *Bothrops asper*. *Toxins* **2012**, *4*, 1500–1516. [CrossRef]
22. Yamane, E.S.; Bizerra, F.C.; Oliveira, E.B.; Moreira, J.T.; Rajabi, M.; Nunes, G.L.; de Souza, A.O.; da Silva, I.D.; Yamane, T.; Karpel, R.L. Unraveling the antifungal activity of a South American rattlesnake toxin crotamine. *Biochimie* **2013**, *95*, 231–240. [CrossRef] [PubMed]
23. Calderon, L.A.; Sobrinho, J.C.; Zaqueo, K.D.; de Moura, A.A.; Grabner, A.N.; Mazzi, M.V.; Marcussi, S.; Nomizo, A.; Fernandes, C.F.; Zuliani, J.P.; et al. Antitumoral activity of snake venom proteins: New trends in cancer therapy. *BioMed Res. Int.* **2014**, *2014*, 203639. [CrossRef] [PubMed]
24. Li, L.; Huang, J.; Lin, Y. Snake venoms in cancer therapy: Past, present and future. *Toxins* **2018**, *10*, 346. [CrossRef] [PubMed]
25. Vyas, V.K.; Brahmbhatt, K.; Bhatt, H.; Parmar, U. Therapeutic potential of snake venom in cancer therapy: Current perspectives. *Asian Pac. J. Trop. Biomed.* **2013**, *3*, 156–162. [CrossRef]
26. Das, I.; Ahmed, N.; Liat, L.B. Venomous terrestrial snakes of Malaysia: Their identity and biology. *Clin. Toxicol.* **2013**, 1–15. [CrossRef]
27. Tweedie, M.W.F. *The Snakes of Malaya*, 3rd ed.; Singapore National Printers: Singapore, 1983; 167p.
28. Lee, M.L.; Tan, N.H.; Fung, S.Y.; Sekaran, S.D. Antibacterial action of a heat-stable form of L-amino acid oxidase isolated from king cobra (*Ophiophagus hannah*) venom. *Comp. Biochem. Physiol. C Toxicol. Pharmacol.* **2011**, *153*, 237–242. [CrossRef] [PubMed]

29. Phua, C.; Vejayan, J.; Ambu, S.; Ponnudurai, G.; Gorajana, A. Purification and antibacterial activities of an L-amino acid oxidase from king cobra (*Ophiophagus hannah*) venom. *J. Venom. Anim. Toxins Incl. Trop. Dis.* **2012**, *18*, 198–207. [CrossRef]
30. Ahn, M.Y.; Lee, B.M.; Kim, Y.S. Characterization and cytotoxicity of L-amino acid oxidase from the venom of king cobra (*Ophiophagus hannah*). *Int. J. Biochem. Cell Biol.* **1997**, *29*, 911–919. [CrossRef]
31. Fung, S.Y.; Lee, M.L.; Tan, N.H. Molecular mechanism of cell death induced by king cobra (*Ophiophagus hannah*) venom L-amino acid oxidase. *Toxicon* **2015**, *96*, 38–45. [CrossRef] [PubMed]
32. Lee, M.L.; Fung, S.Y.; Chung, I.; Pailoor, J.; Cheah, S.H.; Tan, N.H. King cobra (*Ophiophagus hannah*) venom L-amino acid oxidase induces apoptosis in PC-3 cells and suppresses PC-3 solid tumor growth in a tumor xenograft mouse model. *Int. J. Med. Sci.* **2014**, *11*, 593. [CrossRef] [PubMed]
33. Li Lee, M.; Chung, I.; Yee Fung, S.; Kanthimathi, M.S.; Hong Tan, N. Antiproliferative activity of king cobra (*Ophiophagus hannah*) venom L-amino acid oxidase. *Basic Clin. Pharmacol. Toxicol.* **2014**, *114*, 336–343. [CrossRef] [PubMed]
34. Saha, A.; Gomes, A.; Chakravarty, A.; Biswas, A.; Giri, B.; Dasgupta, S. CNS and anticonvulsant activity of a non-protein toxin (KC-MMTX) isolated from king cobra (*Ophiophagus hannah*) venom. *Toxicon* **2006**, *47*, 296–303. [CrossRef] [PubMed]
35. Du, X.-Y.; Clemetson, J.M.; Navdaev, A.; Magnenat, E.M.; Wells, T.N.; Clemetson, K.J. Ophioluxin, a convulxin-like C-type lectin from *Ophiophagus hannah* (king cobra) is a powerful platelet activator via glycoprotein VI. *J. Biol. Chem.* **2002**, *277*, 35124–35132. [CrossRef] [PubMed]
36. Xu, N.; Zhao, H.Y.; Yin, Y.; Shen, S.S.; Shan, L.L.; Chen, C.X.; Zhang, Y.X.; Gao, J.F.; Ji, X. Combined venomics, antivenomics and venom gland transcriptome analysis of the monocoled cobra (*Naja kaouthia*) from China. *J. Proteom.* **2017**, *159*, 19–31. [CrossRef] [PubMed]
37. Kulkeaw, K.; Chaicumpa, W.; Sakolvaree, Y.; Tongtawe, P.; Tapchaisri, P. Proteome and immunome of the venom of the Thai cobra, *Naja kaouthia*. *Toxicon* **2007**, *49*, 1026–1041. [CrossRef] [PubMed]
38. Yap, M.K.K.; Fung, S.Y.; Tan, K.Y.; Tan, N.H. Proteomic characterization of venom of the medically important southeast asian *Naja sumatrana* (equatorial spitting cobra). *Acta Trop.* **2014**, *133*, 15–25. [CrossRef] [PubMed]
39. Petras, D.; Heiss, P.; Sussmuth, R.D.; Calvete, J.J. Venom proteomics of Indonesian king cobra, *Ophiophagus hannah*: Integrating top-down and bottom-up approaches. *J. Proteome Res.* **2015**, *14*, 2539–2556. [CrossRef] [PubMed]
40. Izidoro, L.F.M.; Sobrinho, J.C.; Mendes, M.M.; Costa, T.R.; Grabner, A.N.; Rodrigues, V.M.; da Silva, S.L.; Zanchi, F.B.; Zuliani, J.P.; Fernandes, C.F. Snake venom L-amino acid oxidases: Trends in pharmacology and biochemistry. *BioMed Res. Int.* **2014**, *2014*, 196754. [CrossRef] [PubMed]
41. Augusto-de-Oliveira, C.S.; Stuginski, D.R.; Kitano, E.S.; Andrade-Silva, D.b.; Liberato, T.; Fukushima, I.; Serrano, S.M.; Zelanis, A. Dynamic rearrangement in snake venom gland proteome: Insights into *Bothrops jararaca* intraspecific venom variation. *J. Proteome Res.* **2016**, *15*, 3752–3762. [CrossRef]
42. Shanbhag, V.K.L. Applications of snake venoms in treatment of cancer. *Asian Pac. J. Trop. Biomed.* **2015**, *5*, 275–276. [CrossRef]
43. Essex, H.E.; Priestley, J.T. Effect of rattlesnake venom on Flexner-Jobling's carcinoma in the white rat (*Mus norvegicus* Albinus.). *Proc. Soc. Exp. Biol. Med.* **1931**, *28*, 550–551. [CrossRef]
44. Kurotchkin, T.; Spies, J. Effects of cobra venom on the Fujinami rat sarcoma. *Proc. Soc. Exp. Biol. Med.* **1935**, *32*, 1408–1410. [CrossRef]
45. Kerkkamp, H.; Bagowski, C.; Kool, J.; van Soolingen, B.; Vonk, F.J.; Vlecken, D. Whole snake venoms: Cytotoxic, anti-metastatic and antiangiogenic properties. *Toxicon* **2018**, *150*, 39–49. [CrossRef] [PubMed]
46. Fung, S.Y.; Lee, M.L.; Tan, N.H. Proteomic investigation of the molecular mechanism of king cobra venom L-amino acid oxidase induced apoptosis of human breast cancer (MCF-7) cell line. *Indian J. Exp. Biol.* **2018**, *56*, 101–111.
47. Debnath, A.; Chatterjee, U.; Das, M.; Vedasiromoni, J.R.; Gomes, A. Venom of Indian monocellate cobra and Russell's viper show anticancer activity in experimental models. *J. Ethnopharmacol.* **2007**, *111*, 681–684. [CrossRef] [PubMed]
48. Feofanov, A.V.; Sharonov, G.V.; Astapova, M.V.; Rodionov, D.I.; Utkin, Y.N.; Arseniev, A.S. Cancer cell injury by cytotoxins from cobra venom is mediated through lysosomal damage. *Biochem. J.* **2005**, *390*, 11–18. [CrossRef] [PubMed]

49. Debnath, A.; Saha, A.; Gomes, A.; Biswas, S.; Chakrabarti, P.; Giri, B.; Biswas, A.K.; Gupta, S.D.; Gomes, A. A lethal cardiotoxic-cytotoxic protein from the Indian monocellate cobra (*Naja kaouthia*) venom. *Toxicon* **2010**, *56*, 569–579. [CrossRef] [PubMed]
50. Lipps, B.V. Novel snake venom proteins cytolytic to cancer cells in vitro and in vivo systems. *J. Venom. Anim. Toxins Incl. Trop. Dis.* **1999**, *5*, 172–183. [CrossRef]
51. Zainal Abidin, S.A.; Rajadurai, P.; Hoque Chowdhury, M.E.; Othman, I.; Naidu, R. Cytotoxic, Anti-Proliferative and Apoptosis Activity of L-Amino Acid Oxidase from Malaysian *Cryptelytrops purpureomaculatus* (CP-LAAO) Venom on Human Colon Cancer Cells. *Molecules* **2018**, *23*, 1388. [CrossRef] [PubMed]
52. Zainal Abidin, S.A.; Rajadurai, P.; Chowdhury, M.E.H.; Ahmad Rusmili, M.R.; Othman, I.; Naidu, R. Cytotoxic, Antiproliferative and Apoptosis-inducing Activity of L-Amino Acid Oxidase from Malaysian *Calloselasma rhodostoma* on Human Colon Cancer Cells. *Basic Clin. Pharmacol. Toxicol.* **2018**, *123*, 577–588. [CrossRef] [PubMed]
53. Mauro, V.P.; Adriana, S.P.; Andreimar, M.S.; Juliana, P.Z. An Update on Potential Molecular Mechanisms Underlying the Actions of Snake Venom L-amino Acid Oxidases (LAAOs). *Curr. Med. Chem.* **2018**, *25*, 2520–2530.
54. Zhang, L.; Wei, L.J. ACTX-8, a cytotoxic L-amino acid oxidase isolated from *Agkistrodon acutus* snake venom, induces apoptosis in Hela cervical cancer cells. *Life Sci.* **2007**, *80*, 1189–1197. [CrossRef] [PubMed]
55. Zhang, L.; Wu, W.T. Isolation and characterization of ACTX-6: A cytotoxic L-amino acid oxidase from *Agkistrodon acutus* snake venom. *Nat. Prod. Res.* **2008**, *22*, 554–563. [CrossRef] [PubMed]
56. Zhang, Y.J.; Wang, J.H.; Lee, W.H.; Wang, Q.; Liu, H.; Zheng, Y.T.; Zhang, Y. Molecular characterization of *Trimeresurus stejnegeri* venom L-amino acid oxidase with potential anti-HIV activity. *Biochem. Biophys. Res. Commun.* **2003**, *309*, 598–604. [CrossRef] [PubMed]
57. Guo, C.; Liu, S.; Dong, P.; Zhao, D.; Wang, C.; Tao, Z.; Sun, M.Z. Akbu-LAAO exhibits potent anti-tumor activity to HepG2 cells partially through produced H_2O_2 via TGF-beta signal pathway. *Sci. Rep.* **2015**, *5*, 18215. [CrossRef] [PubMed]
58. Ribeiro, P.H.; Zuliani, J.P.; Fernandes, C.F.; Calderon, L.A.; Stabeli, R.G.; Nomizo, A.; Soares, A.M. Mechanism of the cytotoxic effect of L-amino acid oxidase isolated from *Bothrops alternatus* snake venom. *Int. J. Biol. Macromol.* **2016**, *92*, 329–337. [CrossRef] [PubMed]
59. Suhr, S.M.; Kim, D.S. Identification of the snake venom substance that induces apoptosis. *Biochem. Biophys. Res. Commun.* **1996**, *224*, 134–139. [CrossRef]
60. Rodrigues, R.S.; Izidoro, L.F.; de Oliveira, R.J., Jr.; Sampaio, S.V.; Soares, A.M.; Rodrigues, V.M. Snake venom phospholipases A_2: A new class of antitumor agents. *Protein Pept. Lett.* **2009**, *16*, 894–898. [CrossRef]
61. Zouari-Kessentini, R.; Luis, J.; Karray, A.; Kallech-Ziri, O.; Srairi-Abid, N.; Bazaa, A.; Loret, E.; Bezzine, S.; El Ayeb, M.; Marrakchi, N. Two purified and characterized phospholipases A_2 from *Cerastes cerastes* venom, that inhibit cancerous cell adhesion and migration. *Toxicon* **2009**, *53*, 444–453. [CrossRef]
62. Khunsap, S.; Pakmanee, N.; Khow, O.; Chanhome, L.; Sitprija, V.; Suntravat, M.; Lucena, S.E.; Perez, J.C.; Sanchez, E.E. Purification of a phospholipase A_2 from *Daboia russelii siamensis* venom with anticancer effects. *J. Venom. Res.* **2011**, *2*, 42–51.
63. Liang, Y.J.; Yang, X.P.; Wei, J.W.; Fu, L.W.; Jiang, X.Y.; Chen, S.W.; Yang, W.L. correlation of antitumor effect of recombinant sea snake basic phospholipase A_2 to its enzymatic activity. *Ai Zheng* **2005**, *24*, 1474–1478. [PubMed]
64. Chwetzoff, S.; Tsunasawa, S.; Sakiyama, F.; Menez, A. Nigexine, a phospholipase A_2 from cobra venom with cytotoxic properties not related to esterase activity. Purification, amino acid sequence, and biological properties. *J. Biol. Chem.* **1989**, *264*, 13289–13297. [PubMed]
65. Rudrammaji, L.M.; Gowda, T.V. Purification and characterization of three acidic, cytotoxic phospholipases A_2 from Indian cobra (*Naja naja naja*) venom. *Toxicon* **1998**, *36*, 921–932. [CrossRef]
66. Premzl, A.; Kovacic, L.; Halassy, B.; Krizaj, I. Generation of ammodytoxin-anti-cathepsin B immuno-conjugate as a model for delivery of secretory phospholipase A_2 into cancer cells. *Toxicon* **2008**, *51*, 754–764. [CrossRef] [PubMed]
67. Maity, G.; Mandal, S.; Chatterjee, A.; Bhattacharyya, D. Purification and characterization of a low molecular weight multifunctional cytotoxic phospholipase A_2 from Russell's viper venom. *J. Chromatogr. B Analyt. Technol. Biomed. Life Sci.* **2007**, *845*, 232–243. [CrossRef] [PubMed]

68. Sales, T.A.; Marcussi, S.; Da Cunha, E.F.F.; Kuca, K.; Ramalho, T.C. Can inhibitors of snake venom phospholipases A_2 lead to new insights into anti-inflammatory therapy in humans? A theoretical study. *Toxins* **2017**, *9*, 341. [CrossRef] [PubMed]
69. Quach, N.D.; Arnold, R.D.; Cummings, B.S. Secretory phospholipase A_2 enzymes as pharmacological targets for treatment of disease. *Biochem. Pharmacol.* **2014**, *90*, 338–348. [CrossRef] [PubMed]
70. Chien, C.M.; Chang, S.Y.; Lin, K.L.; Chiu, C.C.; Chang, L.S.; Lin, S.R. Taiwan cobra cardiotoxin III inhibits SRC kinase leading to apoptosis and cell cycle arrest of oral squamous cell carcinoma CA9-22 cells. *Toxicon* **2010**, *56*, 508–520. [CrossRef] [PubMed]
71. Lin, K.L.; Su, J.C.; Chien, C.M.; Chuang, P.W.; Chang, L.S.; Lin, S.R. Down-regulation of the JAK2/PI3k-mediated signaling activation is involved in taiwan cobra cardiotoxin III-induced apoptosis of human breast MDA-MB-231 cancer cells. *Toxicon* **2010**, *55*, 1263–1273. [CrossRef]
72. Chen, K.C.; Lin, S.R.; Chang, L.S. Involvement of mitochondrial alteration and reactive oxygen species generation in Taiwan cobra cardiotoxin-induced apoptotic death of human neuroblastoma SK-N-SH cells. *Toxicon* **2008**, *52*, 361–368. [CrossRef]
73. Paleari, L.; Negri, E.; Catassi, A.; Cilli, M.; Servent, D.; D'Angelillo, R.; Cesario, A.; Russo, P.; Fini, M. Inhibition of non-neuronal α7-nicotinic receptor for lung cancer treatment. *Am. J. Respir. Crit. Care Med.* **2009**, *179*, 1141–1150. [CrossRef] [PubMed]
74. Ortiz, E.; Gurrola, G.B.; Schwartz, E.F.; Possani, L.D. Scorpion venom components as potential candidates for drug development. *Toxicon* **2015**, *93*, 125–135. [CrossRef] [PubMed]
75. Ma, R.; Mahadevappa, R.; Kwok, H.F. Venom-based peptide therapy: Insights into anti-cancer mechanism. *Oncotarget* **2017**, *8*, 100908–100930. [CrossRef] [PubMed]
76. Mahadevappa, R.; Ma, R.; Kwok, H.F. Venom peptides: Improving specificity in cancer therapy. *Trends Cancer* **2017**, *3*, 611–614. [CrossRef] [PubMed]
77. Al-Sadoon, M.K.; Abdel-Maksoud, M.A.; Rabah, D.M.; Badr, G. Induction of apoptosis and growth arrest in human breast carcinoma cells by a snake (*Walterinnesia aegyptia*) venom combined with silica nanoparticles: Crosstalk between Bcl2 and caspase 3. *Cell Physiol. Biochem.* **2012**, *30*, 653–665. [CrossRef] [PubMed]
78. Al-Sadoon, M.K.; Rabah, D.M.; Badr, G. Enhanced anticancer efficacy of snake venom combined with silica nanoparticles in a murine model of human multiple myeloma: Molecular targets for cell cycle arrest and apoptosis induction. *Cell. Immunol.* **2013**, *284*, 129–138. [CrossRef] [PubMed]

toxins

MDPI

Review

Bouganin, an Attractive Weapon for Immunotoxins

Massimo Bortolotti, Andrea Bolognesi * and Letizia Polito

Department of Experimental, Diagnostic and Specialty Medicine-DIMES, General Pathology Section, Alma Mater Studiorum—University of Bologna, Via S. Giacomo 14, 40126 Bologna, Italy; massimo.bortolotti2@unibo.it (M.B.); letizia.polito@unibo.it (L.P.)

* Correspondence: andrea.bolognesi@unibo.it; Tel.: +39-051-209-4700

Received: 18 July 2018; Accepted: 2 August 2018; Published: 8 August 2018

Abstract: Bougainvillea (*Bougainvillea spectabilis* Willd.) is a plant widely used in folk medicine and many extracts from different tissues of this plant have been employed against several pathologies. The observation that leaf extracts of Bougainvillea possess antiviral properties led to the purification and characterization of a protein, named bouganin, which exhibits typical characteristics of type 1 ribosome-inactivating proteins (RIPs). Beyond that, bouganin has some peculiarities, such as a higher activity on DNA with respect to ribosomal RNA, low systemic toxicity, and immunological properties quite different than other RIPs. The sequencing of bouganin and the knowledge of its three-dimensional structure allowed to obtain a not immunogenic mutant of bouganin. These features make bouganin a very attractive tool as a component of immunotoxins (ITs), chimeric proteins obtained by linking a toxin to a carrier molecule. Bouganin-containing ITs showed very promising results in the experimental treatment of both hematological and solid tumors, and one bouganin-containing IT has entered Phase I clinical trial. In this review, we summarize the milestones of the research on bouganin such as bouganin chemico-physical characteristics, the structural properties and de-immunization studies. In addition, the in vitro and in vivo results obtained with bouganin-containing ITs are summarized.

Keywords: antiviral activity; Bougainvillea; bouganin; cancer therapy; immunotherapy; immunotoxins; ribosome-inactivating proteins; rRNA *N*-glycosylase activity; VB6-845

Key Contribution: This manuscript highlights the peculiar features of bouganin, a type 1 RIP, which possesses a high ratio of activity on DNA with respect to rRNA, low systemic toxicity and no cross-reactivity with sera against other RIPs. The availability of a de-immunized form of bouganin led to the construction of immunotoxins showing promising antitumor activity in experimental models and in a Phase I clinical trial.

1. Introduction

Ribosome-inactivating proteins (RIPs) are a family of plant proteins characterized by an enzymatic activity classically identified as rRNA *N*-glycosylase (EC 3.2.2.22). These enzymes, which are widely distributed among plant genera, specifically remove the A4324 adenine residue of the 28S rRNA in rat ribosome thus interfering with the ribosome/elongation factor 2 interaction, damaging ribosomes in an irreversible manner and causing the inhibition of protein synthesis [1]. RIPs also show *N*-glycosylase activity on different other substrates, such as mRNA, tRNA, DNA and poly(A) [2,3]. As regards the structure, RIPs can mainly be divided into two groups: type 1, consisting of a single-chain protein with enzymatic activity, and type 2, consisting of an enzymatic A-chain linked to a B-chain with lectin properties. The presence of the B-chain allows the fast internalization of the toxin into the cell, so conferring to most type 2 RIPs a very high toxicity [4]. RIPs can kill the cells by apoptosis [5,6], even if other cell death mechanisms are involved in the pathogenesis of RIP intoxication [7,8]. It is worth noting that RIPs are also able to depurinate viral nucleic acids. In fact, many RIPs can inhibit animal and plant viruses through mechanisms that have not yet been fully clarified. There is evidence that the antiviral action of RIPs cannot be solely attributed to the inhibition of ribosomes, but also to direct interaction of RIPs with viral RNA or DNA [2,9].

Since 1925, it was known that leaf extracts of some plants were able to prevent the infection of other plant species when mixed with a suspension of tobacco mosaic virus (TMV) [10]. Those observations allowed the partial isolation of the antiviral principle of *Phytolacca americana* [11] and *Dianthus caryophyllus* [12]. Afterwards, the antiviral activity was attributed to specific proteins (named pokeweed antiviral protein (PAP) and dianthin, respectively). Moreover, it was shown that also the protein synthesis inhibition activity, present in the extracts, was due to the same proteins [13–15]. It was subsequently established that most of the tested RIPs, including type 2 ones, were able to prevent infection with TMV in *Nicotiana benthamiana* leaves, albeit at different concentrations. Plant extracts with antiviral activity did not prevent the infection of autologous plants but were effective only on heterologous plants. This led to the conclusion that the antiviral principles acted on the plant, rather than on the viruses. Further studies showed that the in vitro antiviral activity of RIPs could also be directed against animal viruses, both RNA and DNA viruses [9].

In medicine, RIPs found application as toxic moiety of conjugates, chimeric molecules specifically targetable to unwanted cells responsible for pathologic conditions. Conjugates containing RIPs linked to monoclonal antibodies (mAbs) or their fragments are referred to as immunotoxins (ITs). ITs can be obtained both by the chemical linkage of the toxic moiety to mAbs and by genetic engineering to obtain recombinant conjugates [16]. RIP-containing ITs have been included in many clinical trials against various diseases, often achieving promising results, especially in the treatment of hematological neoplasms [17].

2. Purification and Antiviral Properties of Bouganin

Bougainvillea spectabilis Willd., also known as "paper flower" or "Bougainvillea", is a woody vine belonging to Nyctaginaceae family. It is native to South America but spread all over the world for its ornamental characteristics. This plant, in fact, is frequently blooming and its bracts have an intense purple or magenta color.

As for many other RIP-containing plants [18], Bougainvillea has long been used as medicinal plant, mainly in Latin America and Mexico [19]. The extracts from several plant tissues, mainly leaves, flowers and stem barks, are utilized in traditional medicine in forms of infusions, decoctions and tinctures. Drunk as a tea, Bougainvillea extracts are employed against cough, sore throat, flu, fever, diarrhea, diabetes, hepatitis and liver problems, asthma, bronchitis, to reduce stomach acidity, dissolve blood clots, regulate menstruation and stop leucorrhea, and for anemia associated with gastrointestinal bleeding and epigastric pain. Infusion of flowers is drunk as a remedy for low blood pressure [20].

Several studies have been conducted in order to evaluate the pharmacological activities of phytochemical constituents isolated from different Bougainvillea tissues. Experimental evidences showed that such molecules can exert antibacterial, antihyperlipidemic, antidiabetic, antifertility, antioxidant, anti-inflammatory, and antiulcer activities [21].

The first experimental evidences of antiviral effect of Bougainvillea date back to the 80s when it was evidenced that the infection of tobacco plants by TMV was prevented by leaf extracts [22,23]. The prevention of the infection was attributed to protein factors.

In 1997, Bolognesi and co-workers identified for the first time the presence of type 1 RIPs in the leaves of Bougainvillea. At least seven different RIPs were purified by ion-exchange chromatography of leaf extracts. The authors' attention focused on the first eluted pick, corresponding to a protein that was named bouganin. This protein was chosen for further experiments because it had the highest specific inhibitory activity on cell-free protein synthesis and gave the highest yield after purification. Bouganin has the properties of type 1 RIP, in that it: (i) is a single-chain protein with a molecular mass of about 30 kDa and an isoelectric point in the alkaline region; (ii) inhibits protein synthesis in a cell-free system (IC_{50} 10 ng/mL) despite its lower activity on whole cells, compared to other type 1 RIPs; (iii) has *N*-glycosylase activity; and (iv) has antiviral activity. Moreover, it appeared to be homogeneous at 99% by reverse-phase High Performance Liquid Chromatography (HPLC) analysis, and on Sodium Dodecyl Sulphate-PolyAcrylamide Gel Electrophoresis (SDS-PAGE) gave a single band with mobility corresponding to Mr 26200. The RIP released a single adenine residue from rat liver ribosomes but several tens of adenine residues from other substrates as rRNA from E. coli, viral RNA, poly (A) and several hundreds of adenine residues from herring sperm DNA (hsDNA), thus possessing a polynucleotide:adenosine glycosidase activity [24]. As reported in Table 1 and Figure 1, bouganin showed the highest activity ratio compared to other type 1 RIPs and to ricin A chain. This represents an advantage because the RIP can exert its activity on different substrates, triggering cell death through multiple pathways.

Table 1. Adenine: Polynucleotide glycosidase activity, protein synthesis inhibitory activity and toxicity of bouganin compared with different type 1 RIPs and ricin A chain.

RIP	Adenine Released		Protein Synthesis		Toxicity
	hsDNA [1] pmol	Rat Ribosomes [1] pmol	Cell Free [1] 10^3 U/mg *	Raji Cells [2] IC_{50} (nM)	Mouse [3] LD_{50} (mg/kg)
Bouganin	377.7	4.8	75	839	>32 [4,#]
Dianthin 30	239.9	5.7	96	541	14
Momordin I	27.1	3.9	526	n.a. [§]	7.4
PAP-S	503.2	5.1	125	n.a. [§]	2.6
Ricin A chain	48.5	6.2	300	200 [5]	16 [6]
Saporin-S6	376.1	19.1	813	23.6	4

[1] Data from [26]. * One unit of inhibitory activity (U) is defined as the amount of protein causing 50% inhibition in 1 mL of reaction mixture. [2] Data from [27]. [#] No mouse died at the higher tested dose (32 mg/kg). [3] Data from [25]. [4] Datum from [24]. [5] Datum from [28]. [6] Datum from [29]. IC_{50} is the concentration inhibiting 50% of protein synthesis; LD_{50} is the lethal dose for 50% of treated animals. [§] not available.

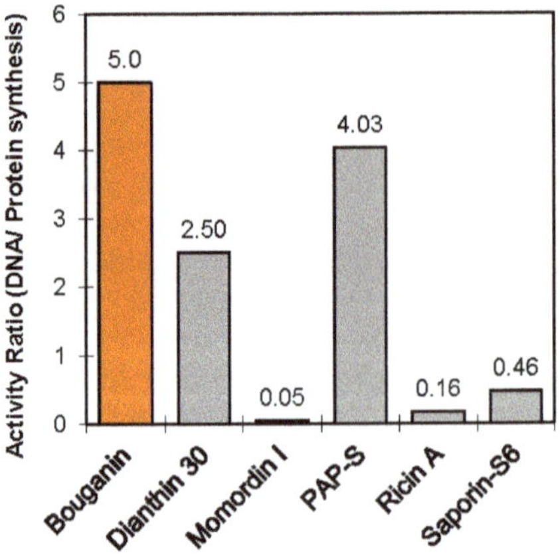

Figure 1. Activity ratio of bouganin compared to other type 1 RIPs and ricin A chain. The bar values represent the ratio between the activity on hsDNA and the activity on cell-free protein synthesis, expressed as 10^3 U/mg, as reported in Table 1.

Bouganin prevented systemic infection by artichoke mottled crinkle virus of *Nicotiana benthamiana* plants, presumably by inhibiting viral replication at the site of infection. Despite the high specific enzymatic activity, bouganin showed a very low toxicity for animals (mice); no mouse was killed by a dosage of 32 mg/kg of bouganin; this dose was very toxic for many other type 1 RIPs. Certainly, bouganin appears to be one of the least toxic RIPs, in comparison with other well-known RIPs (Table 1) [25].

3. Recombinant Bouganin and Structure/Function Studies

In 2002, den Hartog and co-workers cloned, expressed, purified, and characterized the recombinant bouganin. The cDNA, synthetized from total RNA isolated from Bougainvillea leaves, encoded for a precursor protein of 305 amino-acid residues. On the basis of the bouganin primary sequence, it was found that 26 residues at the amino-terminal and 29 residues at the carboxy-terminal are post-translationally removed to produce the mature form, which consists of 250 amino acids. The 26-residue leader portion is a secretory signal sequence rich in hydrophobic amino acids, which probably direct the transport of the nascent polypeptide chain across the endoplasmic reticulum membrane into the endoplasmic reticulum lumen. Bouganin and the other type 1 RIPs have little or no homology in the *C*-terminal cleaved amino-acid sequence. This demonstrates that these sequence motives, of unknown function, are not conserved in nature [30]. The recombinant molecule had similar enzymatic activity in a cell-free protein synthesis assay and had comparable toxicity on cells as compared to native bouganin [30].

Subsequently, the three-dimensional structure of bouganin was solved, demonstrating that its overall structure is like other RIPs, maintaining the typical RIP fold [26]. The *N*-terminal domain (in red in Figure 2a) is made up of a mixed β-sheet of seven strands (β1–β9). The five central strands run antiparallel and the other extern four are parallel to the neighbors. The α2 and the α3 helixes are connected to short structural motifs of β strands that are respectively the first two and the last two of the sheets. In bouganin, the *C*-terminal domain (in green in Figure 2a) is mainly composed of eight α-helices. This region shows a loop, flanking helix α9, composed of two antiparallel β-strands connected by a short helix. By site-directed mutagenesis experiments in RIPs, it has been possible to identify five highly conserved residues, which are involved in *N*-glycosylase activity and correspond to Tyr70, Tyr114, Glu165, Arg168 and Trp198 in bouganin [31]. (Figure 2b). Moreover, Phe169, another highly conserved amino acid near the active site, could play a role in stabilizing the conformation of Arg168 side chain [32].

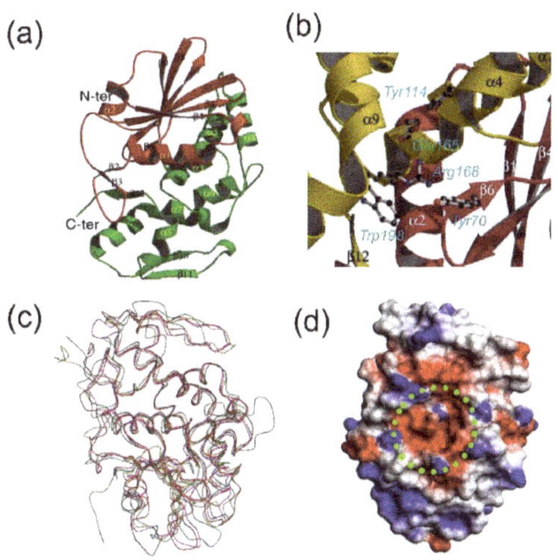

Figure 2. (**a**) Ribbon model of the crystal structure of bouganin (accession number Protein Data Bank 3CTK). The *N*-terminal and the *C*-terminal domains are colored in red and green respectively; (**b**) Catalytic site of bouganin. The conserved important residues are showed in ball-and-sticks; (**c**) Structural comparison between different type 1 RIPs and ricin A chain. Superimposition of the Cα atoms of bouganin (magenta), dianthin 30 (green), PAP-R (red) and ricin A chain (gold). The figures were produced by MOLSCRIPT [33] and rendered by RASTER3D [34]; (**d**) Electrostatic surface potential of bouganin surface at pH 7. The positive regions are represented in blue and the negative ones are colored in red. The active pocket is highlighted by a green circle. The figure was produced by GRASP [35].

Similarly to other RIPs, the active site is in a cleft at the center of the molecule. The authors suggested an explanation for the lower bouganin activity based on one amino acid substitution. In fact, the Asn78 in ricin or the corresponding Asn70 in PAP-R that is proposed to be involved in the interactions with the substrate, probably with the phosphodiester group between the target adenosine and the subsequent guanosine, was conserved in most considered RIPs except in bouganin, which has an aspartate in the corresponding position [26]. The backbone super-imposition of bouganin demonstrates that the central area of the molecule is well super-imposable, while big differences are reported in the peripherical portions (Figure 2c). Analyzing bouganin electrostatic surface potential, it is evidenced that the active site is a wide negative cavity, except for a small positive zone determined by the catalytic residue Arg168 (Figure 2d).

4. Antigenic Properties of Bouganin

Despite the encouraging results obtained with RIP-containing ITs in the treatment of hematological tumors (see Introduction), about 40% of patients respond producing anti-RIP antibodies [36]. This obstacle can be bypassed through multiple treatments with ITs containing different RIPs that do not cross-react each other [37]. Nevertheless, it is crucial to identify the most critical epitopes on RIPs to reduce their immunogenicity and thus improve their therapeutic utility. In this sense, Cizeau and co-workers undertook pioneering work synthetizing 89 peptides of 15 amino acids, each with an overlap of 12 residues, covering the mature bouganin protein. Selecting the peptides that induce T-cell proliferation, through in silico analysis, the residues presented as antigenic determinants for major histocompatibility complex (MHC) class II recognition were identified. The authors expressed a bouganin mutant, called de-bouganin, carrying four mutations (Val123Ala, Asp127Ala, Tyr133Asn, Ile152Ala) that did not affect the catalytic activity of the enzyme but significantly reduced the immune response of the host versus the toxin [38].

Analysis of the primary structure of wild-type bouganin compared to those of some type 1 RIPs (namely, saporin-SO6, tricosanthin and alpha-momorcharin) shows that a Tyr-Tyr-Phe (YYF) or Tyr-Phe-Phe (YFF) sequence is present in all the toxins except bouganin. It has been claimed that YYF/YFF sequences are recognized by different MHC class II alleles, thus justifying the higher antigenicity of saporin-S6, tricosanthin and alpha-momorcharin than bouganin [39]. The Lys-Arg (KR) motif, already identified in trichosanthin as responsible of its immunogenicity, is not significantly represented in conserved positions. Moreover, the bouganin residues inducing immunological responses (Val, Asp, Tyr and Ile) are not conserved amongst this group of type I RIPs [39]. These characteristics make bouganin a RIP with completely different antigenic properties from other type 1 RIPs. In effect, if tested with sera against six other type 1 RIPs, namely saporin-S6, and dianthin 32 (Caryophyllaceae), PAP-R (Phytolaccaceae), momordin I, momorcochin-S and trichokirin (Cucurbitaceae), bouganin gave no cross-reaction with any tested antiserum [24].

5. Bouganin-Containing Immunotoxins, Preclinical Evaluations

Bouganin represents a good candidate as toxic moiety to obtain ITs because of its high activity ratio, very low aspecific toxicity reported in animal models and high stability to derivatization and conjugation procedures [24].

The first bouganin-containing ITs were constructed conjugating bouganin to the M24 mAb (anti-CD80) and to the 1G10 mAb (anti-CD86) and comparing them to ITs built with the same mAbs and two different type 1 RIPs, gelonin, and saporin [27]. Bouganin incremented its toxicity on target cells by 3–4 log upon conjugation, as measured by inhibition of protein synthesis with IC_{50}s ranging from 4.61 to 192 pM as RIPs, comparable with IC_{50} values obtained with gelonin-containing ITs and 1–2 logs higher than saporin-containing ITs built both with the same mAbs and with another anti-CD80 IT [27,40]. All the anti-CD80 ITs resulted in higher cytotoxic effect than anti-CD86 ones, probably reflecting a higher antigen affinity and/or better internalization. This difference was more evident for bouganin-containing ITs. Moreover, bouganin-containing ITs induced apoptosis in target cells and they did not significantly affect the recovery of committed progenitors at concentrations up to 100 nM [27].

Recombinant ITs have the advantage to be stably formulable and less immunogenic than chemically linked ITs, so they are generally more suitable for clinical use. In 2009, Cizeau and co-workers genetically linked de-bouganin to an anti EpCAM Fab moiety via a peptidic linker to create the fusion construct VB6-845. This conjugate bound and selectively killed EpCAM-positive cell lines with a greater potency than many commonly used chemotherapeutic agents. In vivo efficacy was demonstrated using an EpCAM-positive human tumor xenograft model in severe combined immunodeficiency (SCID) mice; the majority of treated mice being tumor free at the end of the study [38]. The same research group described the intracellular trafficking in EpCAM-positive cells of VB6-845. De-bouganin recombinant IT was shown to co-localize along with the EEA1 endosomal and LAMP-2 lysosomal markers after 15 and 45 min whereas it did not traffic via a Golgi/ER pathway in contrast to ricin [41]. The preclinical evaluation of safety and suitability of VB6-845 as a systemically administered drug for the treatment of solid tumors was performed in animal models [42,43]. Efficacy studies in mice bearing human tumors demonstrated that VB6-845 specifically and potently targeted EpCAM-positive tumors. SCID mice bearing subcutaneous ovarian tumor (NIH:OVCAR-3 cells) were treated with 10 and 20 mg/kg VB6-845 IT. All mice were alive at the end of the study and no animal reached the endpoint tumor volume (750 mm^3). In HRLN nu/nu mice bearing subcutaneous MCF-7 VB6-845 at 20 mg/kg gave a 70% survival rate over the study period, with 3 mice achieving a complete tumor regression with no measurable tumor mass. The aspecific toxicity was evaluated in animal models, in dose-ranging studies. In rats, single doses until 100 mg/kg of the IT were well tolerated resulting in no-observable adverse effects, but doses of 200 mg/kg caused mild clinical signs that included excessive licking of forepaws, reddened skin on fore and hind paws, edema of the forepaws, and a slight decrease in activity level. In Cynomolgus monkeys, two treatments (+1 and +8 day) for total doses of the IT until 180 mg/kg were well-tolerated when given as a 3-h infusion mimicking the intended route of administration in the clinic; only mild and transitory clinical side effects were reported. In addition, VB6-845 proved to be minimally immunogenic in monkeys. No immune response was observed against either the humanized Fab or the toxin moiety on day 7; however, antibody responses were detected by day 14, being the reactivity to de-bouganin 3-fold lower than to the Fab fragment.

Several type 1 RIP-containing ITs were designed to target growth factor receptors on human tumors, indicating the validity of the idea [44–47]. In 2016, Dillon and co-workers conjugated de-bouganin to the anti-HER2 mAb trastuzumab. This IT demonstrated greater in vitro potency on HER2-positive cell lines and higher toxicity against tumor cells with cancer stem cell properties than the conjugate containing the tubulin inhibitor mertansine conjugated to trastuzumab. In addition, the authors demonstrated that, unlike for mertansine-trastuzumab IT, T-DM1, the cytotoxic effect of de-bouganin-trastuzumab was not influenced by MK571, an inhibitor of efflux pumps, which is responsible for multidrug resistance [48]. Similarly, ABT-737, a Bcl-2 family inhibitor, modulated T-DM1, but not de-bouganin-trastuzumab cytotoxicity [48]. In the same year, Chooniedass and co-workers described the engineering and biological activity of de-bouganin genetically linked to an anti-HER2 C6.5 diabody (deB-C6.5-diab). On breast cancer cell lines, the DeB-C6.5-diab and de-bouganin-trastuzumab conjugates showed greater cytotoxic activity than auristatin E- and emtansine-trastuzumab conjugates, demonstrating that de-bouganin is effective against tumor cell resistance mechanisms selected in response to immunoconjugates composed by anti-microtubule agents [49]. Main results obtained in preclinical experiments are summarized in Table 2.

Table 2. Characteristics and efficacy of bouganin-containing ITs in pre-clinical experiments.

Bouganin	Carrier	Target	In Vitro Studies		In Vivo Studies		Ref.
			Cell Line	IC_{50} (pM) *	Animals	Survival Rate	
native	M24	CD80	Raji (Burkitt's lymphomaL)	4.61	n.a. #	n.a. #	[27]
native	1G10	CD86	Raji (Burkitt's lymphomaL)	129	n.a. #	n.a. #	[27]
de-bouganin	4D5MOCB	EpCAM	NIH:OVCAR-3 (Ovarian)	700	SCID mice/NIH:OVCAR-3	100%	[38]
de-bouganin	4D5MOCB	EpCAM	MCF-7 (Breast), NIH:OVCAR-3 (Ovarian)	400	HRLN nu/nu/MCF-7	70%	[42]
de-bouganin	Trastuzumab	HER2	HCC1954 (Breast)	45	CB.17 SCID mice/BT-474	83%	[48]
de-bouganin	C6.5 diabody	HER2	HCC202 (Breast)	22	n.a. #	n.a. #	[49]

* The IC_{50} values refer to the cell line resulted the most sensible to the IT in the referenced manuscript. # not available.

6. Bouganin-Containing Immunotoxins, Clinical Evaluations

In 2007, VB6-845 entered a Phase I clinical trial (NCT00481936) sponsored by VIVENTIA Biotech Inc. (Winnipeg, MB, Canada). The purpose was to determine the maximum tolerated dose of VB6-845 and to evaluate its safety and tolerability when administered as intravenous infusion once weekly for 4 weeks to patients with advanced solid tumor of epithelial origin [43]. Fifteen neoplastic patients with cancers affecting kidney, ovary, breast, stomach, pancreas, lung, and colon, were enrolled into the study. The tested dosages were 1.0, 2.0 and 3.34 mg/kg. The maximum treatment duration was 16 weeks. Only one case of dose-limiting toxicity was reported. It was a grade 4 acute infusion reaction, which occurred in a patient with metastatic renal cell carcinoma treated with IT at 2.0 mg/kg that showed hypotension and weakness during the third infusion. These reactions were resolved without consequences after just one day of therapy. Five subjects reported serious adverse events: two of them were reported as related to study treatment, they were infusion reactions consisting of a symptom complex characterized by hypotension, fever, and nausea, weakness, drowsiness, chills, and face and neck hyperemia. The study terminated for corporate reasons unrelated to safety and efficacy of the IT. The adverse events reported in the literature are those available in the clinical database at the moment of the early closure of the trial. Exploratory efficacy data revealed encouraging preliminary results. Seven subjects who completed one full cycle (4 weeks) of treatment showed stable disease using standard imaging techniques. For five of them was reported a stable disease 1 week after the completion of the fourth dose. Amongst the three subjects who continued to receive study treatment after the first cycle, one subject had stable disease at the completion of second (8 weeks) and third (12 weeks) cycles. In addition, one patient with renal cell carcinoma and one patient with breast carcinoma had a reduction of the tumor mass [43].

The validity of the bouganin de-immunization approach was assessed on plasma samples from patients treated with VB6-845 IT, testing their immune responsiveness against both humanized Fab and de-bouganin. After 2 weeks from the treatment, no patient plasma samples showed a detectable immune response. After 3 weeks, only 1 patient showed a moderate anti de-bouganin titer, whereas six of seven patients showed anti-Fab titers. By week 4, anti-Fab titers were detectable in all patients, whilst only two patients had barely measurable anti-de-bouganin titers [43].

7. Conclusions

The plant toxin bouganin can represent an attractive weapon to construct ITs for the experimental therapy of human neoplasia. In fact, it shows interesting features with respect to other RIPs, such as a high ratio of activity on DNA to that on ribosomal RNA and low systemic toxicity. Moreover, the absence of cross-reactivity with sera against RIPs from other taxonomically related or unrelated plants can represent a useful property for circumventing the immune response after repeated administration of ITs. The sequencing of bouganin and the knowledge of its three-dimensional structure allowed to obtain a not immunogenic mutant of bouganin for clinical use. An IT containing modified bouganin showed encouraging results in Phase I clinical trials on patients with advanced carcinoma. The lack of immune responsiveness towards bouganin in patients illustrates the validity of the T cell epitope-depletion approach to dampen the immune response and strongly supports the utility of bouganin as a cytotoxic payload for systemic delivery.

Author Contributions: All the authors collected the literature, wrote, and revised the paper.

Funding: This work was supported by funds for selected research topics from the Alma Mater Studiorum—University of Bologna and by the Pallotti Legacies for Cancer Research.

Conflicts of Interest: The authors declare no conflict of interest.

References

1. Bolognesi, A.; Bortolotti, M.; Maiello, S.; Battelli, M.G.; Polito, L. Ribosome-Inactivating Proteins from Plants: A Historical Overview. *Molecules* **2016**, *21*, 1627. [CrossRef] [PubMed]
2. Barbieri, L.; Valbonesi, P.; Bonora, E.; Gorini, P.; Bolognesi, A.; Stirpe, F. Polynucleotide: Adenosine glycosidase activity of ribosome-inactivating proteins: Effect on DNA, RNA and poly(A). *Nucleic Acids Res.* **1997**, *25*, 518–522. [CrossRef] [PubMed]
3. Battelli, M.G.; Barbieri, L.; Bolognesi, A.; Buonamici, L.; Valbonesi, P.; Polito, L.; Van Damme, E.J.; Peumans, W.J.; Stirpe, F. Ribosome-inactivating lectins with polynucleotide: Adenosine glycosidase activity. *FEBS Lett.* **1997**, *408*, 355–359. [CrossRef]
4. Stirpe, F.; Bolognesi, A.; Bortolotti, M.; Farini, V.; Lubelli, C.; Pelosi, E.; Polito, L.; Dozza, B.; Strocchi, P.; Chambery, A.; et al. Characterization of highly toxic type 2 ribosome-inactivating proteins from Adenia lanceolata and Adenia stenodactyla (Passifloraceae). *Toxicon* **2007**, *50*, 94–105. [CrossRef] [PubMed]
5. Zeng, M.; Zheng, M.; Lu, D.; Wang, J.; Jiang, W.; Sha, O. Anti-tumor activities and apoptotic mechanism of ribosome-inactivating proteins. *Chin. J. Cancer* **2015**, *34*, 325–334. [CrossRef] [PubMed]
6. Polito, L.; Bortolotti, M.; Farini, V.; Battelli, M.G.; Barbieri, L.; Bolognesi, A. Saporin induces multiple death pathways in lymphoma cells with different intensity and timing as compared to ricin. *Int. J. Biochem. Cell Biol.* **2009**, *41*, 1055–1061. [CrossRef] [PubMed]
7. Wei, B.; Huang, Q.; Huang, S.; Mai, W.; Zhong, X. Trichosanthin-induced autophagy in gastric cancer cell MKN-45 is dependent on reactive oxygen species (ROS) and NF-κB/p53 pathway. *J. Pharmacol. Sci.* **2016**, *131*, 77–83. [CrossRef] [PubMed]
8. Polito, L.; Bortolotti, M.; Pedrazzi, M.; Mercatelli, D.; Battelli, M.G.; Bolognesi, A. Apoptosis and necroptosis induced by stenodactylin in neuroblastoma cells can be completely prevented through caspase inhibition plus catalase or necrostatin-1. *Phytomedicine* **2016**, *23*, 32–41. [CrossRef] [PubMed]
9. Parikh, B.A.; Tumer, N.E. Antiviral activity of ribosome inactivating proteins in medicine. *Mini Rev. Med. Chem.* **2004**, *4*, 523–543. [CrossRef] [PubMed]
10. Duggar, B.M.; Armstrong, J.K. The effect of treating the Virus of Tobacco Mosaic with the juices of various plants. *Ann. Missouri Bot. Gard.* **1925**, *12*, 359–366. [CrossRef]
11. Tomlinson, J.A.; Walker, V.M.; Flewett, T.H.; Barclay, G.R. The inhibition of infection by cucumber mosaic virus and influenza virus by extracts from Phytolacca americana. *J. Gen. Virol.* **1974**, *22*, 225–232. [CrossRef] [PubMed]
12. Ragetli, H.W.J.; Weintraub, M. Purification and characteristics of a virus inhibitor from *Dianthus caryophyllus* L.: I. Purification and activity. *Virology* **1962**, *18*, 232–240. [CrossRef]
13. Obrig, T.G.; Irvin, J.D.; Hardesty, B. The effect of an antiviral peptide on the ribosomal reactions of the peptide elongation enzymes, EF-I and EF-II. *Arch. Biochem. Biophys.* **1973**, *155*, 278–289. [CrossRef]
14. Foà-Tomasi, L.; Campadelli-Fiume, G.; Barbieri, L.; Stirpe, F. Effect of ribosome-inactivating proteins on virus-infected cells. Inhibition of virus multiplication and of protein synthesis. *Arch. Virol.* **1982**, *71*, 323–332. [CrossRef] [PubMed]
15. Stirpe, F.; Williams, D.G.; Onyon, L.J.; Legg, R.F.; Stevens, W.A. Dianthins, ribosome-damaging proteins with anti-viral properties from *Dianthus caryophyllus* L. (carnation). *Biochem. J.* **1981**, *195*, 399–405. [CrossRef] [PubMed]
16. FitzGerald, D.J.; Wayne, A.S.; Kreitman, R.J.; Pastan, I. Treatment of hematologic malignancies with immunotoxins and antibody-drug conjugates. *Cancer Res.* **2011**, *71*, 6300–6309. [CrossRef] [PubMed]

17. Kawakami, K.; Nakajima, O.; Morishita, R.; Nagai, R. Targeted anticancer immunotoxins and cytotoxic agents with direct killing moieties. *Sci. World J.* **2006**, *6*, 781–790. [CrossRef] [PubMed]
18. Polito, L.; Bortolotti, M.; Maiello, S.; Battelli, M.G.; Bolognesi, A. Plants Producing Ribosome-Inactivating Proteins in Traditional Medicine. *Molecules* **2016**, *21*, 1560. [CrossRef] [PubMed]
19. Guzmán Gutiérrez, S.L.; Reyes-Chilpa, R.; Jaime, H.B. Medicinal plants for the treatment of "nervios", anxiety, and depression in Mexican Traditional Medicine. *Rev. Bras. Farmacogn.* **2014**, *24*, 591–608. [CrossRef]
20. Argueta, A. *Atlas de las Plantas de la Medicina Tradicional Mexicana*; Instituto Nacional Indigenista: Ciudad de México, México, 1994; p. 1786.
21. Ghogar, A.; Jiraungkoorskul, K.; Jiraungkoorskul, W. Paper Flower, Bougainvillea spectabilis: Update properties of traditional medicinal plant. *J. Nat. Remed.* **2016**, *16*, 82–87. [CrossRef]
22. Murthy, N.S.; Nagarajan, K.; Sastry, A.B. Effect of prophylactic sprays of leaf extracts on the infection of tobacco by tobacco mosaic virus. *Indian J. Agric. Sci.* **1981**, *51*, 792–795.
23. Verma, H.N.; Dwivedi, S.D. Properties of a virus inhibiting agent, isolated from plants which have been treated with leaf extracts from Bougainvillea spectabilis. *Physiol. Plant Pathol.* **1984**, *25*, 93–101. [CrossRef]
24. Bolognesi, A.; Polito, L.; Olivieri, F.; Valbonesi, P.; Barbieri, L.; Battelli, M.G.; Carusi, M.V.; Benvenuto, E.; Del Vecchio Blanco, F.; Di Maro, A.; et al. New ribosome-inactivating proteins with polynucleotide: Adenosine glycosidase and antiviral activities from *Basella rubra* L. and *Bougainvillea spectabilis* Willd. *Planta* **1997**, *203*, 422–429. [CrossRef] [PubMed]
25. Barbieri, L.; Battelli, M.G.; Stirpe, F. Ribosome-inactivating proteins from plants. *Biochim. Biophys. Acta* **1993**, *1154*, 237–282. [CrossRef]
26. Fermani, S.; Tosi, G.; Farini, V.; Polito, L.; Falini, G.; Ripamonti, A.; Barbieri, L.; Chambery, A.; Bolognesi, A. Structure/function studies on two type 1 ribosome inactivating proteins: Bouganin and lychnin. *J. Struct. Biol.* **2009**, *168*, 278–287. [CrossRef] [PubMed]
27. Bolognesi, A.; Polito, L.; Tazzari, P.L.; Lemoli, R.M.; Lubelli, C.; Fogli, M.; Boon, L.; de Boer, M.; Stirpe, F. In vitro anti-tumour activity of anti-CD80 and anti-CD86 immunotoxins containing type 1 ribosome-inactivating proteins. *Br. J. Haematol.* **2000**, *110*, 351–361. [CrossRef] [PubMed]
28. Jaffrézou, J.P.; Levade, T.; Kuhlein, E.; Thurneyssen, O.; Chiron, M.; Grandjean, H.; Carrière, D.; Laurent, G. Enhancement of ricin A chain immunotoxin activity by perhexiline on established and fresh leukemic cells. *Cancer Res.* **1990**, *50*, 5558–5566. [PubMed]
29. Griffin, T.W.; Morgan, A.C.; Blythman, H.E. Immunotoxins therapy: Assessment by animal models. In *Immunotoxins*; Frankel, A.E., Ed.; Kluwer Academic Publishers: Boston, MA, USA, 1988; pp. 433–455. ISBN 0-89838-984-4.
30. Den Hartog, M.T.; Lubelli, C.; Boon, L.; Heerkens, S.; Ortiz Buijsse, A.P.; de Boer, M.; Stirpe, F. Cloning and expression of cDNA coding for bouganin. *Eur. J. Biochem.* **2002**, *269*, 1772–1779. [CrossRef] [PubMed]
31. Fabbrini, M.S.; Katayama, M.; Nakase, I.; Vago, R. Plant Ribosome-Inactivating Proteins: Progesses, Challenges and Biotechnological Applications (and a Few Digressions). *Toxins* **2017**, *9*, 314. [CrossRef] [PubMed]
32. Di Maro, A.; Citores, L.; Russo, R.; Iglesias, R.; Ferreras, J.M. Sequence comparison and phylogenetic analysis by the Maximum Likelihood method of ribosome-inactivating proteins from angiosperms. *Plant Mol. Biol.* **2014**, *85*, 575–588. [CrossRef] [PubMed]
33. Kraulis, P.J. MOLSCRIPT: A program to produce both detailed and schematic plots of protein structures. *J. Appl. Crystallogr.* **1991**, *24*, 946–950. [CrossRef]
34. Merritt, E.A.; Bacon, D.J. Raster3D: Photorealistic molecular graphics. *Methods Enzymol.* **1997**, *277*, 505–524. [PubMed]
35. Nicholls, A.; Sharp, K.A.; Honig, B. Protein folding and association: Insights from the interfacial and thermodynamic properties of hydrocarbons. *Proteins* **1991**, *11*, 281–296. [CrossRef] [PubMed]
36. Pastan, I.; Hassan, R.; FitzGerald, D.J.; Kreitman, R.J. Immunotoxin treatment of cancer. *Annu. Rev. Med.* **2007**, *58*, 221–237. [CrossRef] [PubMed]
37. Kreitman, R.J. Immunotoxins for targeted cancer therapy. *AAPS J.* **2006**, *8*, 532–551. [CrossRef] [PubMed]
38. Cizeau, J.; Grenkow, D.M.; Brown, J.G.; Entwistle, J.; MacDonald, G.C. Engineering and biological characterization of VB6-845, an anti-EpCAM immunotoxin containing a T-cell epitope-depleted variant of the plant toxin bouganin. *J. Immunother.* **2009**, *32*, 574–584. [CrossRef] [PubMed]

39. Giansanti, F.; Flavell, D.J.; Angelucci, F.; Fabbrini, M.S.; Ippoliti, R. Strategies to Improve the Clinical Utility of Saporin-Based Targeted Toxins. *Toxins* **2018**, *10*, 82. [CrossRef] [PubMed]
40. Vooijs, W.C.; Otten, H.G.; van Vliet, M.; van Dijk, A.J.G.; de Weger, R.A.; de Boer, M.; Bohlen, H.; Bolognesi, A.; Polito, L.; de Gast, G.C. B7-1 (CD80) as target for immunotoxin therapy for Hodgkin's disease. *Br. J. Cancer* **1997**, *76*, 1163–1169. [CrossRef] [PubMed]
41. Chaboureau, A.; Ragon, I.; Stibbard, S.; Cizeau, J.; Glover, N.; MacDonald, G.C. Intracellular trafficking of VB6-845, an immunocytotoxin containing a de-immunized variant of bouganin. In Proceedings of the American Association of Cancer Research Annual Meeting, San Diego, CA, USA, 12–16 April 2008.
42. Entwistle, J.; Brown, J.G.; Chooniedass, S.; Cizeau, J.; MacDonald, G.C. Preclinical evaluation of VB6-845: An anti-EpCAM immunotoxin with reduced immunogenic potential. *Cancer Biother. Radiopharm.* **2012**, *27*, 582–592. [CrossRef] [PubMed]
43. Entwistle, J.; Kowalski, M.; Brown, J.; Cizeau, J.; MacDonald, G.C. The Preclinical and Clinical Evaluation of VB6-845: An Immunotoxin with a De-Immunized Payload for the Systemic Treatment of Solid Tumors. In *Antibody-Drug Conjugates and Immunotoxins. Cancer Drug Discovery and Development*; Phillips, G., Ed.; Springer: New York, NY, USA, 2013; pp. 349–367. ISBN 978-1-4614-5455-7.
44. Di Massimo, A.M.; Di Loreto, M.; Pacilli, A.; Raucci, G.; D'Alatri, L.; Mele, A.; Bolognesi, A.; Polito, L.; Stirpe, F.; De Santis, R. Immunoconjugates made of an anti-EGF receptor monoclonal antibody and type 1 ribosome-inactivating proteins from *Saponaria ocymoides* or Vaccaria pyramidata. *Br. J. Cancer* **1997**, *75*, 822–828. [CrossRef] [PubMed]
45. Ricci, C.; Polito, L.; Nanni, P.; Landuzzi, L.; Astolfi, A.; Nicoletti, G.; Rossi, I.; De Giovanni, C.; Bolognesi, A.; Lollini, P.L. HER/erbB receptors as therapeutic targets of immunotoxins in human rhabdomyosarcoma cells. *J. Immunother.* **2002**, *25*, 314–323. [CrossRef] [PubMed]
46. Zhou, X.; Qiu, J.; Wang, Z.; Huang, N.; Li, X.; Li, Q.; Zhang, Y.; Zhao, C.; Luo, C.; Zhang, N.; et al. In vitro and in vivo anti-tumor activities of anti-EGFR single-chain variable fragment fused with recombinant gelonin toxin. *J. Cancer Res. Clin. Oncol.* **2012**, *138*, 1081–1090. [CrossRef] [PubMed]
47. Yip, W.L.; Weyergang, A.; Berg, K.; Tønnesen, H.H.; Selbo, P.K. Targeted delivery and enhanced cytotoxicity of cetuximab-saporin by photochemical internalization in EGFR-positive cancer cells. *Mol. Pharm.* **2007**, *4*, 241–251. [CrossRef] [PubMed]
48. Dillon, R.L.; Chooniedass, S.; Premsukh, A.; Adams, G.P.; Entwistle, J.; MacDonald, G.C.; Cizeau, J. Trastuzumab-deBouganin Conjugate Overcomes Multiple Mechanisms of T-DM1 Drug Resistance. *J. Immunother.* **2016**, *39*, 117–126. [CrossRef] [PubMed]
49. Chooniedass, S.; Dillon, R.L.; Premsukh, A.; Hudson, P.J.; Adams, G.P.; MacDonald, G.C.; Cizeau, J. DeBouganin Diabody Fusion Protein Overcomes Drug Resistance to ADCs Comprised of Anti-Microtubule Agents. *Molecules* **2016**, *21*, 1741. [CrossRef] [PubMed]

MDPI
St. Alban-Anlage 66
4052 Basel
Switzerland
Tel. +41 61 683 77 34
Fax +41 61 302 89 18
www.mdpi.com

Toxins Editorial Office
E-mail: toxins@mdpi.com
www.mdpi.com/journal/toxins

www.ingramcontent.com/pod-product-compliance
Lightning Source LLC
LaVergne TN
LVHW071951080526
838202LV00064B/6723
* 9 7 8 3 0 3 9 2 1 1 8 9 0 *